Flow Cytogenetics

ANALYTICAL CYTOLOGY SERIES

Flow Cytogenetics

Edited by

Joe W. Gray

Biomedical Sciences Division
Lawrence Livermore National Laboratory
Livermore, California, USA

ACADEMIC PRESS

(Harcourt Brace Jovanovich, Publishers)

London   San Diego   New York   Berkeley
Boston   Sydney   Tokyo   Toronto

This book is printed on acid free paper ∞

ACADEMIC PRESS LIMITED
24–28 Oval Road
LONDON NW1 7DX

United States Edition published by
ACADEMIC PRESS, INC.
San Diego, CA 92101

British Library Cataloguing in Publication is available

ISBN 0-12-296110-2

Typeset by EJS Chemical Composition, Bath, England
and printed in Great Britain by Thomson Litho Ltd, East Kilbride, Scotland

Contents

Contributors

K. L. ALBRIGHT, Los Alamos National Laboratory, Life Sciences Division, Los Alamos, New Mexico 87545, USA

N. A. ALLEN, Biomedical Sciences Division, Lawrence Livermore National Laboratory, Livermore, California 94550, USA

J. A. ATEN, Laboratory for Radiobiology, Academic Medical Center, University of Amsterdam, Amsterdam, The Netherlands

MARTY F. BARTHOLDI, Los Alamos National Laboratory, US Department of Energy, Los Alamos, New Mexico 87545, USA

J. G. J. BAUMAN, TNO Radiobiological Institute, Lange Kleiweg 151, 2288 GJ Rijswijk, The Netherlands

N. C. BROWN, Los Alamos National Laboratory, Life Sciences Division, Los Alamos, New Mexico 87545, USA

G. BRUNS, The Children's Hospital, Department of Pediatrics, Harvard Medical School, Harvard University, Cambridge, Massachusetts 02138, USA

E. W. CAMPBELL, Los Alamos National Laboratory, Life Sciences Division, Los Alamos, New Mexico 87545, USA

A. V. CARRANO, Biomedical Sciences Division, Lawrence Livermore National Laboratory, Livermore, California 94550, USA

MARY CASSIDY, Los Alamos National Laboratory, US Department of Energy, Los Alamos, New Mexico 87545, USA

M. CHRISTENSEN, Biomedical Sciences Division, Lawrence Livermore National Laboratory, Livermore, California 94550, USA

L. M. CLARK, Los Alamos National Laboratory, Life Sciences Division, Los Alamos, New Mexico 87545, USA

J. G. COLLARD, The Netherlands Cancer Institute (Antoni van Leeuwenhoekhuis), Department of Experimental Cytology, H6 121 Plesmanlaan, 1066 CX Amsterdam, The Netherlands

L. SCOTT CRAM, Los Alamos National Laboratory, US Department of Energy, Los Alamos, New Mexico 87545, USA

P. N. DEAN, Biomedical Sciences Division, Lawrence Livermore National Laboratory, Livermore, California 94550, USA

L. L. DEAVEN, Los Alamos National Laboratory, Life Sciences Division, University of California, Los Alamos, New Mexico 87545, USA

T. DONLON, The Children's Hospital, Department of Pediatrics, Harvard Medical School, Harvard University, Cambridge, Massachusetts 02138, USA

G. J. VAN DEN ENGH, Biomedical Sciences Division, Lawrence Livermore National Laboratory, Livermore, California 94550, USA

JOHN H. EVANS, MRC Human Genetics Unit, Western General Hospital, Crewe Road, Edinburgh , EH4 2XU, Scotland, UK

JUDITH A. FANTES, MRC Human Genetics Unit, Western General Hospital, Crewe Road, Edinburgh EH4 2XU, Scotland, UK

J. J. FAWCETT, Los Alamos National Laboratory, US Department of Energy, Los Alamos, New Mexico 87545, USA

A. FLINT, The Children's Hospital, Department of Pediatrics, Harvard Medical School, Harvard University, Cambridge, Massachusetts 02138, USA

J. C. FUSCOE, Center for Environmental Health, U-40, The University of Connecticut, Storrs, Connecticut 06268, USA

JOE. W. GRAY, Biomedical Sciences Division, Lawrence Livermore National Laboratory, Livermore, California 94550, USA

DARYLL K. GREEN, MRC Human Genetics Unit, Western General Hospital, Crewe Road, Edinburgh EH4 2XU, Scotland, UK

P. HARRIS, The Children's Hospital, Department of Pediatrics, Harvard Medical School, Harvard University, Cambridge, Massachusetts 02138, USA

C. E. HILDEBRAND, Los Alamos National Laboratory, US Department of Energy, Los Alamos, New Mexico 87545, USA

P. J. JACKSON, Los Alamos National Laboratory, US Department of Energy, Los Alamos, New Mexico 87545, USA

J. W. G. JANSSEN, Deutsches Rotes Kreuz, Blutspendezentrale, Ulm, Oberer Easelberg 10, Postfach 1564, 7900 Ulm/Donau, FRG

J. H. JETT, Los Alamos National Laboratory, US Department of Energy, Los Alamos, New Mexico 87545, USA

P. DE JONG, Biomedical Sciences Division, Lawrence Livermore National Laboratory, Livermore, California 94550, USA

S. KOLLA, Los Alamos National Laboratory, Life Sciences Division Los Alamos, New Mexico 87545, USA and Stanford University Graduate School, Stanford, California, USA

PAUL M. KRAEMER, Los Alamos National Laboratory, US Department of Energy, Los Alamos, New Mexico 87545, USA

L. KUNKEL, The Children's Hospital, Department of Pediatrics, Harvard Medical School, Harvard University, Cambridge, Massachusetts 02138, USA

D. KURNIT, The Children's Hospital, Department of Pediatrics, Harvard Medical School, Harvard University, Cambridge, Massachusetts 02138, USA

M. LALANDE, The Children's Hospital, Department of Pediatrics, Harvard Medical School, Harvard University, Cambridge, Massachusetts 02138, USA

RICHARD G. LANGLOIS, Biomedical Sciences Division, Lawrence Livermore National Laboratory, Livermore, California 94550, USA

S. A. LATT,† The Children's Hospital, Department of Pediatrics and Genetics, Harvard Medical School, Harvard University, Cambridge, Massachusetts 02138, USA

ROGER V. LEBO, Department of Obstetrics, Gyneacology and Pediatrics, University of California, San Francisco, California 94143, USA

J. L. LONGMIRE, Los Alamos National Laboratory, US Department of Energy, Los Alamos, New Mexico 87545, USA

CHARLOTTE LOZES, Biomedical Sciences Division, Lawrence Livermore National Laboratory, Livermore, California 94550, USA

JOE LUCAS, Biomedical Sciences Division, Lawrence Livermore National Laboratory, Livermore, California 94550, USA

J. S. McNINCH, Biomedical Sciences Division, Lawrence Livermore National Laboratory, Livermore, California 94550, USA

M. L. MENDELSOHN, Biomedical Sciences Division, Lawrence Livermore National Laboratory, Livermore, California 94550, USA

† Deceased.

L. J. MEINKE, Los Alamos National Laboratory, US Department of Energy, Los Alamos, New Mexico 87545, USA

J. MEYNE, Los Alamos National Laboratory, US Department of Energy, Los Alamos, New Mexico 87545, USA

DAN H. MOORE II, Biomedical Sciences Division, Lawrence Livermore National Laboratory, Livermore, California 94550, USA

R. K. MOYZIS, Los Alamos National Laboratory, US Department of Energy, Los Alamos, New Mexico 87545, USA

U. MULLER, The Children's Hospital, Department of Pediatrics, Harvard Medical School, Harvard University, Cambridge, Massachusetts 02138, USA

J. MULLIKIN, Biomedical Sciences Division, Lawrence Livermore National Laboratory, Livermore, California 94550, USA

A. C. MUNK, Los Alamos National Laboratory, US Department of Energy, Los Alamos, New Mexico 87545, USA

R. NEVE, The Children's Hospital, Department of Pediatrics, Harvard Medical School, Harvard University, Cambridge, Massachusetts 02138, USA

L. PEDERSON, 4421 Vine Street, Vancouver, British Columbia V4P5W6, Canada

J. PERLMAN, Center for Environmental Health, U-40, The University of Connecticut, Storrs, Connecticut 06268, USA

DONALD C. PETERS, Biomedical Sciences Division, Lawrence Livermore National Laboratory, Livermore, California 94550, USA

D. PINKEL, Biomedical Sciences Division, Lawrence Livermore National Laboratory, Livermore, California 94550, USA

M. VAN DER PLOEG, Sylvius Laboratory, 72 Wassenaarseweg, Leiden, The Netherlands

F. ANDREW RAY, Los Alamos National Laboratory, US Department of Energy, Los Alamos, New Mexico 87545, USA

A. J. SILVA, Biomedical Sciences Division, Lawrence Livermore National Laboratory, Livermore, California 94550, USA

U. TANTRAVAHI, The Children's Hospital, Department of Pediatrics, Harvard Medical School, Harvard University, Cambridge, Massachusetts 02138, USA

BARB J. TRASK, Biomedical Sciences Division, Lawrence Livermore National Laboratory, Livermore, California 94550, USA

A. TULP, The Netherlands Cancer Institute (Antoni van Leeuwenhoekhuis), Department of Experimental Cytology, H6 121 Plesmanlaan, 1006 CX Amsterdam, The Netherlands

M. A. VAN DILLA, Biomedical Sciences Division, Lawrence Livermore National Laboratory, University of California, Livermore, California 94550, USA

BRYAN D. YOUNG, Medical Oncology Laboratory, Imperial Cancer Research Fund, St Bartholomew's Hospital, 45 Little Britain, London EC1, UK

Preface

Flow Cytogenetics describes the application of flow cytometry and sorting to the special tasks of mammalian chromosome classification and purification. The first publications on flow cytogenetics, which appeared in 1975; promised accurate, statistically precise chromosome classification using flow cytometry (called flow karyotyping) as well as high purity chromosome fractionation. The 17 chapters in this book illustrate that the early promise of flow cytogenetics has been realized and exceeded. Application areas for flow karyotyping now include; quantification of radiation induced chromosome damage for biological dosimetry, determination of chromosome normality in cell cultures to suport somatic cell genetic studies and detection of numerical and structural aberrations associated with genetic diseases such as Down's syndrome. Applications for sorted chromosomes include gene mapping and the construction of chromosome-specific recombinant DNA libraries. This book covers both the technical aspects and application areas of flow cytogenetics.

The first chapter serves as an introduction and overview and should be consulted by readers not familiar with the field.

Chapters 2 to 6 are devoted to the technical aspects of flow cytogenetics; specifically, instrumentation, cell culture, chromosome isolation, chromosome staining and flow karyotype analysis. These chapters cover the technical developments that have allowed flow cytogenetics to become a practical tool in numerous laboratories.

Chapters 7 and 8 cover the use of univariate and bivariate flow karyotyping for chromosome classification and for detection of homogeneously occurring aberrations (i.e. aberrations that occur in most or all of the cells from which chromosomes were isolated). These chapters should be consulted for examples of the use of flow karyotyping for clinical detection of disease-linked chromosome aberrations or for characterization of cultured cell lines (including human-rodent hybrids).

Chapters 9 and 10 cover the progress that has been made on the use of flow cytometry for detection of the random structural aberrations produced by clastogenic agents such as ionizing radiation.

The next three chapters cover specialized developments in flow cytogenetics. Chapter 11 describes the use of silt-scan flow cytometry for chromosome classification and aberration detection based on chromosome shape. Chapter 12 summarizes the use of velocity sedimentation for chromosome enrichment prior to sorting. Chapter 13 describes the use of high speed sorting for chromosome purification.

Chapters 14 through 16 summarize the applications of flow sorted chromosomes. Chapter 14 is devoted to gene mapping using sorted chromosomes. Chapter 15 focuses on techniques for construction of recombinant DNA libraries using sorted chromosomes and on the use of these libraries in genetics research. Chapter 16 continues the theme of the previous chapter; summarizing the current status of the human-chromosome-specific libraries produced by the National Laboratory Gene Library Project.

The last chapter focuses on the use of fluorescence *in situ* hybridization to stain specific chromosomes, genes or gene transcripts in interphase cells or nuclei so that their presence and normality can be assessed flow cytometrically. The techniques covered here are rapidly developing and represent one of the important future directions in flow cytogenetics.

The chapters in this book describe the substantial progress that has been made in flow cytogenetics during the past 15 years. In addition, they suggest areas where development is continuing. At least three forces are acting to drive future work: 1. The availability of commercial instrumentation that is capable of high resolution chromosome analysis and sorting. 2. The need for chromosome characterization and for chromosome specific DNA generated by the International Genome effort. 3. The increasing availability of chromosome-specific probes to support chromosome specific staining using *in situ* hybridization. This book is aimed at investigators just entering the field as well as those interested in the continued development or application of Flow Cytogenetics.

JOE W. GRAY
Livermore, California
July 1989

Human Chromosome Analysis by Flow Cytometry

BRYAN D. YOUNG

Medical Oncology Laboratory
Imperial Cancer Research Fund
St Bartholomew's Hospital
45 Little Britain, London EC1, UK

I. INTRODUCTION

The application of banding techniques to the study of metaphase chromosomes has revealed many chromosomal aberrations associated with genetic disease and cancer. As techniques have improved, the cytogenetic analysis of the human karyotype has assumed greater diagnostic value. In a routine service, it is generally reckoned that a cytogeneticist can analyze about 250 samples per year. However, in cancer cytogenetics, where more complex karyotypes are often encountered, this figure may be considerably less. This area of diagnosis is not only highly labor intensive but is dependent on the skill and objectivity of the cytogeneticist.

The automation of karyotype analysis would bring great benefits to the study of human disease, both in terms of efficiency and objectivity. Automatic slide-based computer analysis of the karyotype has been in development for a number of years. Although these efforts have not been entirely successful, they have shown that measurement of the DNA content of chromosomes provides a reliable basis for karyotype analysis because, unlike

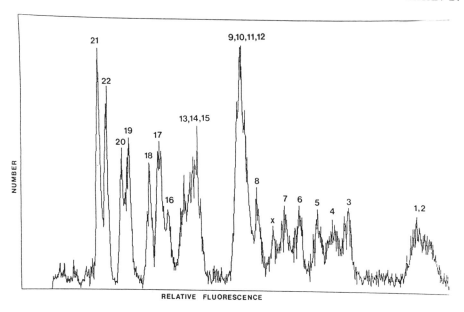

Fig. 1. Flow karyotype measured for chromosomes isolated from a peripheral blood sample of a normal individual. Chromosomes were prepared from 10 ml of blood using a previously described protocol (*49*). Chromosomes were stained with ethidium bromide at 4°C and analyzed on a FACS 440 cell sorter with the laser tuned to 488 nm. Background noise was reduced by electronic gating (*38*). The peaks due to each chromosome are indicated.

length or banding measurements, DNA content measurements are independent of the degree of contraction (*32,33,34*). Flow chromosome analysis is an alternative to conventional analytical techniques that are applied to chromosomes in metaphase spreads mounted on microscope slides. In the flow approach, chromosomes are isolated from mitotic cells, stained with one or more fluorescent dyes (usually DNA specific) and analyzed as they flow one by one through the beam(s) of a flow cytometer. Instrumentation for chromosome analysis is discussed in detail in Chapter 2. The results of analyzing several hundred thousand chromosomes are accumulated to form a flow karyotype like that shown in Fig. 1 for human chromosomes stained with ethidium bromide. Each peak is produced by chromosomes that yield the same fluorescent signals. The mean of the peak is proportional to the relative chromosome DNA content. The area of the peak is proportional to the frequency of occurrence of that chromosome type in the population. The coefficient of variation (CV) of the peak is an indication of the measurement precision. Flow cytometry offers the advantages of high resolution (measurement CVs of less than 2% are common) and statistical precision (measurement of 10^5 chromosomes can be accomplished in a few minutes). Measurement of chromosomal DNA content by flow cytometry has a number of advantages over slide-based systems and could, in the future, form the basis of an alternative system for abnormality detection.

Flow sorting offers the possibility of purifying chromosomes according to their fluorescence intensity. With proper care, this technology allows collection of microgram quantities of DNA from one or a few chromosome types. Purities above 90% are often achieved. The DNA collected in this manner has proved useful for gene mapping (*28,29*) and for production of recombinant DNA libraries (*21,45*).

It is intended in this chapter to provide an overview of the extent to which flow systems can be applied to the detection and analysis of chromosomal abnormalities and to chromosome purification. Both chromosome preparation, crucial to obtaining high resolution, and the application of flow analysis and sorting to particular problems are discussed. The details of these and other aspects of chromosome analysis and sorting are covered in subsequent chapters and will be referenced as appropriate.

II. FACTORS AFFECTING RESOLUTION

The first step in flow chromosome analysis is the production of a suspension of isolated, fluorescently stained chromosomes. This process requires cell culture and collection of mitotic cells, mechanical shearing to release the chromosomes into an aqueous buffer designed to stabilize the isolated chromosomes and addition of one or more fluorescent dyes.

A. SOURCE OF CELLS

The first human flow karyotypes were measured for chromosomes isolated from fibroblast cells (11,12). Subsequently, high-resolution flow karyotypes were obtained from short-term PHA-stimulated peripheral blood lymphocytes (49) and lymphoblastoid cell lines (21). Sufficient chromosomes for a flow karyotype can be obtained from as little as 2 ml of human peripheral blood. The use of short-term cultures has the advantage that there is little opportunity for karyotype alterations that can occur in established cell lines. Lymphoblastoid cell lines, on the other hand, are an ideal source, where large numbers of chromosomes are required, such as for sorting experiments. Also there are many lymphoblastoid lines with known aberrant chromosomes that may be sorted to facilitate analysis of the genotype of the aberration(s) or to simplify subchromosome gene mapping. In practice, flow karyotypes can be measured for most types of cells whether grown in suspension or monolayers. The most important factor for good resolution seems to be that the cell culture should be growing optimally and be free of dead cells or bacterial or mycoplasma contamination. Cell culture for chromosome isolation is covered in more detail in Chapter 3.

B. PREPARATIVE TECHNIQUES

Careful preparation of a chromosome suspension is crucial to obtaining high-resolution flow karyotypes. Damage to chromosomes not only alters their DNA content but may cause aggregation and loss of resolution. As a result, chromosome fragments or changes may be generated that cannot be distinguished from normal chromosomes. These "debris particles" produce a slowly varying continuum in the flow karyotype that underlies the peaks produced by intact chromosomes. Hence, much work has been devoted to development of techniques in which the chromosome isolation procedure is optimized to minimize chromosome damage.

Although several different approaches have been developed over the last few years, they all have certain common features. A culture of growing cells is usually treated with an agent such as colcemid or vinblastine in order to arrest cells in metaphase. The period of treatment may depend on the cell cycle time, but extensive culture in the presence of such agents should be avoided since this may lead first to a high degree of chromosome contraction and ultimately to cell death. If the cells grow as an attached layer, mitotic shake-off can be used to obtain a population enriched in metaphase cells. Suspension cell lines may also be used if sufficient metaphase cells are present and if care is taken not to lyse nuclei of the interphase cells which predominate in the population. The cell population is usually subjected to hypotonic swelling, treatment with a detergent, and lysis by mechanical disruption. Typically, disruption is caused by passage through a fine needle or by vigorous vortexing. During the lysis process it is helpful to monitor the disruption of the mitotic cells on a fluorescence or phase-contrast microscope since insufficient shearing will not lyse sufficient cells or disrupt chromosome clumps whereas excessive treatment can cause chromosome breakage. Differential centrifugation may be used to remove interphase nuclei, if present.

Several methods for chromosome isolation are summarized in Table I. The first flow karyotypes (11,12,41) were obtained using the hexylene glycol procedure of Wray and Stubblefield (48). This approach has the advantage that the banding structure of chromosomes is sufficiently preserved to allow direct analysis of chromosomes after sorting. Carrano et al. (5) investigated the influence of several factors on this protocol, such as shearing forces and cell concentration. The addition of sodium dodecyl sulfate, ribonuclease and trypsin were found to have a negative effect. An alternative approach for the bulk isolation of chromosomes using polyamines to stabilize the DNA was developed for flow analysis by Sillar and Young (38) from the method of Blumenthal et al. (3). A direct comparison of the two methods using the same CHO cell culture indicated that the polyamine method yielded lower background levels and superior measurement resolution. In addition, the molecular weight of the DNA in the isolated chromosomes was very high. A disadvantage of the polyamine method is that the isolated

chromosomes are highly condensed so that direct banding analysis of the sorted chromosomes is difficult. However, successful quinacrine banding of sorted chromosomes prepared in this way has been described (10). This technique has been used successfully in our laboratory on a variety of cell lines and peripheral blood samples and has been adopted by a number of other groups. A comparative study (42), which did not include flow analysis, concluded that the polyamine method (3,38) was superior to the hexylene glycol based method for the preparation of DNA for study by micrococcal nuclease digestion. Yu et al. (50) investigated the use of 4′-aminomethyl-4,5′,8-trimethyl-psoralen to stabilize the chromosomes, but their data did not indicate that this method was superior to the hexylene glycol method.

The use of the intercalating fluorochrome propidium iodide to stabilize chromosome structure has been reported by Buys et al. (4). This study demonstrated that chromosomes isolated in this way maintain their banding structure. A similar protocol but with 0.1% sodium citrate and the addition of ethanol to 20% prior to lysis has been reported by Matsson and Rydberg (31). This method is based on a procedure developed for staining human lymphoblasts (20), and was claimed to cause less damage to interphase cell nuclei than other methods. Stoehr et al. (40) have reported a

Table I

METHODS FOR CHROMOSOME ISOLATION

pH	Detergent	Lysis	Stabilizing agent	Fluorochrome	Reference
7.5	—	S	Hexylene glycol	EB	(11,12,39)
7.5	Triton X-100	S	Mg^{2+}	EB + M	(6)
7.2	Digitonin	V	Polyamines	EB	(36)
—	—	V	PI	PI	(29)
—	Triton X-100	S	Psoralen (AMT)	Ho	(47)
—	Triton X-100	S	PI	PI	(4)
1% Acetic acid	—	U	Hexylene glycol	DAPI	(38)
7.5	Triton X-100	S	Mg^{2+}	PI	(2)
8.0	—	S	Mg^{2+}	PI, Ho	(41,42)

S = syringe; V = vortexing; U = ultrasound; EB = ethidium bromide; PI = propidium iodide; M = mithramycin; Ho = Hoechst 33258.

hexylene glycol based protocol which involves mild fixation with 1% acetic acid and sonication to disrupt the mitotic cells. Although acidic treatment (50% acetic acid) was shown (42) to result in extensive damage to the chromosomal DNA, the relatively mild treatment in this protocol did not prevent DNA analysis on sorted chromosomes. A low pH protocol has been devised (7) for the velocity sedimentation of chromosomes followed by flow analysis. Mitotic cells are lysed in the presence of 2% citric acid which gives superior separation during 1g velocity sedimentation. However, since chromosomes sedimented at low pH are not suitable for DNA analysis, Collard et al. (8) have subsequently modified the sucrose gradient to neutral pH.

A comparative study of the influence of several ions and detergents has been reported (2). The optimal chromosome buffer contained magnesium (3–5 mM), sodium (10 mM), Tris (10 mM) and Triton X-100 (0.4%) as judged by coefficient of variation measured by flow analysis. Magnesium ions were found to be necessary for the stability of the chromosomes, whereas sodium and potassium had no effect, and calcium had a negative effect. Van den Engh et al. (43) have described an isolation buffer at pH 8.0 which uses $MgSO_4$ to stabilize the chromosomes. Another feature of this method is the use of RNase to reduce the background fluorescence due to binding of fluorochromes to RNA. Furthermore, it has been reported that the addition of sodium citrate to 10 mM can improve the resolution obtainable by this method (44). Chromosome isolation is treated in more detail in Chapter 4.

C. CHOICE OF DNA-SPECIFIC STAINS

There are several criteria that a DNA-specific stain must fulfill to be useful in flow cytogenetics. Firstly, the wavelength of excitation must match the available light source. Most flow instruments use laser illumination which is restricted to wavelengths

in the range 458–514 nm, and to the ultraviolet at ~360 nm. The relatively small DNA content of most chromosomes requires that the stain has the highest possible quantum efficiency after binding to DNA. This is important in reducing statistical variation in measurement of chromosomal fluorescence. Also, the stain should be insensitive to the degree of chromosomal contraction, which can be quite variable in a chromosomal preparation.

The first flow karyotypes were obtained using ethidium bromide as the DNA-specific stain (11,12,41), and this has remained one of the principal stains used. Jensen et al. (19) investigated the suitability of ethidium bromide, Hoechst 33258, and chromomycin A3 for chromosome staining. Hoechst 33258 and chromomycin A3 resulted in flow karyotypes that were different from that obtained with ethidium bromide. It has been shown that Hoechst 33258 has an AT-binding preference (25), chromomycin A3 has a GC-binding preference (1), and ethidium bromide and propidium iodide have no base sequence preference (30). It has therefore been concluded that the differences in flow karyotypes observed with such stains are due to differences in chromosomal base composition or accessibility. A comparative study of the use of ethidium bromide, Hoechst 33258, chromomycin A3, DAPI and propidium iodide for flow cytometry of metaphase chromosomes has been reported (23). DAPI binds preferentially to AT-rich DNA and the relative intensities of DAPI-stained chromosomes are similar to those of Hoechst-33258-stained chromosomes. Similarly, propidium iodide, which is not thought to have a base sequence preference, yields flow karyotypes that are similar to those obtained using ethidium bromide. It was also shown by analysis of doubly stained chromosomes that bound stains interact by energy transfer with little or no binding competition.

Conventional fluorescence microscopy was used (26) to study interactions between a number of DNA-binding dyes on metaphase chromosomes. It was shown that contrast in donor fluorescence can be enhanced if an energy acceptor dye (e.g. actinomycin D or methyl green) is used that has a binding specificity opposite to that of the donor (e.g. Hoechst 33258, quinacrine *vs* chromomycin A3). Use of such dyes produces patterns selectively highlighting standard or reverse chromosome bands. Recently, the counterstaining of human chromosomes with the non-fluorescent AT-specific DNA-binding agent, netropsin has been used in flow analysis (*35*). The use of netropsin alters the fluorescence signal from the C-band heterochromatin of certain human chromosomes and thus can increase the degree to which some chromosomes can be distinguished. For example, counterstaining with either netropsin or distamycin A facilitated identification of an abnormal chromosome 15 (*22*).

One of the most promising innovations, particularly for the analysis of human chromosomes, has been the introduction of dual-beam flow cytometry. In this approach, chromosomes are typically double-stained with Hoechst 33258 and chromomycin A3, each of which is excited independently. The resultant fluorescence values are presented in bivariate distributions. It has been demonstrated that the Hoechst 33258 and the chromomycin A3 fluorescence can be each measured almost independently (*13*). With this technique, called bivariate flow karyotyping, it has been possible to resolve all the human chromosomes, except chromosomes 9 through 12, and 14 and 15 (*24*). A disadvantage of this approach that has prevented its widespread application is that expensive lasers are required to produce the ultraviolet light needed for Hoechst 33258 illumination.

Cytochemical considerations for flow cytogenetics are considered in more detail in Chapter 5.

D. Flow Instrumentation

The accurate analysis of chromosomes in flow systems required that the systems be operated at the highest possible resolution. Therefore, it is most important to ensure that the flow cytometer is performing optimally. The analysis of fluorescent microspheres with a low coefficient of variation (CV) is a valuable check on the performance of the machine. Misalignment of the laser or problems with the flow system are corrected at this stage so that the CV of the microspheres is minimal. It is also important that the laser is performing optimally and that its output is stable. The final instrument optimization for flow karyotyping is best performed as the chromosomes pass through the machine. A substantial improvement in resolution can often be obtained by reducing the flow rate of chromosomes to 100–200/sec. Although a longer period for data acquisition is required, the chromosomes are more centered in the stream and therefore receive a more uniform illumination. The large number and close spacing of the peaks in a human karyotype is such that 1000 channels of resolution are recommended for data acquisition. A lower number of channels may result in loss of resolution.

The early flow karyotypes were obtained using specially designed instrumentation. More recently, commercially available machines have been used to generate high-resolution flow karyotypes. The use of dual-beam cytometers has clearly improved the resolution of some human chromosomes. Other aspects of instrumentation which have improved resolution include both slit-scanning and optimal collection of the emitted fluorescence. Instrumentation for chromosome analysis is discussed in more detail in Chapters 2, 11 and 15.

III. APPLICATION OF FLOW SYSTEMS TO KARYOTYPE ANALYSIS

A. Chromosome Classification

Flow karyotypes are sensitive both to chromosomal DNA content and to the frequency of occurrence of the various chromosome types. The peak means are remarkably reproducible, usually varying by less than 0.5% from day to day for chromosomes from the same cell line or person. The peak means and areas are usually estimated by least squares fitting the sum of a number of normal distributions to the experimental data as described in Chapter 6. The means and areas of fitted distributions are taken as estimates of the means and frequencies of occurrence of the various chromosome types. In practice, some chromosome types may have the same mean and thus cannot be distinguished (e.g. human chromosomes 9–12 in Fig. 1).

.

B. ABERRATION DETECTION

The measurement of CVs obtained during flow karyotyping are routinely between 1 and 2% (49). With this level of resolution it is possible to detect differences in chromosomal DNA content of 1/2000 of the male genome. This is less than the DNA content of one prometaphase band. As a result, flow karyotyping is potentially useful for the detection of aberrations that cause small alterations in the DNA content of one or more chromosome types. It is therefore appropriate to consider the application of flow analysis to human chromosome classification and to the detection and quantitation of abnormalities. In principle, it should be relatively simple to detect certain aberrations. For example, a trisomy should cause a 50% increase in area of the peak of the trisomic chromosome. A balanced reciprocal translocation between two chromosomes which results in a net change in their DNA contents should result in the appearance of two new peaks whose areas are each equivalent to that of a chromosome present at one copy in each cell. The areas of the peaks due to the two unaffected homologs should be reduced to that of a single chromosome. An aberration such as a small marker chromosome may result in a peak in a position where no peaks normally exist.

In practice, the detection of such abnormalities using flow karyotyping without the support of conventional cytogenetics is complicated by the high level of chromosomal polymorphism in the general population. Certain chromosomes have regions of centric heterochromatin which can vary considerably in DNA content from person to person resulting in microscopically visible differences in size and changes in the flow karyotypes. Variations in flow karyotypes have been correlated with specific C and Q band heteromorphisms (49,24,16). As a result, there can be no standard normal flow karyotype when based solely on DNA content. In order to investigate these variations further, Harris et al. (17) determined the DNA contents of chromosomes from 20 normal humans using high-resolution flow cytometry. In this study, ethidium bromide was used as the DNA-specific fluorochrome. The mean and standard deviation of the relative fluorescence of each chromosome are shown in Table II. The mean relative fluorescence value for each chromosome was in close agreement with the most recent cytophotometric values of DNA content (32). Chromosomes 1, 9, 16 and Y, showed the largest variations, in accord with their large centric heterochromatic regions. The other chromosomes also showed significant variations, with the X chromosome being the least variable. The peak due to chromosomes 10–12 was used for standardization. Therefore, the variability of these chromosomes was not assessed in this study.

Interpretation of flow karyotypes would be greatly simplified if the fluorescence from the polymorphic chromosome regions could be eliminated. Netropsin, which is a non-fluorescent, AT-specific, DNA-binding agent, has been used to reduce the fluorescence from the C-band heterochromatin of certain chromosomes (35). It is possible that this approach may be developed to "normalize" flow karyotypes so that all normal profiles would be identical.

An alternative approach, which is immediately applicable, is to compare the flow karyotype in question with flow karyotypes of the parents of the individual being examined. Chromosome polymorphisms usually appear to

Table II

ETHIDIUM BROMIDE RELATIVE
FLUORESCENCE VALUES
(RELATIVE DNA CONTENT)

Chromosome	Mean	SD
1	1014.7	18.681
2	989.8	7.002
3	829.1	5.421
4	788.8	3.465
5	754.1	3.076
6	709.8	2.803
7	667.0	2.528
X	640.4	2.520
8	604.8	2.522
9	570.2	19.545
10–12[a]	560.0	—
13	449.8	9.219
14	429.3	9.613
15	409.76	12.467
16	382.8	10.731
17	348.8	3.433
18	332.0	6.569
20	274.7	3.371
19	256.6	5.815
Y	227.6	15.259
22	211.6	83.744
21	183.8	7.134

[a] Used as the basis of karyotype normalizations.

be inherited unaltered in size (37). Hence any feature that cannot be seen in either of the parental flow karyotypes can be assumed to be of *de novo* occurrence. This approach has been used by Harris *et al.* (18) to study a series of families with dysmorphic children. Although no chromosome abnormality was detected, the usefulness of this approach was clearly demonstrated.

Another approach to improving the accuracy with which flow karyotypes can be interpreted is to measure different chromosomal features so that the differences between the various chromosomes are large compared to polymorphic variability. DNA content and base composition (see Chapter 5), and DNA content and centromeric index (see Chapter 12), have been used to enhance chromosomal identification.

Additional details concerning univariate and

bivariate flow karyotyping and the applications of these techniques are presented in Chapters 7 and 8, respectively.

C. Flow Analysis of a Human "Microchromosome"

Despite the difficulties of applying flow analysis to the routine detection of abnormalities, there are certain aberrations which can be well studied by this approach. The occurrence of human "microchromosomes" represents such an aberration. These small accessory marker chromosomes are associated with dysmorphic features and have been detected by conventional cytogenetics (39). Since they are usually smaller than the smallest human chromosome, they produce an abnormal peak in a flow karyotype in the region between the origin and the peak for chromosome 21. An example of the flow cytogenetic analysis of cells from such a patient is shown in Fig. 2. This patient had previously been shown by conventional cytogenetics to have a constitutional microchromosome in an otherwise normal karyotype (see Fig. 3). This flow karyotype was obtained by increasing the laser power and electronic gain to expand this region of the flow karyotype and by electronically gating on light scatter to minimize the background due to debris. Analysis suggested that the minichromosome had about 40% of the DNA content of a normal chromosome 21 and thus it could be estimated to be composed of 2×10^7 bp of DNA. The precise origin of this type of aberration is unknown and this minichromosome has been sorted and cloned to allow the determination of its chromosomal identity.

It is uncertain whether this represents the lowest limit of detection in a flow karyotype. Most flow karyotypes have a rising continuum near the origin due to debris, and this could be a limiting factor. The magnitude of this continuum must be sufficiently small that it does not overwhelm the peak of interest. Improvements in preparative techniques may

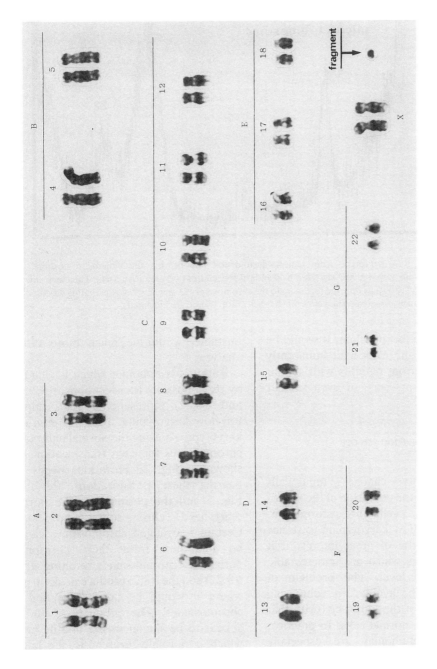

Fig. 3. Conventional Geimsa-banded karyotype of a lymphoblastoid cell line established from a dysmorphic child. The "microchromosome" is indicated and was present in all cells.

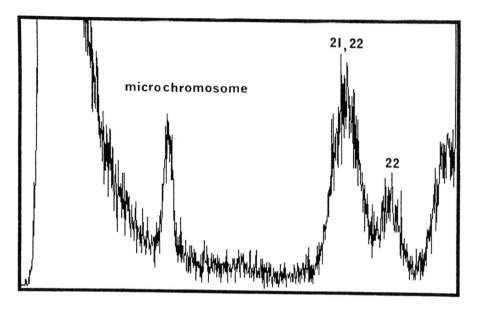

Fig. 2. Partial flow karyotype from lymphoblastoid cell line bearing the "microchromosome". Chromosomes were prepared and analyzed by standard procedures (*49*) on a FACS 440. The electronic gain was increased to display only the region up to the chromosomes 21 and 22. Electronic gating was used to minimize the rising baseline. The peaks due to each chromosome are indicated.

eventually eliminate this problem. It would be of interest to examine cells with apparently normal karyotypes from patients with similar syndromes for the presence of even smaller microchromosomes.

D. FLOW KARYOTYPE ANALYSIS OF TUMOR CELLS

The study of karyotype changes in tumor cells has been essential in the definition of the role of oncogene activation in cellular transformation. The application of flow karyotyping to tumor cytogenetics would be of great interest if it could reveal features which are undetectable at the microscopic level. The problem of polymorphism can be avoided by comparing the flow karyotype of the tumor cell with that of the patient's own normal cells. In practice, leukemias and lymphomas are especially suitable to this form of analysis, since they tend to have relatively simple chromosomal alterations. In contrast, solid tumours, which are often aneuploid and karyotypically

unstable, would be much more difficult to analyze.

Burkitt's lymphoma, which is characterized by three possible translocations, t(8;14), t(2;8) and t(8;22) represents a good opportunity for flow karyotyping. Partial Geimsa-banded karyotypes showing the normal and rearranged chromosomes for each translocation type are shown in Fig. 4. Flow karyotypes of cells bearing these translocations are shown in Fig. 5 and the positions of the normal and rearranged chromosomes are indicated. Certain rearranged chromosomes can readily be identified from their positions. The derivative chromosomes resulting from the t(8;22) and the t(8;2) produce peaks in positions where it would be unusual to find normal chromosomes. The products of the t(8;14) appear to be similar in size and lie in the gap which is normally present to the left of the chromosome 9–12 peak. This analysis demonstrates how the flow karyotype of a tumor cell can be interpreted in the light of the knowledge of its Geimsa-banded

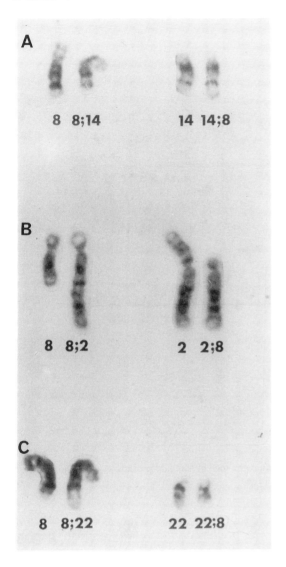

Fig. 4. Geimsa-banded partial karyotype of Burkitt's lymphoma cell lines. Each cell line bears one of the typical translocations: A, t(8;14); B, t(2;8); C, t(8;22). The relevant normal and rearranged chromosomes are illustrated for each cell.

resolution when compared with diploid cells. This is not unexpected since such progressive changes, including the acquisition of marker chromosomes, will gradually distort the flow karyotype. This phenomenon has been used by Otto and Oldiges (36) to monitor the clastogenic (chromosome breaking) effects of certain agents. These experiments require careful controls to demonstrate that the peak broadening observed is due to chromosome damage caused by the clastogenic agent and is not an artifact of the chromosome isolation process. More details about the applications of flow karyotyping to detection of random induced chromosome damage are presented in Chapters 9 and 10.

IV. FLOW SORTING

Flow sorting allows purification of any chromosome group that can be distinguished flow cytometrically. However, the rate at which chromosomal DNA can be accumulated is low. For example, ~24 h are required to purify 1 μg of chromosomal DNA from an average-sized human chromosome if 2000 chromosomes/sec are processed by the system. As a result, chromosomes are usually processed at the highest rate consistent with the desired sorting purity (sorting purity decreases with increasing processing rate). High-speed sorting (Chapter 15) and velocity sedimentation (Chapter 14) have been developed to increase the rate at which chromosomes can be sorted. In spite of these limitations, sorting allows collection of usable quantities of chromosomes at higher purity than by any other method.

V. APPLICATIONS OF CHROMOSOME SORTING

Chromosome sorting has facilitated chromosomal localization of cloned DNA probes. In this application ~20 000 chromosomes of each type are sorted onto nitrocellulose filters and hybridized to radioactively labeled probe DNA. If the probe

karyotype. A similar analysis has been performed on lymphoma cells by Wirchubsky (46) and been extended to murine plasmacytomas (47).

The flow karyotypes of cell lines that have accumulated many changes in culture show a broadening of their peaks and a general loss of

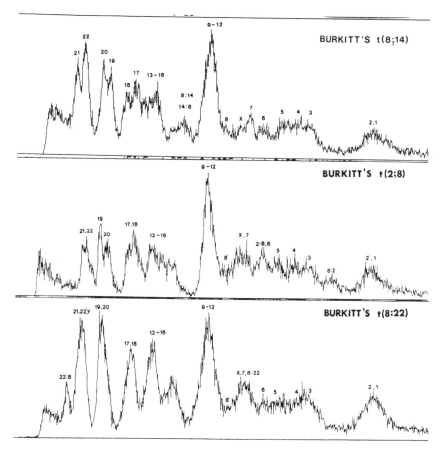

Fig. 5. Flow karyotypes of the Burkitt's lymphoma cell lines obtained by standard procedures (*46*). The positions of normal and rearranged chromosomes are indicated.

is uniquely present in haploid DNA, it will hybridize only to the chromosomal DNA in which the sequence is located (i.e. only to one chromosome type). The use of this procedure for gene mapping is described in more detail in Chapter 13.

Sorted chromosomes of a single type have also been used for the production of chromosome-specific recombinant DNA libraries. This technique was first applied for production of complete digest phage libraries for human chromosomes 21 and 22 (*21*). Since that time, libraries have been constructed for a broad range of normal and abnormal chromosome types from several species. In addition, two complete phage libraries have

been produced for each normal human chromosome type (*45*) and are available through The American Type Culture Collection (Camden, NJ). The construction and characteristics of these libraries are discussed in more detail in Chapters 15 and 16.

VI. FUTURE PROSPECTS

It is clear that with the technology currently available, flow karyotyping is best used in close conjunction with conventional cytogenetic analysis for chromosome analysis. Indeed it may have a role as a backup to conventional analysis for cases in which there is doubt about a specific abnormality. The occurrence of small marker chromosomes, or microchromosomes, is an area in which the resolving power of flow analysis could be immediately exploited.

Although flow cytometry can now yield high-resolution flow karyotypes, the problem of polymorphisms remains an obstacle to its routine application to automatic karyotype analysis. Although family studies can circumvent the problem of polymorphism to some extent, they will not always be possible or appropriate. Even if it were possible to limit the variability in normal profiles by quenching the fluorescence from the centric heterochromatin, difficulties of interpretation would remain. For example, a balanced reciprocal translocation in which there was no net change in DNA content would remain undetectable.

The development of flow karyotyping into a generally applicable form of automatic karyotype analysis may be hastened by new techniques. The slit-scanning technique has the potential to identify individual chromosomes in flow by their centromeric indices (*14,15*; Chapter 11). Also, the *in situ* hybridization by biotinylated DNA probes to chromosomes (*27*; Chapter 17) and detection by fluorescence shows considerable promise for flow analysis. It is possible that DNA libraries of flow-sorted chromosomes could provide probes to form the basis of chromosome-specific "stains" (Chapter 17). It is thus possible to envisage a flow system which, combining such advances, could attain the goal of complete analysis of the human karyotype.

Increased resolution in flow karyotypes would also improve chromosome sorting. The speed of sorting is currently limited by the need to maintain very high resolution during sorting to keep the purity high. Staining procedures that make selected chromosomes easier to distinguish will allow sorters to be run faster while maintaining high purity.

ACKNOWLEDGEMENTS

I am grateful to Dr P. Harris and Professor M. Ferguson-Smith for valuable collaborations and discussions, to Miss B. Gibbons and Mrs D. Rowe for cytogenetic analysis and to Miss L. Davis for FACS analysis. The microchromosome analysis is part of a collaboration with Dr V. Sgaramella and Dr L. Feretti of Pavia University, Italy. This work has been supported by the Cancer Research Campaign, the Imperial Cancer Research Fund, the National Fund for Research into Crippling Diseases and the Scottish Home and Health Department.

REFERENCES

1. Behr, W., Honikel, K., and Hartmann, G. (1969). Interaction of the RNA polymerase inhibitor chromomycin with DNA. *Eur. J. Biochem.* **9**, 82–92.
2. Bijman, Th. J. (1983). Optimization of mammalian chromosome suspension preparation employed in a flow cytometric analysis. *Cytometry* **3**, 354–358.
3. Blumenthal, A. B., Dieden, J. D., Kapp, L. N., and Sedat, J. W. (1979). Rapid isolation of metaphase chromosomes containing high molecular weight DNA. *J. Cell Biol.* **81**, 255–259.
4. Buys, C. H. C. M., Koerts, T., and Aten, J. A. (1982). Well-identifiable human chromosomes isolated from mitotic fibroblasts by a new method. *Hum. Genet.* **61**, 157–159.
5. Carrano, A. V., Van Dilla, M. A., and Gray, J. W. (1979). Flow cytogenetics: a new approach to chromosome analysis. *In* "Flow Cytometry and Sorting" (M. Melamed, P. F. Mullaney and M. L. Mendelsohn, eds). Wiley, New York.
6. Carrano, A. V., Gray, J. W., Langlois, R. G., Burkhart-Schultz, K. J., and Van Dilla, M. A. (1979). Measurement and purification of human chromosomes by flow cytometry and sorting. *Proc. Natl Acad. Sci. USA* **76**, 1382–1384.
7. Collard, J. G., Tulp, A., Stegeman, J., Boezeman, J., Bauer, F. W., Jongkind, J. F., and Verkerk, A. (1980). Separation of large quantities of Chinese hamster chromosomes by velocity sedimentation at unit gravity followed by flow sorting (FACS II). *Exp. Cell Res.* **130**, 217–227.
8. Collard, J. G., Tulp, A., Stegeman, J., and Boezeman, J. (1981). Separation of human metaphase chromosomes at neutral pH by velocity sedimentation at 20 times gravity. *Exp. Cell Res.* **133**, 341–346.
9. Cram, L. S., Bartholdi, M. F., Ray, F. A., Travis, G. L., and Kraemer, P. M. (1983). Spontaneous neoplastic evolution of Chinese hamster cells in culture: multistep progression of karyotype. *Cancer Research* **43**, 4828–4837.
10. Fantes, J. A., Green, D. K., and Cooke, H. J. (1983).

Purifying human Y chromosomes by flow cytometry and sorting. *Cytometry* **4**, 88–91.

11. Gray, J. W., Carrano, A. V., Steinmetz, L. L., Van Dilla, M. A., Moore II, D. H., Mayall, B. H., and Mendelsohn, M. L. (1975). Chromosome measurement and sorting by flow systems. *Proc. Natl Acad. Sci. USA* **72**, 1231–1234.

12. Gray, J. W., Carrano, A. V., Moore II, D. H., Steinmetz, L. L., Minkler, J., Mayall, B. H., Mendelsohn, M. L., and Van Dilla, M. A. (1975). High-speed quantitative karyotyping by flow microfluorometry. *Clin. Chem.* **21**, 1258–1262.

13. Gray, J. W., Langlois, R. G., Carrano, A. V., Burkhart-Schultz, K., and Van Dilla, M. A. (1979). High resolution chromosome analysis: one and two parameter flow cytometry. *Chromosoma (Berl.)* **73**, 9–27.

14. Gray, J. W., Peters, D., Merrill, J. T., Martin, R., and Van Dilla, M. A. (1979). Slit-scan flow cytometry of mammalian chromosomes. *J. Histochem. Cytochem.* **27**, 441–444.

15. Gray, J. W., Lucas, J., Pinkel, D., Peters, D., Ashworth, L., and Van Dilla, M. A. (1980). Slit-scan flow cytometry: analysis of Chinese hamster M3-1 chromosomes. *In* "Flow Cytometry IV," pp. 249–255. Universitetsforlaget, Norway.

16. Green, D. K., Fantes, J. A., and Spowart, G. (1984) Radiation dosimetry using the methods of flow cytogenetics. *In* "Biological Dosimetry" (W. G. Eisert and M. L. Mendelsohn, eds). Springer-Verlag, Berlin.

17. Harris, P., Boyd, E., Young, B. D., and Ferguson-Smith, M. A. (1986). Determination of the DNA content of human chromosomes by flow cytometry. *Cytogenet. Cell Genet.* **41**, 14–21.

18. Harris, P., Cook, A., Boyd, E., Young, B. D., and Ferguson-Smith, M. A. (1987). The potential of family flow karyotyping for the detection of chromosome abnormalities. *Hum. Genet.* **76**, 129–133.

19. Jensen, R. H., Langlois, R. G., and Mayall, B. H. (1977). Strategies for choosing a deoxyribonucleic acid stain for flow cytometry of metaphase chromosomes. *J. Histochem. Cytochem.* **25**, 954–964.

20. Krishan, A. (1977). Flow cytometry: long term storage of propidium iodide/citrate-stained material. *Stain Technol.* **52**, 339–343.

21. Krumlauf, R., Jeanpierre, M., and Young, B. D. (1982). Construction and characterization of genomic libraries of specific human chromosomes. *Proc. Natl Acad. Sci. USA* **79**, 2971–2975.

22. Laland, M., Schreck, R. R., Hoffman, R., and Latt, S. A. (1985). Identification of inverted duplicated 15 chromosomes using bivariate flow cytometric analysis. *Cytometry* **6**, 1–6.

23. Langlois, R. G., Carrano, A. V., Gray, J. W. and Van Dilla, M. A. (1980). Cytochemical studies of metaphase chromosomes by flow cytometry. *Chromosoma* **77**, 229–251.

24. Langlois, R. G., Yu, L. C., Gray, J. W., and Carrano, A. V. (1982). Quantitative karyotyping of human chromosomes by dual beam flow cytometry. *Proc. Natl Acad. Sci. USA* **79**, 7876–7880.

25. Latt, S. A., and Wohlleb, J. (1975). Optical studies of the interaction of 33258 Hoechst with DNA, chromatin and metaphase chromosomes. *Chromosoma* **52**, 297–316.

26. Latt, S. A., Sahar, E., Eisenhard, M. E., and Juergens, L. A. (1980). Interactions between pairs of DNA binding dyes: results and implications for chromosome analysis. *Cytometry* **1**, 2–12.

27. Lau, Y. F. (1985). Detection of Y-specific repeat sequences in normal and variant human chromosomes using *in situ* hybridization with biotinylated probes. *Cytogenet. Cell Genet.* **39**, 184–187.

28. Lebo, R., Carrano, A., Burkhart-Schultz, K., Dozy, A., Yu, L.-C., and Kan, Y. W. (1979). Assignment of human β-, γ- and δ-globin genes to the short arm of chromosome 11 by chromosome sorting and DNA restriction analysis. *Proc. Natl Acad. Sci. USA* **76**, 5804–5808.

29. Lebo, R. (1982). Chromosome sorting and DNA sequence localization. *Cytometry* **5**, 145–154.

30. Le Pecq, J. B., and Paoletti, C. (1967). A fluorescent complex between ethidium bromide and nucleic acids. Physical-chemical characterization. *J. Mol. Biol.* **27**, 87–94.

31. Matsson, P., and Rydberg, B. (1981). Analysis of chromosomes from human peripheral lymphocytes by flow cytometry. *Cytometry* **1**, 369–372.

32. Mayall, B. H., Carrano, A. V., Moore II, D. H., Ashworth, L., Bennett, D., and Mendelsohn, M. (1984). The DNA based human karyotype. *Cytometry* **9**, 376–385.

33. Mendelsohn, M. L., and Mayall, B. H. (1974). Chromosome identification by image analysis and quantitative cytochemistry. *In* "Human Chromosome Methodology," (J. Yunis, ed.), pp. 311–346. Academic Press, New York.

34. Mendelsohn, M. L., Mayall, B. H., Bogart, E., Moore, D. H., and Perry, B. H. (1973). DNA content and DNA-based centromeric index of the 24 human chromosomes. *Science* **179**, 1126–1129.

35. Meyne, J., Bartholid, M., Travis, G., Cram, L. S. (1984). Counterstaining human chromosomes for flow karyology. *Cytometry* **5**, 580–583.

36. Otto, F. J., Oldiges, H. (1980). Flow cytogenetic studies in chromosomes and whole cells for the detection of clastogenic effects. *Cytometry* **1**, 13–17.

37. Robinson, J. A., Buckton, K. E., Spowart, G., Newton, M., Jacobs, P. A., Evans, H. J., and Hill, R. (1986). The segregation of human chromosome polymorphisms. *Am. Hum. Genet.* **40**, 113–121.

38. Sillar, R., Young, B. D. (1981). A new method for the preparation of metaphase chromosomes for flow analysis. *J. Histochem. Cytochem.* **29**, 74–78.

39. Steinbach, P., Djalali, M., Hansmann, I., Kattner, E., Meisel-Stosiek, M., Probeck, H. D., Schmidt, A., and Wolf, M. (1983). The genetic significance of accessory bisatellited marker chromosomes. *Hum. Genet.* **65**, 155–164.

40. Stoehr, M., Hutter, K. J., Frank, M., Goerttler, K. (1982). A reliable preparation of mono dispersed chromosome suspensions for flow cytometry. *Histochemistry* **74**, 57–61.

41. Stubblefield, E., Cram, S., and Deaven, L. (1975). Flow microfluorometric analysis of isolated Chinese hamster chromosomes. *Exp. Cell Res.* **94**, 464–468.

42. Tien Kuo, M. (1982). Comparison of chromosomal structures isolated under different conditions. *Exp. Cell Res.* **138**, 221–229.

43. van den Engh, G., Trask, B., Cram, S., and Bartholdi, M. (1984). Preparation of chromosome suspensions for flow cytometry. *Cytometry* **5**, 108–123.

44. van den Engh, G. J., Trask, B., Gray, J. W., Langlois, R. G., and Yu, L.-C. (1985). Preparation and bivariate analysis of suspensions of human chromosomes. *Cytometry* **6**, 92–100.

45. Van Dilla, M. A., Deaven, L. L., Albright, K. L., Allen, N. A., Aubuchon, M. R., Bartholdi, M. F., Browne, N. C., Campbell, E. W., Carrano, A. V., Clark, L. M., Cram, L. S., Fuscoe, J. C., Gray, J. W., Hildebrand, C. E., Jackson, P. J., Jett, J. H., Longmire, J. L., Lozes, C. R., Luedemann, M. L., Martin, J. C., McNinch, J. S., Meincke, L. J., Mendelsohn, M. L., Meyne, J., Moyzis, R. K., Munk, A. C., Perlman, J., Peters, D. C., Silva, A. J., and Trask, B. J. (1986). Human chromosome-specific DNA libraries: construction and availability. *Biotechnology* **4**, 537–552.

46. Wirchubsky, Z., Perlmann, C., Lindsten, J., and Klein, G. (1983). Flow karyotype analysis and fluorescence activated sorting of Burkitt-lymphoma-associated translocation chromosomes. *Int. J. Cancer* **32**, 147–153.

47. Wirchubsky, Z., Weiner, F., Spira, J., Sumegi, J., Perlmann, C., and Klein, G. (1984). The role of specific chromosomal translocations and trisomies in the genesis of certain plasmacytomas and lymphomas in mice, rats and humans. *Cancer Cells/Oncogenes Viral Genes* **2**, 253–260.

48. Wray, W., Stubblefield, E. (1970). A new method for the rapid isolation of chromosomes, mitotic apparatus or nuclei from mammalian fibroblasts at near neutral pH. *Exp. Cell. Res.* **59**, 469–478.

49. Young, B. D., Ferguson-Smith, M. A., Sillar, R., Boyd, E. (1981). High resolution analysis of human peripheral lymphocyte chromosomes by flow cytometry. *Proc. Natl Acad. Sci. USA* **78**, 7727–7731.

50. Yu, L.-C., Aten, J., Gray, J., and Carrano, A. V. (1981). Human chromosome isolation from short term lymphocyte culture for flow cytometry. *Nature* **293**, 154–155.

Instrumentation for Chromosome Analysis and Sorting

JOE W. GRAY and G. J. VAN DEN ENGH

Biomedical Sciences Division
Lawrence Livermore National Laboratory
Livermore, CA 94550, USA

I. INTRODUCTION

Flow cytometry applied to isolated chromosomes stained with one or more fluorescent DNA-specific dyes allows measurement of chromosomal dye (or DNA) content with precision approaching 1% (5,9,18). As a result, chromosomes differing by as little as 1 fg in DNA content can be distinguished using flow cytometry. Furthermore, hundreds or thousands of chromosomes can be measured each second so that a statistically precise measurement of the distribution of DNA (or dye) content among the chromosomes of a population can be made in only a few minutes. This capability allows accurate chromosome classification, detection of disease-linked numerical and structural aberrations and analysis of chromosomal changes that may be important in biological dosimetry (6; Chapters 8–11). The principles of flow cytometric chromosome classification, hereafter called flow karyotyping, are illustrated in Fig. 1. Isolated mitotic chromosomes stained with one or more fluorescent dyes flow one by one through one or two laser beams that are adjusted to efficiently excite the chromosomal dye(s). Light resulting from the passage of chromosomes through each laser beam is collected by a lens that projects the light through a spectral filter and onto a photomultiplier. The spectral filter is selected to reject scattered laser light and pass chromosomal fluorescence. Each photomultiplier emits electrical pulses that are proportional in amplitude to the intensities of the light flashes reaching it. These electrical pulses are then amplified, digitized and stored as a histogram in the memory of a pulse height analyzer. The histograms accumulated in this fashion typically are composed of a number of peaks, each produced by one or a few chromosome types in the population that

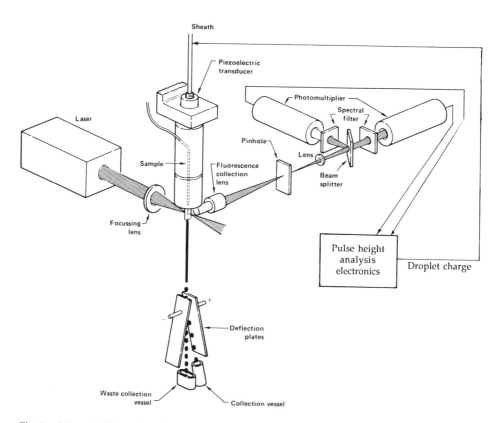

Fig. 1. Schematic illustration of a dual-beam flow sorter. Chromosomes flow one at a time through two laser beams. One beam is adjusted to emit 351–364 nm light to excite HO. The other is tuned to 458 nm to excite CA3. Chromosomes are classified and sorted according to their HO and CA3 fluorescence intensities. The chromosomes exit the measurement chamber in a thin (~75 μm) liquid jet. The jet is forced to break into regular droplets by vibrating piezoelectric crystal attached to the flow chamber. The droplets containing the chromosomes of interest are charged as they separate from the liquid jet. The charged droplets are separated from the others as they pass through a high-voltage electric field.

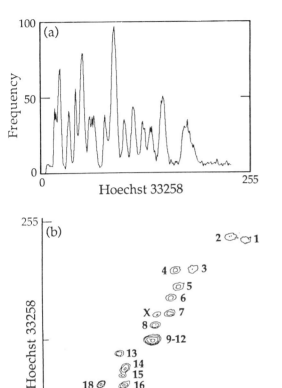

Fig. 2. Flow karyotypes measured for human chromosomes. (a) A univariate flow karyotype measured for chromosomes stained with the dye HO. (b) A bivariate flow karyotype measured for chromosomes stained with HO and CA3.

somes responsible for each peak are indicated in the figure.

Chromosomes of one type can be purified to an unprecedented degree using flow sorting. For example, many human chromosome types can be sorted with high purity [e.g. 70–90% (*3,4,19*; Chapters 13–16)]. Chromosomes are selected for purification according to their location in the flow karyotype (e.g. in one peak). The chromosomes flow from the measurement region in a liquid jet in air as illustrated in Fig. 1 until they reach the end of the jet. When a chromosome that has been selected for sorting reaches the point where the jet breaks into droplets, an electric charge is induced on the jet for a few microseconds. As a result, that chromosome becomes enclosed in a charged liquid droplet. The charged droplets are separated from the uncharged droplets as they pass through a high-voltage electric field. Two classes of chromosomes can be separated simultaneously by putting a positive charge on the droplets carrying one class and a negative charge on the droplets carrying the other class.

This chapter is devoted to the specialized instrumental techniques and considerations that are required to achieve the high measurement resolution that is required for chromosome classification and sorting. Techniques used to achieve the high chromosome sorting rates and purities that are needed to accumulate chromosomes for molecular biological studies are also discussed.

II. MEASUREMENT PRECISION

A central requirement for chromosome flow cytometry is that the precision with which chromosome fluorescence intensity is measured be as high as possible to allow maximal discrimination between chromosomes of similar DNA content and DNA base composition. The variation in dye content among chromosomes of the same type is less than 0.01 in good chromosome suspensions (*18*), and the measurement precision should be of the same order to take full advantage of the precision

contain the same amount of fluorescent dye. Chromosome classification and aberration detection is accomplished by analysis of the location and area of these peaks. Thus, these histograms are called flow karyotypes. Figure 2(a) shows a flow karyotype measured for human chromosomes stained with the DNA-specific fluorescent dye Hoechst 33258 (HO). Figure 2(b) shows a bivariate flow karyotype for chromosomes stained with Hoechst 33258 (HO) and chromomycin A3 (CA3). The chromo-

of chromosome staining. In practice, measurement coefficients of variation (CVs) around 0.02 are usually required for human flow karyotyping. However, CVs of 0.03–0.06 may be tolerated for sorting by judicious choice of cell lines in which the chromosome to be sorted is well separated in the flow karyotype from other chromosomes (*3,4,19*).

One important determinant of measurement precision is the variation in the number of photoelectrons measured for chromosomes of one type as they pass through a flow cytometer. The number of photoelectrons depends on a number of factors, as indicated by the equation

$$P = k \{N\}$$
$$\times \left\{ \varepsilon_\lambda V(\varrho)^{-1} \int I(\chi,\varrho)\, d\chi \right\}$$
$$\times \left\{ \phi\alpha(\varrho) \int q_{PM}(\lambda)\, F(\lambda)\, T(\lambda)\, d\lambda \right\} \quad (1)$$

$= k$ {Number of excitable dye molecules} × {absorption of laser light} × {conversion to photoelectrons},

where P is the number of photoelectrons detected, k is a constant, ε_λ is the molar extinction of the fluorescent dye at the wavelength of the laser beam, N is the number of excitable dye molecules per chromosome [decreases with increasing beam intensity and time in the excitation beam due to bleaching and saturation (*1,11*)], $I(\chi,\varrho)$ is the laser beam intensity along the chromosome flow axis χ at a distance ϱ from the center of the sample stream, $V(\varrho)$ is the velocity of the chromosomes through the laser beam at ϱ, ϕ is the fluorescence quantum efficiency of the stain, $\alpha(\varrho)$ is the fraction of fluorescence light collected by the optical system for a chromosome at ϱ, $q_{PM}(\lambda)$ is the quantum efficiency of the photomultiplier as a function of fluorescence wavelength λ, $F(\lambda)$ is the relative intensity of the fluorescence and $T(\lambda)$ is the transmission efficiency of the spectral filters and the other optical components. We discuss in the following sections aspects of flow cytometry that are important in achieving high measurement

precision. Specific examples are presented indicating the steps required to keep the variation in each component of Eq. (1) below 0.02. However, variations well below this level would be required if measurement CVs of 0.01 are to be achieved.

A. Measurement Uniformity

High measurement resolution requires that all chromosomes are exposed to nearly the same illumination intensity, $I(\chi,\varrho)$, as they pass through the flow chamber and that the efficiency, $\alpha(\varrho)$, with which fluorescence is collected is approximately the same for all chromosomes. That is, $I(\chi,\varrho)$ and $\alpha(\varrho)$ must be nearly independent of ϱ across the entire sample stream. In addition, the number of photons, P, collected for each chromosome must be sufficiently high that the total chromosome fluorescence intensity can be estimated with high precision.

The maximum diameter of the sample stream, M_d, for which the illumination intensity varies by less than 2% across the stream width varies linearly with laser beam width:

$$M_d = 0.5\,(-2 \ln 0.98)^{1/2} FW1/e^2. \quad (2)$$

The maximum sample stream diameter for which this is true is only about 2 μm at a laser beam diameter (full width at $1/e^2$ intensity; FW1/e^2) of 20 μm. However, a sample stream diameter of 10 μm is allowable if the laser beam diameter is 100 μm. In general, the beam width should be kept as small as possible since the chromosome illumination intensity (very important for high-resolution chromosome analysis) decreases linearly with laser beam width.

The constraints on chromosome illumination uniformity discussed above can be relaxed considerably if the chromosome illumination is sufficiently high that substantial dye saturation (*1*) and/or photobleaching (*11*) occurs. If either of these phenomena occur, the number of photons emitted by each chromosome begins to become independent of the illumination intensity and is determined mostly by the number of dye molecules in the chromosome. Figure 3 shows that the fluorescence intensity measured for human chromosomes stained with Hoechst 33258 is *not* linearly related to the laser beam intensity at high laser power densities. In the Lawrence Livermore National Laboratory (LLNL) dual-beam cytometer, for example, the fluorescence emitted per chromosome becomes increasingly independent of laser power when the laser power is increased beyond 0.5 W. This is advantageous for high-resolution chromosome analysis since slight variations in chromosome illumination (due to laser instability or position in sample stream) do not cause variations of the same magnitude in emitted fluorescence. Figure 3 shows that at 1 W, a 7% variation in illumination intensity causes only a 2% change in fluorescence intensity. This phenomenon allows use of larger sample streams (advantageous for sorting where high throughput is needed).

The collection efficiency of fluorescence collection lenses depends on where in the sample stream the chromosomes flow. However, the collection efficiency of commonly used lenses varies much more slowly with stream diameter than the intensity of the illumination. Figure 4(a), for example, shows the fluorescence collection efficiency *vs* distance from the point of focus of the collection

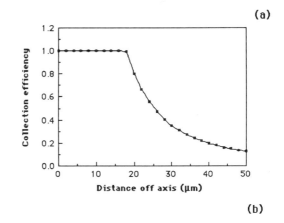

(a)

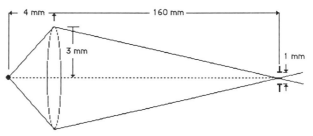

(b)

Fig. 4. Fluorescence collection efficiency considerations. (a) This curve shows the theoretical fluorescence collection efficiency at the edge of the sample stream nearest the fluorescence collection lens and the sample stream diameter. (b) A schematic representation of the fluorescence collection optics assumed for the calculation in part (a). The actual calculations assume a 0.6 NA lens located 4.209 mm from the sample stream, a focal length of 4.1 mm, and a 1 mm pinhole located 160 mm from the lens.

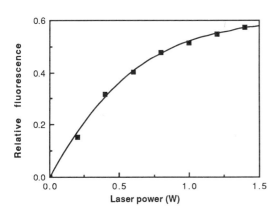

Fig. 3. Fluorescence intensities measured for human chromosomes stained with HO and excited at laser beam intensifies from 0.2 to 1.4 W.

lens for the optical system shown in Fig. 4(b) [similar to that employed in the LLNL dual-beam flow cytometer (2)]. In this calculation, the collection efficiency was calculated for chromosomes flowing at the edge of the sample stream as near as possible to the collection lens. In this system, sample stream diameters greater than 15 μm can be tolerated with little change in collection efficiency.

B. PHOTON STATISTICS

High-resolution flow karyotyping requires that the statistical fluctuations in the number of photoelectrons, P, is small compared to the desired measurement precision. Poisson statistics predict that 2500 photoelectrons must be generated if 2% measurement precision is required. The number of photoelectrons generated depends on the intensity of chromosome illumination, on the dye quantum efficiency, on the absolute efficiency with which fluorescence is collected and transferred to the face of the photomultiplier and on the efficiency with which this light is converted to photoelectrons as described in Eq. (1). In the case of chromosomes stained with ethidium bromide (EB), it is reasonable to assume that the smaller human chromosomes bind a few million dye molecules. The absorption coefficient of EB is $\sim 5 \times 10^3/M/cm$ (10). Assuming this absorption coefficient and dye density, a 1 W laser beam focused to a spot $100 \times 10 \mu m$, and a $1 \mu sec$ beam residence time and no photobleaching or dye saturation; Eq. (1) predicts that approximately 10^7 photons would be emitted by each small chromosome. In fact, photobleaching and/or saturation *do* occur so that only $\sim 2 \times 10^6$ photons may be emitted. The efficiency of fluorescence collection is determined to a large extent by the lens system used for fluorescence collection. The lens used on the LLNL dual-beam cytometer has a numerical aperture of 0.6 and collects only about 10% of the total fluorescence emitted by each chromosome. The efficiency with which fluorescence is transmitted through the filters and other optical components may be as low as 0.5 so that only 10^5 photons reach the face of the photomultiplier. The quantum efficiency of most photomultipliers is only about 0.2 for 600 nm fluorescence so that, in the worst case, only about 2×10^4 photoelectrons are generated for the smaller chromosomes. This is barely adequate for high-resolution chromosome analysis. As a result, the flow cytometer whose specification was used in this exercise requires a high power laser (e.g. several hundred milliwatts). High-speed sorters require even higher-power lasers to compensate for the fact that the residence time of the chromosomes in the laser beam is about five times shorter than in conventional systems so that correspondingly

fewer photons are produced for each chromosome. Of course, these requirements can be eased by using dyes with higher absorption coefficients and/or more binding sites per chromosome and by selection of spectral filters and optical components that pass fluorescence at high efficiency while rejecting scattered laser light.

III. FLUIDICS

The diameter of the region in which chromosomes are constrained to flow is critical in chromosome analysis and sorting. Small sample streams are desirable because they increase the precision of chromosome analysis by minimizing variations in chromosome illumination and fluorescence collection efficiency as described above. In addition, small sample streams minimize variations in the velocity of chromosome flow. The flow velocity, $V(\varrho)$, depends somewhat on the distance, ϱ, from the center of the sample stream. In systems where the flow is laminar (e.g. in systems where illumination occurs in long quartz capillaries), the flow velocity can be expressed as

$$V(\varrho) = 2V_a(1 - \varrho^2/R^2), \qquad (3)$$

where V_a is the average flow velocity and R is the diameter of the flow chamber (13). This equation indicates that differences in velocity between chromosomes that flow along the edge or down the center of the sample stream are negligible for systems in which the diameter of the flow channel is $\sim 250 \mu m$; even for sample streams $20 \mu m$ in diameter.

Unfortunately, small sample streams also reduce the rate at which chromosomes can be processed. The highest chromosome concentration that can be achieved reliably using current chromosome isolation procedures (4; Chapter 3) is about 10^8/ml and concentrations around 10^7 are common. Figure 5 shows the theoretical measurement rates *vs* sample stream diameter, assuming a chromosome concentration of 10^8 for both

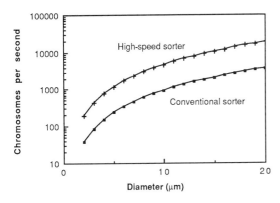

Fig. 5. Relationship between sample stream diameter and chromosome throughput for a conventional (1.2 m/sec flow velocity) and a high-speed sorter (6 m/sec flow velocity). These calculations assume a chromosome concentration of 10^8/ml.

conventional and high-speed sorters. These measurement rate calculations assume that measurements are made in a 250 μm square cross-section quartz flow channel in both systems and the jet diameter is $\sim$70 μm so that the flow velocities in the quartz channels in conventional and high-speed sorters are 1.2 and 6 m/sec, respectively. This graph suggests that analysis rates of only a few hundred chromosomes per second are possible in a conventional sorter at a sample stream diameter of 3 μm (a sample stream diameter that can be reliably achieved in a conventional sorter). A sample stream diameter of 10 μm yields a maximum analysis rate of $\sim$1000/sec in conventional systems. An analysis rate of $\sim$16 000/sec (typical for high-speed sorting) can be achieved at a sample stream diameter of 6 μm in the high-speed sorter if the chromosome concentration is 10^8. However, a sample stream diameter of almost 20 μm is required in the high-speed sorter if the chromosome concentration is only 10^7. Thus, two modes of chromosome analysis are suggested: (1) low-speed analysis (a few hundred chromosomes per second) at the smallest possible stream diameter for high-resolution flow karyotyping and (2) high-speed sorting at the largest sample stream (e.g. 5–10 μm) that allows sufficient resolution for sorting at the desired purity.

IV. ELECTRONICS

Chromosomal fluorescence striking the photocathode of a photomultiplier tube (PMT) is converted into an electric current. This current is amplified, digitized by an analog-to-digital converter (ADC) and added to an accumulating flow karyotype. There are several ways to process the signals prior to digitization. The optimal method depends on several aspects of the chromosome analysis process including the analysis rate, the intensity of the background fluorescence, the size of the chromosomes relative to the laser beam diameter and the speed with which they move through the measurement area. The details of pulse processing are discussed using the LLNL dual-beam cytometer (2) and the high-speed sorter (6,12) as examples. More general discussions of the electronics for flow cytometry are presented elsewhere (7,8,14,15,16).

A. DETECTORS

The photons that strike the photocathode of the PMT liberate electrons, the number of which depends on the photocathode material and the wavelength of the photon. Hoechst 33258, DAPI, mithramycin and chromomycin A3 all emit light in the region between 400 and 600 nm. Photomultipliers with bialkali or S11 cathodes have a good sensitivity in this range [q_{PM}(200–600 nm) $\simeq$ 0.1–0.2]. However, their sensitivity decreases rapidly above 550 nm. PMTs with S20 photocathodes [q_{PM}(600–700 nm) $\simeq$ 0.05] should be used for dyes such as ethidium bromide and propidium iodide that fluoresce at longer wavelengths.

The photoelectrons accelerate from the photocathode through a series of dynodes at increasingly higher electric potential. Several electrons are emitted for each incident electron at each dynode, the number of which is proportional to the accelerating potential. Thus, the electron amplification increases exponentially with the PMT voltage and may be as high as 10^6 (e.g. if each of 10 dynodes produces a four-fold amplification). Assuming an amplification of 10^6 and a quantum efficiency of 0.1, 10^5 photons striking the photocathode for $1\,\mu sec$ will induce an anode current of 10^{10} electrons or 1.6 mA.

PMTs should be used to amplify the signals as much as possible since they add little noise to the signal. However, at high anode currents, the PMT response will deviate from linearity. Depending on the PMT design, the maximum permissible anode current may vary between 0.2 and 10 mA. This current is determined both by the fluorescence from the chromosomes and from the free dye molecules in the sample stream. An increase in the chromosome sample stream diameter or in the free dye concentration may drive the PMT into a nonlinear mode of operation. Since the pulse-processing electronics remove the DC component from the signal, changes in the free dye signal may easily be overlooked. Thus, monitoring the total cathode current is recommended.

B. PREAMPLIFICATION

The cathode currents of the PMTs are converted to voltage pulses by preamplifiers and boosted so they can be transported over cables to the signal-processing electronics. In most systems, the measurement precision is determined by the preamplifiers. They are composed of integrated circuits that have, at best, a precision and long-term stability of $\sim 1\,mV$. Additional circuits, such as a baseline restoration circuit, may further degrade the performance. Thus, measurement of CVs of 0.01 require $\sim 100\,mV$ signals from the smallest particles. The high voltage on the PMT should be adjusted to achieve signals of this magnitude. The largest and smallest chromosomes in some cell populations differ in fluorescence intensity by a factor of 10. Thus, the full-scale output of the preamplifiers should be at least 1 V. There are two types of current to voltage converters: linear and integrating. Both are illustrated by the preamplifier circuit shown in Fig. 6. It consists of an operational amplifier (op-amp) with a resistor (R) and a capacitor (C) in its feedback loop.

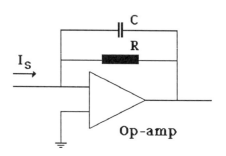

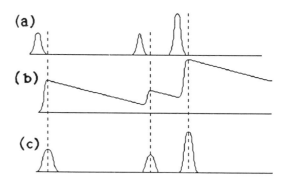

Fig. 6. Diagram of a current-to-voltage converter. The values of the resistor (R) and the capacitor (C) in the feedback loop of the operational amplifier determine the characteristics of the circuit. If the RC product is smaller than the rise time of the current pulse (I_s), the output voltage is inversely proportional to the input current [trace (a); for illustrative purposes all signals are drawn positive]. If the RC product is much larger than the pulse width, the height of the output pulse is proportional to the area of the input pulse. The integrated pulse decays with the time constant of the RC product [trace (b)]. To remove this tail from the signal, the RC integrating preamplifier should be followed with a differentiating amplifier [trace (c)]. The heights of the resulting pulses are proportional to the areas of the input signals. Their widths are typically 2–5 times those of the input pulses. RC integration followed by "pole zero" cancellation does not require gate or trigger signals.

Linear preamplification is accomplished by making C small. In this case, the signal current flowing into the negative op-amp input (I_s) is diverted through R. The op-amp tries to balance the voltages at the two inputs so that the inverting (negative) input is at virtual ground and the output voltage (V_{lin}) is

$$V_{lin} = - I_s R. \qquad (4a)$$

Thus, output voltage is proportional to the input current [trace (a); Fig. 6]. The amplification is determined by R (e.g. $R = 1\,k\Omega$ gives $-1\,V/mA$). Linear pre-amplification is used when chromosome analysis rates are high (high-speed cell sorting) or when the pulse shape needs to be conserved (e.g. for slit scanning). Pulses from linear pre-amplifiers are usually integrated during a later step.

The preamplifier can be configured as an integrator by increasing R and C. A current pulse at the inverting input of the op-amp now accumulates as a charge on C. The op-amp keeps the inverting input at ground so the output voltage of the preamplifier is

$$V_{int} = -1/C \int I_s\,dt + \text{offset}. \qquad (4b)$$

The charge on C dissipates through R [trace (b); Fig. 6] with a decay time that is determined by the RC product. The RC product should be chosen so it is long compared to the widths of the pulses to be processed. Pulse integration by an RC network in the preamplifier feedback loop has several advantages: (1) the network precision and signal-to-noise ratio are high; (2) the rise time of the output is much slower than the unintegrated input signal so that the electronics that follow are less critical than with a linear preamplifier; (3) RC integration does not require external gating signals. Therefore, RC integration can be achieved by replacing an existing preamplifier combination as long as the subsequent amplifiers can shape the pulse properly for presentation to the analog-to-digital convertor (ADC). However, RC integration is not suitable for high pulse rates or for samples with a high background fluorescence. Since the circuit operates as a linear amplifier for signals that change slower than the RC time and since R is very high, a small DC or low-frequency component will saturate the amplifier. For example, an average input current of $0.1\,mA$ into a preamplifier with $R = 100\,k\Omega$ will drive the output to $-10\,V$ and thus saturates the preamplifier. Thus, while RC integration cannot be used in instruments that require a high throughput rate such as the high-speed cell sorter, it has been used for high-resolution flow karyotyping on the LLNL dual-beam cytometer ($1\,\mu sec$ pulse width, $RC = 60\,\mu sec$) where analysis rates are less than $1000/sec$. Flow karyotypes with peak CVs of $1-2\%$ are achieved routinely.

C. AMPLIFICATION

The $0.1-1\,V$ signals from the preamplifiers must be amplified since ADCs typically require pulses in the range of $0-10\,V$. Amplification of the voltage pulses from preamplifiers is becoming increasingly easy with the development of monolytic high-speed op-amps. The AD744 (Analog Devices, $300\,nsec$ settling time to 0.1%, $30\,pA$ bias current, $<0.25\,mV$ noise, drift and offset) is an example of a modern high-performance op-amp that can be used in most applications.

Pulses from the linear preamplifiers used for high-speed sorting must be "actively" integrated prior to conversion by the ADC. Several schemes for such integration are in use. In one, the input signal is integrated while it is above a fixed threshold. High-precision analysis is difficult with amplifiers of this type. The integration time will depend on the pulse width so that variable amounts of any DC component are included in the integral. In another scheme, integration occurs for a fixed period following arrival of a gate pulse. This approach is accurate only if the DC component is constant and properly subtracted during the entire run (difficult for chromosome analyses where the sample stream diameter and thus the DC level may change considerably over time). Thus, the high analysis rates achieved by active integration may lead to lower measurement precision. This is observed on the LLNL high-speed sorter where a chromosome measurement precision better than $3-4\%$ is rarely achieved.

Signals from an RC integrating preamplifier require additional pulse shaping prior to digitization [trace (c); Fig. 6] to remove the decay tail and the offset introduced by the integrating capacitor. This is achieved with a differentiating amplifier. Differentiating amplifiers with good linearity and accurate base line restoration are commercially available. The high-resolution karyotypes from the LLNL dual-beam cytometer, for example, are generated using Ortec 450 research amplifiers. These amplifiers handle fluorescence pulses that are ~1 μsec wide. The time constants are set at a rise time of 0.5 μsec and a decay time of 5 μsec.

D. Sample and Hold Circuits

Signals into the ADC must be held constant during the conversion process. Typically, this takes 5–30 μsec depending on the pulse amplitude. This is accomplished by pulse sample and hold circuits (PSH). These circuits hold the maximum voltage that occurs at its input terminals during a gate period. The gate signal for the PSH may be generated internally for pulses that exceed a fixed threshold or may be supplied externally. PSH circuits are likely to be a weak link in the measurement electronics. To maintain the full measurement precision the PSH must have an accuracy and linearity that is ~10 mV over a full range of 10 V. If offsets or nonlinearities are suspected in chromosome measurements (i.e. if the relative peak means in the flow karyotype vary by more than 1% from run to run on the same sample) the characteristics of the PSH should be checked by testing its performance with calibrated pulses as described below.

E. Analog-to-Digital Conversion

ADCs calculate for each pulse a binary number proportional to the amplitude of the pulse. ADCs can be characterized in terms of their speed, precision and linearity. ADC speed is defined as the longest time needed to perform a conversion. The precision of the ADC conversion is expressed as the number of bits that the ADC uses to code the pulse amplitude. For example, an eight-bit ADC maps pulse amplitudes into 256 channels while a ten-bit ADC maps into 1024 channels. The ADC precision should be selected to preserve the precision of the input electronics. A wide variety of cheap monolytic ten-bit ADCs are available. Most of these are successive approximation ADCs or flash converters. Some care should be taken in using these for high-resolution flow karyotyping since they may have poor channel width characteristics leading to nonlinearities in the flow karyotype. Wilkinson ADCs have superior channel width characteristics. However, these ADCs are relatively slow, require many components and are bulky and expensive. Conversion also can be done with successive approximation ADCs but this requires a careful analysis of the ADCs properties. The ADCs used should have no missing codes and a nondifferential linearity of 1% or better. To achieve this performance successive approximation ADCs with a precision that exceeds the resolution requirements should be used. A successive approximation ADC that has a resolution that is four bits higher than the signal precision may be used. Thus, 12-bit successive approximation ADCs should be used to generate 256-channel flow karyotypes in systems where a measurement precision of 0.01 is required. Detailed discussions of ADCs can be found in the texts by Shapiro (14) and Hiebert and Sweet (8). An excellent overview of AD conversion techniques and ADC characteristics is presented by Sheingold (15).

F. Additional Processing

Flow cytometers may contain additional circuits for signal modification. For example, many flow cytometers contain circuits that compensate for detector cross talk. Such circuits should not be used in chromosome measurements since they change the relative peak means upon which chromosome classification is based. Logarithmic amplification has been found useful in flow karyotyping since it facilitates histogram interpretation (see Chapter 7). The logarithmic signal compression achieved using such amplifiers leads to flow karyotypes in which peaks with the same coefficients of variation have the same width irrespective of where they are located in the flow karyotype. A flow karyotype of human chromosomes covers a range of approximately one decade. Expressed logarithmically, an eight-bit histogram (256 channels) is sufficient to cover all chromosomes with a constant channel separation of 1%. If the fluorescence is expressed linearly, a ten-bit histogram has to be used to obtain the same resolution for the smallest chromosomes. Unfortunately, logarithmic signal amplification is relatively slow, often introduces signal distortion and has poor temperature and long-term stability. The logarithmic representation is also very sensitive to small offset voltages. Therefore, the offset voltages in all signal-processing steps have to be adjusted very accurately.

G. CALIBRATION

Many flow karyotypes are generated using commercial instruments where modification of the design for optimal performance of chromosome analysis may be difficult. In addition, the exact details of pulse processing may be obscure. Thus, regular measurement of the overall instrument performance is important. The precision of the electronics can be assessed by analyzing pulses of known amplitudes and areas and whose shapes are similar to real fluorescence signals. A design of a simple calibration circuit that can be used for linear and logarithmic circuits has been reported by van den Engh *et al.* (*17*). This circuit produces a repeating series of pulses whose amplitudes increase linearly. Analysis of the locations of the histogram peaks produced by these pulses reveals system nonlinearities and zero offsets that may compromise flow karyotype interpretation.

V. DATA ACQUISITION

Parameters that are commonly estimated from flow karyotypes include: peak means,

areas, CVs and the magnitude of the continuum underlying the peaks. The accuracy with which these parameters can be estimated depends heavily on the manner in which the data are accumulated.

A. HISTOGRAM DIMENSION

Closely spaced peaks in flow karyotypes can usually be resolved and accurately analyzed if their means are separated by ~2.5 times their standard deviations (*4*). For example, peaks near channel 100 whose CVs are 0.02 can be resolved as distinct peaks if their means are separated by ~5 channels. This relationship is valid, however, only if the peak means are sufficiently large that each peak comprises several channels. Simulation studies suggest that accurate peak parameter estimation can be accomplished if the channel to CV ratio is 0.8 or larger. Thus, if CVs of 0.02 are to be achieved for all peaks, the gain should be adjusted so that the mean of the lowest peak is in channel 40 or higher. Since the fluorescence signals from the largest human chromosome is ~5.5 times larger than that for the smallest, the peak for the largest chromosome should be near channel 225. This can be achieved in 256-channel univariate flow karyotypes. However, it cannot be achieved in 64×64-channel bivariate flow karyotypes. This suggests the use of 256×256 histograms for high-resolution bivariate flow karyotypes.

B. MEASUREMENT PRECISION

The precision with which the mean and area (or volume) of a peak can be estimated depends on the number of events measured for each peak. An estimate of the number of events required in flow karyotypes can be gained by considering the number of events required for estimation of the mean and area of a single isolated peak. The precision with which the mean of a peak can be estimated depends on the peak CV and the number of events, n, measured for that peak. The fractional error in peak mean estimation, e_μ, can be written as

$$e_\mu = 1.96CV/\sqrt{n}, \qquad (5)$$

assuming a 95% confidence interval. Peak means rarely need to be estimated to better than 0.1%. Equation (5) suggests that this can be accomplished by measuring ~1500 events for each peak (i.e. ~40 000 events per human flow karyotype). Accumulation of this number of events takes only a few minutes even at the low processing rates required for high-resolution flow karyotyping. The fractional error in peak area estimation, e_A, can be expressed as

$$e_A = 1/\sqrt{n}. \qquad (6)$$

Peak areas rarely need to be measured with a precision better than 10%. This can be accomplished by measuring only 100 events per peak and thus is not an issue in flow karyotypes. Of course, peak mean and area estimation becomes much more difficult and more events should be measured in flow karyotypes where several of the peaks overlap with each other or with a debris continuum or where the peaks are not truly normal distributions. Thus, the foregoing numbers should be used as lower limits and complex high-resolution flow karyotypes like those measured for human chromosomes should not be measured with less than ~50 000 total events.

C.   DRIFT CORRECTION

The locations of the peaks in flow karyotypes depend on the concentration of the dye(s) in the buffer in which the chromosomes are suspended, on the gain of the detectors and amplifiers used for data acquisition, on the laser power and on the trajectory of the sample stream through the flow chamber. Unfortunately, some of these factors may change during analysis. The effective dye concentration changes in the beginning of a run as a result of mixing between sample and sheath fluid in the tube used to transport sample to the flow chamber. The effective laser power may change as a result of mode changes in the laser or in

beam pointing. The trajectory of the sample stream through the measurement region may be altered by bubbles or solid deposits. The changes produced by these phenomena cause a loss of resolution or peak skewing in flow karyotypes. Most of the changes introduced by these system fluctuations affect the mean of each peak proportionately (i.e. they act to change the overall system gain). Such changes can be readily detected in list mode data and corrected.

VI.   DATA DISPLAY

A.   UNIVARIATE FLOW KARYOTYPES

Univariate flow karyotypes accumulated using linear amplification typically show several peaks of equal area increasing in width and decreasing in height with increasing fluorescence intensity. This phenomena occurs because the measurement CV is roughly constant for all peaks so that the peak standard deviation increases with increasing peak mean. Ideally, the height of each peak should be proportional to the area of the peak. This can be accomplished partially by accumulating the flow karyotype using logarithmic instead of linear amplification. Application of this technique is illustrated in Chapter 10. This technique is effective, however, only if the peak CVs are constant. This is often not the case since the peak CV may increase slightly with decreasing fluorescence intensity since fewer photoelectrons are generated for the smaller chromosomes, thereby reducing the measurement precision. Thus, flow karyotypes measured using logarithmic amplification may show peaks increasing in height with increasing fluorescence intensity.

B.   BIVARIATE FLOW KARYOTYPES

Bivariate flow karyotypes may be displayed as contour (or gray level) plots, scatter plots or isometric displays. It is important that such displays carry information both about the

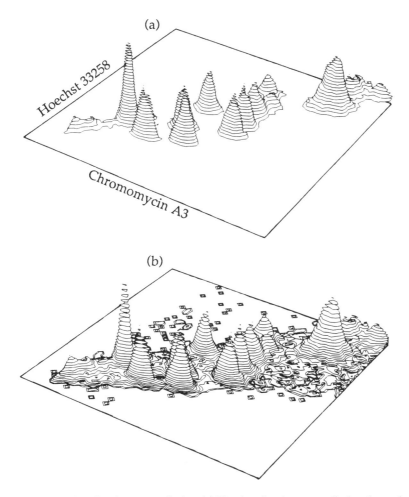

Fig. 7. Human bivariate flow karyotype display. (a) Bivariate flow karyotype displayed as an isometric contour plot with the contours spaced evenly and with all contours below 5% of the distribution maximum suppressed. (b) The same data displayed with no data suppression and the contours spaced such that the distance between contours increases with increasing frequency.

peaks (their means, CVs and areas) and about the magnitude of any debris continuum that underlies the peaks. Information about the debris continuum is usually lost in plots in which information is displayed linearly along the frequency axis or where the lower contours are not displayed. Thus, these plots may mislead sorting experiments by suggesting that the sorts will be more pure than they actually are. Figure 7(a), for example, shows an isometric contour plot of a human flow karyotype displayed linearly and with all histogram element frequencies lower than 5% of the histogram maximum not displayed. Little can be learned from this plot about the debris continuum. Figure 7(b), on the other hand, shows the same data displayed such that the contour spacing increases with increasing peak height. This and similar nonlinear display schemes emphasize both the peak and low-frequency components of the flow karyotype.

VII. CHROMOSOME SORTING

Chromosome sorting yields chromosomes of a single type with purity unequaled by any other means. Such chromosomes have proved valuable for gene mapping (see Chapter 14) and for production of chromosome-specific recombinant DNA libraries (see Chapters 15 and 16). This section focuses on the factors that determine the rate, efficiency and purity of chromosome sorting.

A. Sort Purity

The purity with which chromosomes can be sorted depends on the measurement resolution in the sorter, on the degree of overlap between the peak from which chromosomes are to be sorted and nearby peaks for other chromosomes, on the magnitude of the debris continuum underlying the peak from which chromosomes are to be sorted, on the sorting efficiency and on the frequency of DNA fragments that are below the detection limit of the sorter.

The resolution achieved during sorting is probably the single most important instrumental determinant of sorting purity. High-resolution sorting allows small windows to be established for sorting so that the fraction of events in the window due to debris fragments and chromosomes from adjacent peaks is minimal. Figure 8(c), for example, shows the sort of purity to be expected at constant sorting efficiency as a function of measurement resolution for the hypothetical situation where two peaks exist whose means are separated by 5% [Fig. 8(a) and 8(b)]. The purity achieved at 75% sorting efficiency (i.e. with the sort window centered on the peak and including 75% of all chromosomes of the type to be sorted) is almost twice as high at a precision of 2% as the purity at a precision of 6%. Decreasing the size of the sort window to exclude more of the chromosomes from the adjacent peak (to a sorting efficiency of 50%) increases the purity only slightly. This example illustrates a general phenomenon: measurement resolution is more important to sorting purity than window size. Thus, increased sorting efficiency (important when sorting large

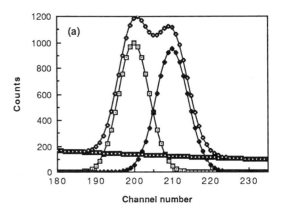

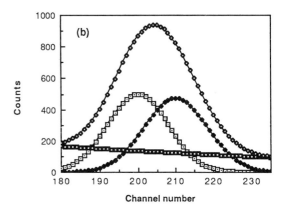

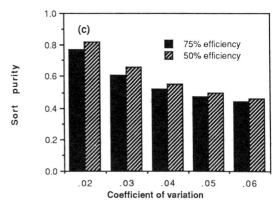

Fig. 8. Dependence of the sorting purity on peak separation and window width. (a) A univariate distribution showing two peaks superimposed on a continuum. Each peak has an area of 10 000. The peak means are separated by 5% and the peak CVs are 0.02. (b) The same univariate distribution as in part (a) except that the peak CVs are 0.04. (c) This plot shows the dependence of the sorting purity on the peak CV and window width.

numbers of chromosomes for library production) sometimes can be obtained with only a small loss in purity by increasing the size of the sorting window.

The quality and chromosomal composition of the chromosomal suspension from which chromosomes are to be sorted also play important roles in determining the sort purity. High-purity sorting can be accomplished much more easily in preparations in which the chromosome of interest is well resolved from surrounding chromosomes. Thus, human cell lines are often useful sources of the smaller human chromosomes but are less useful as sources for the larger chromosomes. Human × hamster hybrid cells have been used as sources for the larger human chromosomes since hybrid cell lines exist in which the larger human chromosomes are well resolved from other human chromosomes and from the hamster chromosomes. High-purity sorting is also more readily achieved in preparations relatively free of DNA fragments and chromosome clumps. Debris and clumps can contaminate a sort if these particles cannot be distinguished from the chromosome being sorted. Small debris particles can also contaminate a sort if they are sufficiently small that they are not sensed by the sorter during their passage through the measurement region. Since they are not sensed, they will not be rejected by the sorter anti-coincidence circuitry. These particles occur at very high frequency in some preparations. DNA in debris particles is most likely derived from all parts of the genome while DNA in chromosome clumps is more likely to be from only a few chromosome types.

B. Sort Rate

The rate at which chromosomes can be sorted depends on the rate at which droplets are produced in the sorter jet, on the deadtime of the sorter electronics, on the fraction of fluorescent objects comprised by the chromosome to be sorted and on the maximum

sample stream diameter that is consistent with high-purity sorting. The droplet production rate determines the maximum rate at which objects can be processed for sorting; assuming that sorting is aborted to avoid sorting multiple objects together. The actual sorting rate, S, achieved can be expressed as

$$S_r = f_i \, re^{-r\tau}, \qquad (7)$$

where f_i is the fraction of the total objects in the chromosome suspension comprised by the chromosome of interest, r is the rate at which objects (chromosomes, debris, nuclei, etc.) pass through the system and τ is the minimum time between objects that would not lead to a sort abort. In systems where three drops are sorted for each event, $\tau = 3/\omega$ where ω is the frequency of droplet production in Hz. These equations predict that the sort rate increases with throughput rate, reaching a maximum at $r = 1/\tau$ and decreasing thereafter. Figure 9 illustrates the sorting rates expected for conventional ($\omega = 30\,000/\text{sec}$) and high-speed sorters ($\omega = 200\,000/\text{sec}$) assuming three drop sorts in all instruments and that $f_i = 1$. The maximum sort rate for the 30 kHz sorter is thus 10 000/sec while the maximum for the 200 kHz sorter is 67 000/sec. However, the efficiency at these rates is only 37%. This efficiency is usually unacceptable because preparation of high-quality chromosome suspensions is time

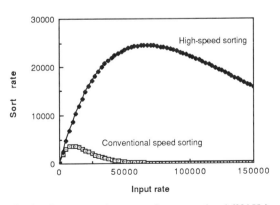

Fig. 9. Sort rate *vs* input rate for conventional (30 kHz) and high-speed (200 kHz) sorters. These calculations assume that all objects passing through the system fall within the sorting window.

consuming and labor intensive. Most sorts are run at much lower input rates so that the loss of chromosomes due to coincidence is less than 25%. This occurs at input rates of $\sim$3000/sec for 30 kHz sorters and $\sim$20 000/sec for 200 kHz sorters.

The fraction, f_i, of the total fluorescent objects in the chromosome suspension comprised by the desired chromosome type is also a critical determinant of the actual sort rate. It is tempting to use Eq. (4) to predict that the sort rate for a single human chromosome type should be $\sim$650/sec for high-speed sorters ($f_i = 1/23$; $r = 20\,000$; $\tau = 1.5 \times 10^{-5}$) and $\sim$100/sec in conventional sorters ($f_i = 1/23$; $r = 3000$; $\tau = 1 \times 10^{-4}$). However, this assumes that all of the objects in the chromosome suspension are chromosomes. In fact, it is not uncommon to find that as many as 75% of the objects in a chromosome preparation are nuclei, clumps and small debris particles so that, in actuality, high-speed sorting rates are around 200/sec and conventional sorting rates are around 25/sec. The true chromosome sort rate can be increased by raising the lower-level discriminator in the sorter so that the system does not "see" the small debris particles, thereby raising f_i. However, this comes at the expense of reduced purity since the small debris fragments are no longer rejected by the anticoincidence circuitry.

VIII.   FUTURE DIRECTIONS

Developments in several areas seem likely to occur in the next few years to increase the efficiency and purity of chromosome sorting. Sorting efficiency may be increased by increasing the number of chromosomes sorted simultaneously (e.g. by sorting four chromosome types simultaneously instead of two). The sorting efficiency also may be increased by pre-enriching for the chromosomes of interest prior to sorting (e.g. using velocity sedimentation; see Chapter 12) and by using systems capable of bit-mapped sorting so that the most efficient sort window can be used. Sorting purity can be improved by increasing the distinctness with which the chromosomes of interest are stained. Improved staining may come from the use of improved DNA-specific dyes (see Chapter 5), from chromosome-specific immunofluorescence staining (see Chapter 5) or from chromosome-specific fluorescence *in situ* hybridization (see Chapter 17). In addition, sorting purity may be improved by measuring additional chromosome parameters such as size and/or shape to reject clumps and debris fragments (see Chapter 11). Sorting according to chromosome shape using slit-scanning is particularly interesting in this regard. Sorting purity may also be improved by developing sorting instrumentation that adjusts for small changes in system gain (e.g. due to laser beam intensity changes) that occur during extended sorts. This may be accomplished by adjusting amplifier gains to account for the drift or by adjusting the sort windows to track the peaks as they move in the flow karyotype.

ACKNOWLEDGEMENTS

This work was performed under the auspices of the US Department of Energy by the Lawrence Livermore National Laboratory under contract number W-7405-ENG-48 with support from USPHS grant CA17665. The authors thank Ms Lil Mitchell for assistance in preparation of this manuscript.

REFERENCES

1. Bartholdi, M. F., Sinclair, D. C., and Cram, L. S. (1983). Chromosome analysis by high illumination flow cytometry. *Cytometry* **3**, 395–401.

2. Dean, P., and Pinkel, D. (1978). High resolution dual laser flow cytometry. *J. Histochem. Cytochem.* **26**, 622–627.

3. Deaven, L., Van Dilla, M., Bartholdi, M., Carrano, A., Cram, L., Fuscoe, J., Gray, J., Hildebrand, E., Moyzis, R., and Perlman, J. (1986). Construction of human-chromosome specific DNA libraries from flow sorted chromosomes. *Cold Spring Harbor Symp. Quantit. Biol.* **51**, 159–168.

4. Gray, J., Dean, P., Fuscoe, J., Peters, D., Trask, B., van den Engh, G., and Van Dilla, M. (1987). High speed chromosome sorting. *Science* **238**, 323–329.

5. Gray, J. W., Trask, B., van den Engh, G., Silva, A., Lozes, C., Grell, S., Schoenberg, S., and Golbus, M. (1988). Application of flow karyotyping in prenatal detection of chromosome aberrations. *Am. J. Hum. Genet.* **42**, 49–59.

6. Gray, J. W., and Cram, L. S. (1989). Flow karyotyping and sorting. *In* "Flow Cytometry and Sorting" (M. Melamed, T. Lindmo and M. Mendelsohn, eds). (In press.)

7. Hiebert, R. D., Jett, J. H., and Salzman, G. C. (1981). Modular electronics for flow cytometry and sorting: the LACEL system. *Cytometry* **1**, 337–341.

8. Hiebert, R. D., and Sweet, R. G. (1985). Electronics for flow cytometers and sorters. *In* "Flow cytometry: Instrumentation and Data Analysis" (M. A. Van Dilla, P. N. Dean, O. D. Laerum and M. R. Melamed, eds). Academic Press, London.

9. Langlois, R. G., Yu, L.-C., Gray, J. W., and Carrano, A. V. (1982). Quantitative karyotyping of human chromosomes by dual beam flow cytometry. *Proc. Natl Acad. Sci. USA* **79**, 7876–7880.

10. Le Pecq, J. B., and Paoletti, C. (1967). A fluorescent complex between ethidium bromide and nucleic acids. *J. Mol. Biol.* **27**, 87–106.

11. Mathies, R. A., and Stryer, L. (1986). Single-molecule detection: a feasibility study using phycoerythrin. *In* "Applications of Fluorescence in the Biomedical Sciences" (D. L. Taylor, A. S. Waggoner, F. Lanni, R. F. Murphy and R. R. Birge, eds). Alan R. Liss, New York.

12. Peters, D., Branscomb, E., Dean, P., Merrill, T., Pinkel, D., Van Dilla, M., and Gray, J. W. (1985). The LLNL high speed sorter: design features, operational characteristics and biological utility. *Cytometry* **6**, 290–301.

13. Pinkel, D., and Stovel, R. (1985). Flow chambers and sample handling. *In* "Flow Cytometry: Instrumentation and Data Analysis" (M. Van Dilla, P. Dean, O. Laerum and T. Lindmo, eds). Academic Press, London.

14. Shapiro, H. (1985). "Practical Flow Cytometry." Alan R. Liss, New York.

15. Sheingold, D. H. (1986). "Analog–Digital Conversion Handbook." Prentice-Hall, Englewood Cliffs, NJ.

16. van den Engh, G. J., and Stokdijk, W. (1989). A parallel processing data acquisition system for multi-laser flow cytometry and cell sorting. *Cytometry* **10**, 282–293.

17. van den Engh, G. J., Stokdijk, W., Trask, B. J., and Gray, J. W. (1988). A calibration circuit for determining the offset and non-linearity of flow cytometer electronics. *Cytometry* (submitted).

18. van den Engh, G., Trask, B., Lansdorp, P., and Gray, J. W. (1988). Improved resolution of flow cytometric measurements of Hoechst- and chromomycin-A3-stained human chromosomes after addition of citrate and sulfite. *Cytometry* **9**, 266–270.

19. Van Dilla, M., Deaven, L., *et al.* (1986). Human chromosome-specific DNA libraries: construction and availability. *Biotechnology* **4**, 537–552.

Cell Culture for Chromosome Isolation

CHARLOTTE LOZES

Biomedical Sciences Division
Lawrence Livermore National Laboratory
Livermore, CA 94550, USA

I. INTRODUCTION

Flow cytogenetic analyses require suspensions of isolated chromosomes. Preparation of such suspensions begins with the collection of mitotic cells from the cell population to be studied. The quality of chromosome preparations for flow cytogenetics depends critically on the fraction of mitotic cells in the cell population from which the chromosomes are to be isolated and on the viability and biochemical characteristics of the cell culture from which the chromosomes are isolated. Several different criteria can be applied to determine the quality of the chromosome preparation, including: (1) The extent of chromosome clumping and the fraction of DNA-containing debris fragments in the population. Minimization of chromosome clumping and DNA fragment production is important to flow karyotyping since these objects often complicate quantitative analysis of the flow karyotype. These objects are also undesirable during sorting since they reduce the effective sorting rate (their presence reduces the frequency of the chromosomes to be sorted in the population) and the purity of the sort (if they cannot be distinguished from the chromosomes to be sorted). (2) The absolute number of chromosomes obtained from a fixed number of cultured cells. High efficiency of chromosome isolation is important for sorting to minimize the amount of cell culture required for sorter purification of a large number of chromosomes (e.g. as required for production of recombination DNA libraries). (3) The molecular weight of the DNA in the isolated chromosomes. High-molecular-weight DNA is important for sorting if the DNA from the sorted chromosomes is to be used for production of large-insert recombinant DNA libraries.

This chapter focuses on cell culture factors that affect chromosome preparation quality. These include (1) the culture environment, (2) the density at which cells are seeded initially and (3) the techniques for collection of mitotic cells. Also discussed are specific techniques that have been successfully applied to selected cell types. The details of chromosome isolations are covered in Chapter 4.

II. CULTURE ENVIRONMENT

A. CULTURE CONDITIONS

One measure of the efficiency of cell culture for chromosome isolation is the number of mitotic cells that can be recovered from a culture. This number is a function of the number of cells in the culture and the growth rate (or alternately, the fraction of cells passing through mitosis per unit time). Figure 1 illustrates the dynamics of a hypothetical cell population with a minimum doubling time of 25 h. The cell number mitotic index and growth rates are initially low as the cells adapt to the new environment created by cell subculture [Fig. 1(a)]. Thus, the mitotic cell yield [Fig. 1(b); cell number times mitotic index] is low at this time. Later, the growth rate and the cell number increase as the population enters exponential growth. The mitotic cell yield is maximal near the end of this period. Later, the mitotic cell yield falls as the growth rate decreases, even though the cell number is high. Thus, mitotic cells should be collected near the end of the exponential growth phase for maximum efficiency. The time and magnitude of maximum efficiency can be adjusted by changing the initial seeding density of the culture. This is usually accomplished

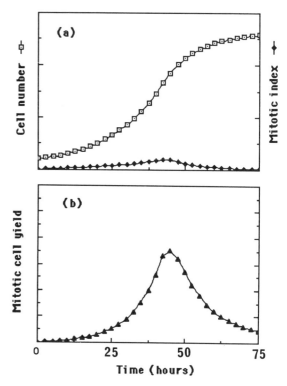

Fig. 1.   Growth dynamics of a hypothetical cell population with a doubling time of 25 h in which all cells proliferate. (a) Plots of the relative cell number and the fraction of cells in mitosis as a function of time after subculturing. (b) Relative mitotic cell yield per culture as a function of time after subculturing.

empirically (e.g. cell cultures are diluted 1:1, 1:2, 1:4 and 1:8 and used to establish cultures from which chromosomes will be isolated) and harvested at several times after culture set up (e.g. at 24, 36 and 48 h) to maximize the probability of successful flow karyotyping.

Cultures of established cell lines such as Chinese hamster ovary cells are ideal for production of chromosome suspensions since they have adapted for rapid growth in the *in vitro* environment. Their rapid growth (doubling times around 12 h) allows a large fraction of the cells to pass through mitosis during a 3–4 h period. Primary cell strains and human solid tumors are usually more difficult to deal with since they have not adapted for growth in artificial media and thus may proliferate slowly. Human cells, for example,

Table I

SUGGESTED MEDIA AND CELL TYPES
SUITABLE FOR GROWTH IN THEM

Cell type	Suggested medium	Reference
Human fibroblast	MEM	(3)
	F-12	(9)
Human lymphoblastoid	RPMI 1640	(6)
Amniotic	MEM-α	(1)
	Chang	(12)
Chinese hamster × human hybrid	MEM-α	(15)
Chinese hamster	MEM-α	(15)
Human tumor	F-12	(2)

rarely have doubling times less than 24 h. It is important that the nutrient requirements for these cells are identified and supplied without altering the osmolarity of the medium (4). Table I suggests several media and additives and the cell types that are suited to growth in them. It is also essential that growth additives such as fetal calf serum be selected to optimize the growth of the cell type(s) from which chromosomes are to be isolated.

B. Suspension vs Monolayer Cultures

Suspension cultures are appealing as sources of mitotic cells for chromosome isolation because large numbers of cells can be carried with minimal effort in each culture. A single 1 liter suspension culture may contain well over 10^8 cells. In addition, establishment of new cultures requires only that cells from an established culture be transferred into the new cultures. The disadvantages of suspension cultures are: (1) Not all cells may be carried in suspension. Some transformed fibroblasts can be adapted to growth in suspension with considerable effort. However, most primary (nontransformed) cells cannot adapt at all. Cells that normally grow in suspension *in vivo* are easiest to grow in suspension *in vitro*. Transformed human lymphoblastoid cells, for example, grow readily. (2) Mitotic cells cannot

be separated easily from nonmitotic cells. Thus, it is almost impossible to obtain populations of cells for chromosome isolation in which the mitotic index is higher than 0.5. This is a disadvantage during chromosome isolation since interphase nuclei may be disrupted to produce debris. This phenomenon is minimal, however, if the nonmitotic cells are healthy.

Monolayer cultures are appealing as sources of mitotic cells for chromosome isolation because most fibroblast cell types grow readily in monolayer culture. In addition, it is possible to collect, from monolayer cultures, populations of mitotic cells in which the mitotic cell fraction is very high. This is accomplished by shaking the cultures vigorously (7). The mitotic cells are less tightly attached to the monolayer than the interphase cells. As a result, the mitotic cells are removed selectively by shaking. Some cells, for example Chinese hamster ovary cells, are sufficiently well suited to mitotic selection that mitotic indices greater than 0.95 can be attained. Mitotic cells from human cells grown in monolayer can also be shake-selected. Mitotic indices around 0.75 can be readily attained for human amniocytes and fibroblasts using this technique. Shake-selection procedures often can be improved by shaking prior to the beginning of the mitotic cell collection process (i.e. prior to addition of colcemid; see below) to eliminate weakly attached interphase cells and dead and dying cells. Monolayer cultures suffer from two disadvantages: (1) The cells grown in monolayer must be removed by enzymatic treatment* to establish new cultures. This is time consuming and labor intensive. In addition, the trypsin-treated cells may require several hours to resume active proliferation. (2) Monolayer cultures are expensive. Numerous culture flasks are required for production of large numbers of mitotic cells.

---

* Trypsinization is accomplished in our laboratory by aspirating the medium from the flask, rinsing for a few seconds to a few minutes with Puck's EDTA, removing the Puck's EDTA, incubating with 3 ml trypsin solution (0.025% trypsin in Puck's EDTA) at 37°C for about 3 min, and tapping to loosen the cells. Medium with fetal calf serum is added to deactivate the trypsin.

For example, 40 T150 culture flasks may be required to produce the same number of mitotic cells as are contained in a single 1 liter spinner culture.

III. MITOTIC CELL ACCUMULATION

The efficiency of mitotic cell collection is usually increased by treatment with colcemid or another mitotic inhibitor for an extended period so that the fraction of mitotic cells in the population increases. The duration of treatment with inhibitors has been extended as long as 12–14 h for some cultures. However, the health of the culture tends to deteriorate as the colcemid duration increases since the cells may either escape the block and divide (often by nondisjunction) or die as a result of unbalanced growth. Some studies have shown that these events may begin to occur after only 3–4 h (13). Nuclei and chromosomes from the dying cells are usually fragile and thus may be broken to produce DNA-containing debris during the chromosome isolation step. Free DNA from these cells may also contribute to subsequent chromosome clumping. It is important to minimize the duration of the mitotic inhibition period in experiments where lightly condensed chromosomes are required (e.g. when the isolated chromosomes are to be banded or slit-scanned). The condensation of chromatin in chromosomes increases continuously while the cells are blocked in mitosis (14).

Several chemicals may be used for mitotic cell accumulation. These act by inhibiting the formation of the spindle fibers required for mitotic cell division. Agents of this type include colchicine and its N-deacetyl-N-methyl derivative, colcemid, vincristine and vinblastine. Colcemid has been used in most chromosome isolation efforts to date. The optimal treatment duration and concentration depends on the cell type. Human fibroblasts usually are treated with 0.3 μg/ml colcemid for 12–15 h while Chinese hamster ovary cells may be treated with 0.03 μg/ml colcemid for 4–6 h. There is some evidence in the literature that these agents also affect cell processes such as the rate of DNA synthesis and chromatin structure (11). However, there is no evidence that these phenomena are important to chromosome isolation for flow karyotyping.

The rate at which mitotic cells can be accumulated in mitosis can be maximized (and hence the mitotic inhibition period can be minimized) by synchronizing the cell population. Cell synchronization can be accomplished by physically selecting cells from a limited part of the cell cycle or by chemically blocking cells for a short period in a limited part of the cell cycle (7). Once the cell population has been synchronized, the mitotic inhibition treatment is timed to span the period when the synchronous cohort of cells is passing through mitosis. Elutriation and other velocity sedimentation techniques can be used to select cells according to size. This effectively selects cells in a limited part of the cell cycle since cell size increases continuously as cells progress from the G1-phase of the cell cycle toward mitosis. Thus, the smallest cells in the population will be in the G1-phase and the largest will be in the G2-phase plus mitosis. These techniques are especially well suited to synchronization of cells grown in suspension. Amethopterin has been used to synchronize lymphocytes by acting as a folic acid antagonist thereby blocking the de novo synthesis of thymidine. Cells treated in this fashion are reversibly blocked in S-phase and at the G1-phase/S-phase boundary. Cells accumulated in these portions of the cell cycle can be released by treatment with thymidine to reverse the block so that they move in a partly synchronous wave through mitosis. Other procedures for cell synchronization are described by Stubblefield et al. (13) and Grdina et al. (7). These procedures may lead to an increase in the mitotic index by as much as four- to five-fold (5).

IV. SPECIFIC EXAMPLES

A. FIBROBLASTS

LL811 is a cell strain (i.e. nontransformed) established from human placental material. It is grown in monolayer culture in Minimum essential medium-alpha (MEM-α) with 20% fetal calf serum (FCS), penicillin and streptomycin. The cells have an average doubling time of 28 h. The cultures are established at a density of 1.6×10^6/T150 culture flask and colcemid (0.1 μg/ml) is added 48 h later. The colcemid treatment period is 10–16 h. Chromosomes can be easily isolated from about 2.5×10^5 mitotic cells collected by shake-selection. The mitotic index on these cells is about 0.60. Production of large numbers of mitotic cells (e.g. as required for production of bulk quantities of chromosomes for chromosome sorting) requires multiple T150 flasks. For example, about 50 T-150s yield 7 to 8×10^7 mitotic cells.

M3-1 is the Chinese hamster cell line used in many early flow karyotyping studies. These cells are transformed and have been grown continuously in culture for many generations. They are usually grown in monolayer in MEM-α with 10% FCS, penicillin and streptomycin. These cells have an average doubling time of about 12 h. Cultures for mitotic cell harvest are established by using cells from a nearly confluent culture diluted 1:4 in fresh medium. Colcemid is added for 4 h to cultures that have grown for about 8–12 h. Mitotic cells are collected by shake selection. The mitotic index of these cells is usually around 0.90. Chinese hamster $\times$ human hybrid cell lines generally require culture conditions similar to those for hamster cells.

The cell line LOVO, developed from human adenocarcinoma cells in the laboratory of Dr B. Drewinko (2) also has been successfully cultured for flow karyotyping. These cells are grown as monolayer cells in MEM-α plus 10% FCS, penicillin and streptomycin. This cell line is propagated by subculturing every 2 weeks. The cells are split about 1:5 at each subculturing. The LOVO cells are removed from the flask for subculturing by treating with hyaluronidase for 5 min followed by 2.5% trypsin for 10 min at room temperature. The viability of the cells removed by this method is approximately 0.90. Cultures are established for mitotic cell collection from a 1:1 or 1:2 dilution of cells from a nearly confluent culture (2).

Cells taken for prenatal diagnosis are usually amniocytes or chorionic villus cells. Human amniocytes established from 2–4-week-old primary amniocyte cultures have been routinely cultured for flow karyotyping in our laboratory (8). These cultures may contain several cell types. One type resembles normal fibroblasts like the LL811 strain. A second type is probably derived from placental trophoblasts (termed AF cells). These cells form colonies that grow in a loosely organized formation with a growth zone on the periphery of the colony. The third cell type appears to be epitheloid. These cells are difficult to remove by trypsinization. Of approximately 170 amniotic cultures analyzed in our laboratory, 37% contained fibroblast clones, 31% contained epitheloid clones and 90% contained AF clones. Cultures of amniotic cells are established from a single 2–5 week old T-25 culture of amniotic fluid cells. The cells are removed from the T-25 cultures by rinsing with Puck's EDTA and treating for 15–30 sec with 0.5 ml of 0.025% trypsin in EDTA. The flask is then trapped to loosen the cells and 4.5 ml of MEM-α plus FCS is added to deactivate the trypsin. The cells are split 1:2 or 1:3 and grown for 48 h in MEM-α plus 20–30% FCS supplemented with L-glutamine prior to addition of colcemid for mitotic cell collection. The colcemid blocking period is typically 10 h.

B. LYMPHOCYTES AND LYMPHOBLASTOID CELLS

Flow karyotypes have been measured for chromosomes isolated from whole blood cultures or from lymphocyte preparations of human peripheral blood cultures. Whole blood obtained by venipuncture is collected in a green-top heparin-containing tube and is cultured in a T-25 flask containing 5 ml of RPMI 1640 plus 20% FBS, supplemented with L-glutamine penicillin–streptomycin, 10 units per ml heparin (Elkins-Sim, Inc.), phytohemagglutinin-M form (4 ml per 100 ml medium), and 25 mM Hepes buffer by innoculating the flask with 300 λ whole blood and incubating at 37°C in a controlled 5% CO_2 environment for 72 or 96 h. Colcemid (0.1 μg/ml) is then added and the flask is re-incubated at 37°C for 10–12 h. Flow karyotypes from such cultures usually contain debris (B. Trask, unpublished observations).

Higher-quality flow karyotypes are obtained by separating lymphocytes from whole blood (12 ml) obtained by venipuncture and placed in a sterile bottle containing 120 units of heparin. It is mixed with an equal volume of Hank's balanced salt solution (HBSS). This mixture is layered over 18 ml of lymphocyte separation medium (LSM, Litton Bionetics) in a plastic centrifuge tube and centrifuged at 400g for 30 min at room temperature. The visible white cell layer is collected and resuspended in a 50-ml tube, filled to the top with 4:1 HBSS to culture medium (RPMI 1640 plus 20% FBS). It is centrifuged at 340g for 10 min at room temperature. The pellet is resuspended in a 15-ml tube containing 4:1 HBSS to culture medium and centrifuged for 6 min at 340g. The pellet is then loosened and 4 ml of culture medium, which is the same as that used for whole blood above, is added. The cultures are established at a density of 2 × 10^5 cells/ml and grown under the same conditions as whole blood for 72 h. Colcemid (0.1 μg/ml) is added and the culture is grown for a further 10 h. The mitotic index is usually 10–40% (16).

Transformed lymphoblastoid cultures (e.g. GM131) have been grown using similar culture conditions. That is, they are usually established at a density of 2.5 × 10^5/ml in RPMI 1640 plus 20% FCS and antibiotics. These cells are grown for about 30 h before addition of colcemid (0.1 μg/ml). The colcemid treatment duration is

typically 10 h. Mitotic indices average about 35% with this procedure. Lymphoblastoid cells can also be grown in spinner cultures for isolation of large numbers of mitotic cells (e.g. to facilitate isolation of very large numbers of cells required for chromosome sorting). Lymphoblastoid cells should be allowed to adjust to growth in spinner flasks before attempting to set up cultures for chromosome isolation.

C. SOLID TUMORS

Chromosomes also have been prepared from cells isolated from solid tumors growing *in vivo*: Kooi *et al.* (*10*), for example, have isolated cells from rat R-1 rhabdomyosarcomas grown in the flanks of female rats. In this study, the tumor-bearing rats were injected intraperitoneally with the stathmokinetic agent, vindesin. Twenty-four hours later, the tumor was removed, rinsed in PBS and treated with 0.05% trypsin for 15 min to produce a suspension of single cells. These cells were then treated with DNase for 10 min to remove free DNA. Chromosomes were isolated from this cell suspension and flow karyotyped. The flow karyotypes clearly showed several peaks superimposed on a substantial debris continuum. The magnitude of the debris continuum was reduced by culturing cells isolated from untreated tumors for 13 h prior to addition of 0.03 μg/ml of vindesin. Chromosomes were isolated 7 h after addition of vindesin and flow karyotyped.

V. SUMMARY

The success of flow karyotyping depends critically on the production of populations enriched in healthy mitotic cells containing minimal free DNA, DNA fragments or cells susceptible to disruption during chromosome isolation. Production of such cell populations requires culture conditions that maintain high cell viability, rapid cell growth and a high growth fraction. Unfortunately, a universally useful set of culture conditions has not been devised. Thus, careful attention must be given to optimizing the culture of each cell type to be flow karyotyped. Such optimization has already been accomplished for a variety of cell types including human fibroblasts, amniocytes, lymphocytes and lymphoblastoid cells. Near optimal culture conditions have also been established for Chinese hamster and hamster × human cell lines. Continued work will be required, however, to establish cell culture conditions that allow routine application of flow karyotyping to solid tumors.

ACKNOWLEDGEMENTS

I wish to thank Kerry Brookman for pointing out the needs of other technologists and Richard Langlois and Irene Jones for many helpful comments. I thank Joe Gray for final editing of this chapter.

This work was performed under the auspices of the US Department of Energy by the Lawrence Livermore National Laboratory under contract number W-7405-ENG-48 with support from USPHS grant 17665.

REFERENCES

1. Barker, P., and Lawce, H. (1980). Amniotic fluid cell culture. *In* "Association of Cytogenetic Technologists Laboratory Manual" (M. Hack and H. Lawce, eds), pp. 223–247, University of California, San Francisco, CA.
2. Drewinko, B., Romsdahl, M. M., Yang, L. Y., Ahearn, M. J., and Trujillo, J. (1976). Establishment of human carcinoembryonic antigen-producing colon adenocarcinoma cell line. *Cancer Res.* **36**, 467–475.
3. Eagle, H. (1959). Amino acid metabolism in mammalian cell cultures. *Science* **130**, 432.
4. Freshney, R. (1983) Culture of animal cells. *In* "A Manual of Basic Technique." Alan R. Liss, New York.
5. Gallo, J. H., Ordong, J. V., Brown, G. E., and Testa, J. R. (1984). Synchronization of human leukemic cells: relevance for high-resolution chromosome banding. *Human Genet.* **66**, 220–224.
6. Glick, J. (1980). *In* "Fundamentals of Human Lymphoid Cell Culture." Dekker, New York.
7. Grdina, D., Meistrich, M., Meyn, R., Johnson, T., and White, A. (1986). Cell synchrony techniques: a comparison of methods. *In* "Techniques in Cell Cycle Analysis." (J. W. Gray and Z. Darzynkiewicz, eds), Humana Press, Clifton, NJ.
8. Gray, J. W., Trask, B., van den Engh, G., Silva, A., Lozes, C., Grell, S., Schoenberg, S., and Golbus, M. (1988). Applications of flow karyotyping in prenatal detection of chromosome aberrations. *Am. J. Hum. Genet.* **42**, 49–59.
9. Ham, R. G. (1965). Clonal growth of mammalian cells in a chemically defined synthetic medium. *Proc. Natl Acad. Sci. USA* **53**, 288–293.
10. Kooi, M. W., Aten, J. A., Stap, J., Kipp, J. B. A., and Barendsen, G. W. (1984). Preparation of chromosome suspensions from cells of a solid experimental tumor for measurement by flow cytometry. *Cytometry* **5**, 547–549.
11. Mueller, G. A., Gauldren, M. E., and Drane, W. (1972). The effects of varying concentrations of colchicine on the progression of grasshopper neuroblasts into metaphase. *In* "Experimental Control of Mitosis II", p. 113. MSS Information, New York.
12. Salk, D., Disteche, C., Stenchever, M. R., Mitchell, L., and Rodrigez, M. (1984). Use of Chang medium for prenatal diagnosis. Presented at the "1984 Meeting of the Association of Cytogenetic Technologists."
13. Stubblefield, E. (1968). Synchronization methods for mammalian cell culture. *In* "Methods in Cell Physiology IV", Chapt. 2. Academic Press, New York.
14. Terasima, T., and Tolmach, L. (1961). Changes in X-ray sensitivity of Hela cells during the cell cycle. *Nature* **190**, 1210–1211.
15. Thompson, L. H., Carrano, A. V., Sato, K., Salazar, E. P., White, B. F. and Stewart, S. A. (1987). Identification of nucleotide-excision-repair genes on human chromosomes 2 and 13 by functional complementation in hamster–human hybrids. *Somat. Cell Molec. Genet.* **13** (5), 539–552.
16. Yu, L.-C., Aten, J., Gray, J., and Carrano, A. (1981). Human chromosome isolation from short-term lymphocyte culture for flow cytometry. *Nature* **293**, 154–155.

Chromosome Isolation Procedures

BARB TRASK

Biomedical Sciences Division
Lawrence Livermore National Laboratory
Livermore, CA 94550, USA

I. INTRODUCTION

Preparation of chromosomes for flow karyotyping or sorting requires that they be released from mitotic cells, stabilized in suspension and stained with fluorescent dyes. The resolution that can be obtained in a flow karyotype is dependent on the isolation procedure as well as the measuring instrument (see Chapter 2). Ideally, each chromosome type should produce a distinct peak in the flow karyotype. The fluorescent staining should be homogeneous, stable and reproducible. The relative peak positions in a flow karyotype should be the same in repeated measurements for the same individual. Peak areas should be proportional to the relative frequency of chromosomes in the original cells. In other words, chromosomes should not be selectively lost, and the number of clumps and debris particles in the chromosome suspension should be low. It should be possible to fluorescently label the isolated chromosomes with a variety of DNA and protein stains. Storage of chromosomes before flow analysis should not result in degradation of the flow karyotype, loss of proteins or decreased DNA molecular weight. The yield of chromosomes from mitotic cells should be high. Finally, the method should be simple.

Further requirements are placed on the chromosome preparation, and thus on the isolation procedure, by specific applications such as prenatal cytogenetics, biological dosimetry and preparative chromosome sorting for DNA cloning. The importance of certain chromosome characteristics, such as their bandability, size or protein content, or the molecular weight of the DNA they contain, may vary with the application. Examples of applications and their specific requirements follow below.

Flow karyotypic detection and study of chromosomal abnormalities require preparations of chromosome suspensions from normal and abnormal cells. Peak resolution must be high and the contribution of clumps and debris must be low to allow the detection of small changes in peak position or peak volume. For clinical studies, the isolation procedure should perform well on a variety of cell types, despite low numbers of available mitotic cells.

The level of chromosomal debris found between peaks in a flow karyotype is sometimes taken as a measure of the exposure of

Table I

PROCEDURES FOR THE ISOLATION OF CHROMOSOMES

Hexylene Glycol	Propidium Iodide	Polyamine	Citrate/Ethanol
[Wray and Stubblefield (81); modified by Yu et al. (84)]	[Aten et al. (4); Buys et al. (11); Cram et al. (20)]	[Blumenthal et al. (10); Sillar and Young (67); Lalande et al. (45)]	[Matsson and Rydberg (60)]
Centrifuge 150g, 8 min	Centrifuge 150g, 8 min	Centrifuge 150g, 10 min	Centrifuge 150g, 10 min
Wash in Hank's BSS + cult med + 0.037 μg/ml colchicine	—	(Optional: Wash in Hank's BSS)	Wash in PBS
Centrifuge 150g, 10 min	—	Centrifuge 150g, 10 min	Centrifuge 150g, 10 min, 4°C
Resuspend in 75 mM KCl + 0.037 μg/ml colchicine	Resuspend in 1.0 ml 75 mM KCl + 5–50 μg/ml PI	Resuspend in 75 mM KCl	Resuspend in 0.1% sodium citrate, 50 μg/ml PI at 10^6 cells/ml
4°C, 7 min	25°C, 10 min or 37°C, 25 min	4°C, 10 min, or 25°C, 25 min	25°C, 20 min
Centrifuge 200g, 10 min	—	Centrifuge 150g, 8 min	
Resuspend in buffer: 25 mM Tris-HCl, pH 7.5 0.75 M hexylene glycol 0.5 mM CaCl₂ 1.0 mM MgCl₂ 0.037 μg/ml colchicine	Add 0.5 ml 75 mM KCl + 5–50 μg/ml PI + 1.0–1.5% Triton X-100 (+ optional 1.0 mg/ml RNase)	Wash in buffer: 15 mM Tris-HCl, pH 7.2, + 2 mM EDTA + 0.5 mM EGTA + 80 mM KCl + 20 mM NaCl + 14 mM β-mercaptoethanol + 0.2 mM spermine + 0.5 mM spermidine[a]	Add ethanol to 20% final concentration while vortexing
Incubate 37°C, 10 min	Incubate 25°C, 4 min	Centrifuge 1000 rpm, 8 min	
Incubate 4°C, 30 min	—	Resuspend in above buffer + 0.1% digitonin	
Shear with needle or Virtis homogenizer	Shear with needle	Vortex 10–20 sec	
	(Optional: Incubate 37°C, 30 min)		

individuals to mutagenic agents. Obviously, for this type of analysis to yield consistent and reliable information about chromosome damage, chromosomes must not be damaged during the isolation procedure itself.

By measuring the fluorescence intensity profile of a chromosome as it passes a small laser beam (slit scan), the position of the centromere can be determined and used to identify chromosomes. For these measurements, it is desirable that the chromosomes be swollen or stretched so that the centromeric constriction is evident. For conventional flow karyotyping, on the other hand, the total fluorescence intensity of a chromosome is measured. Compact chromosomes are acceptable, and may be desirable.

Flow sorting can be used to purify chromosomes for gene mapping and for

Table I

CONTINUED

Citric Acid	Tris/MgCl$_2$/Triton X-100	Hepes/MgSO$_4$
[Collard *et al.* (*17*)]	[Otto *et al.* (*65*); similar procedures described by Disteche *et al.* (*27*); Bijman (*9*)]	[van den Engh *et al.* (*75–77*)]
Centrifuge 150*g*, 10 min, 4°C	Centrifuge 150*g*, 5 min	Centrifuge 150*g*, 10 min
Wash twice in PBS	—	—
150*g*, 10 min, 4°C	—	—
Resuspend in 75 m*M* KCl	Resuspend in 20 m*M* Tris-HCl, 40 m*M* MgCl$_2$	Resuspend in 50 m*M* KCl, 5 m*M* Hepes, 10 m*M* MgSO$_4$ (opt for PI stain: 0.15 mg/ml RNase), 3 m*M* dithiothreitol, pH 8.0
15°C, 5–30 min	4°C, 10 min	22°C, 10 min
150*g*, 10 min, 4°C	—	—
Remove all but 0.1 ml KCl Add cold buffer dropwise, while vortexing: 2% citric acid + 1 m*M* MgCl$_2$ + 1 m*M* CaCl$_2$ + 0.1 *M* sucrose at pH 2.4; 10^7 cells/ml	Add Triton X-100 to final concentration: 0.25%	Add Triton X-100 to final concentration: 0.25%
—	Incubate 22°C, 10 min	Incubate 4°C, 10 min
Shear with needle at 4°C	Disrupt with homogenizer or needle	Shear with needle or vortex
Dilute with 0.1 *M* Tris-HCl, 0.1 *M* NaCl, 5 m*M* MgCl$_2$ + EB, pH 7.4	Dilute with 150 m*M* NaCl, 40 m*M* MgCl$_2$, 25 m*M* Tris-HCl, pH 7.5	(Optional for PI stain: incubate 37°C, 30 min)
		Add 10 m*M* sodium citrate + 25 m*M* sodium sulfite before analysis

[a] Alternatively, pellet may be resuspended directly in 4°C isol buffer + 0.1% digitonin. (M. Luedemann, unpublished results; *77*.)

the construction of chromosome-enriched recombinant DNA libraries. The ability to identify sorted chromosomes by conventional banding techniques or by fluorescence *in situ* hybridization to check the purity of sorted fractions is desirable. Use of sorted chromosomes for construction of large-insert libraries requires that the DNA in the isolated chromosomes be of high molecular weight. In addition, large numbers of sorted chromosomes are required. Thus, efficient recovery of chromosomes from mitotic cells is important. A low recovery efficiency means that large numbers of cells must be cultured for the production of a sufficient number of chromosomes. In addition, highly concentrated chromosome suspensions are required for high-speed sorting.

Many procedures have been developed specifically for the isolation of chromosomes for flow cytometric analysis. These procedures have several features in common. The frequency of mitotic cells in a cell population is increased by the addition of a spindle poison such as colcemid for several hours. The cells are pelleted by centrifugation and are then swollen in a hypotonic solution. During the cell swelling, chromosomes spread from each other within the confines of the plasma membrane, facilitating the ultimate preparation of a monodispersed chromosome suspension. Detergents and physical disruption are used to break open the swollen cells and to release the chromosomes into the isolation buffer. A stabilizing agent must be present in the buffer to maintain intact chromosomes. Before flow cytometric analysis, the chromosomes are stained with fluorescent dyes.

Within this framework of basic steps, the chromosome isolation procedures vary in the mode of chromosome stabilization, number of mitotic cells needed for optimal results, applicability to different cell types, fluorescent stains that can be used, suitability for production of banding patterns in sorted chromosomes and size of DNA fragments that can be extracted from isolated chromosomes. Eight procedures listed in Table I are compared

in the discussion of chromosome preparation presented in the following pages.

II. PREPARATION DETAILS

A. CELL TYPES

Chromosomes can be isolated from mitotic cells obtained from fibroblast, peripheral blood lymphocyte, amniocyte, lymphoblast and leukemic blast cultures. The quality of the flow karyotype and the amount of small fluorescent debris in a suspension depends on the health of the cell culture. High-resolution flow karyotypes can be obtained from cultures with mitotic indices as low as 5% but, in general, cultures with many dead cells and long doubling times will result in karyotypes with prominent debris continua. Figure 1 shows flow karyotypes of human chromosomes isolated from several different cell types using the Hepes/MgSO$_4$ isolation procedure.

A survey of the literature reveals karyotypes of chromosomes isolated from human fibroblast cells using the hexylene glycol, citric acid, polyamine, Hepes/MgSO$_4$, modified propidium iodide (PI), and Tris/MgCl$_2$ procedures and variants of these procedures (4,18,34,41,44,50,61,67,76). Chromosomes can be successfully isolated from lymphoblast suspension cultures using the polyamine buffer or the Hepes/MgSO$_4$ buffer (8,38,43,45,46,50,52,67,77,78). Cultures of tumor cells, such as Lovo, Daudi, K562, and other leukemia and retinoblastoma cell lines, have been shown to yield analyzable chromosome preparations using the Hepes/MgSO$_4$, polyamine, citric acid, Tris/MgCl$_2$ or hexylene glycol procedures (76; B. Trask and G. van den Engh, unpublished observations; Fig. 1; *15,26,41,50,80*; L.-C. Yu, personal communication).

For flow karyotyping to be useful in clinical studies of chromosome abnormalities, chromosomes isolated from clinical samples must yield flow karyotypes with the same high resolution as is obtained with fibroblast or

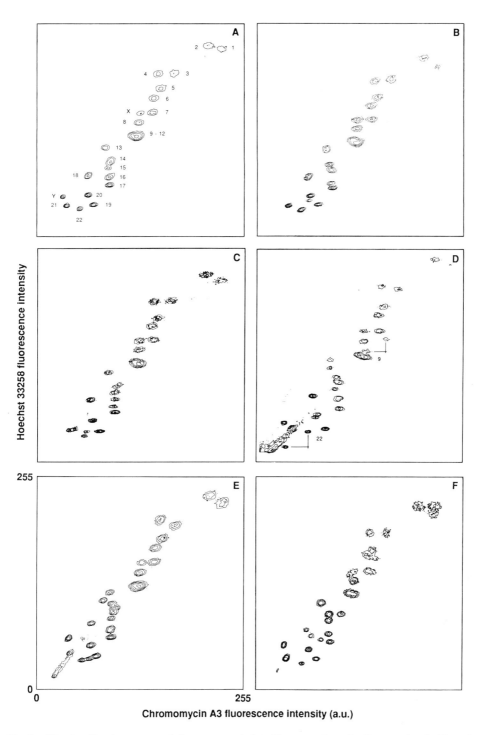

Fig. 1. Bivariate flow karyotypes of chromosomes isolated from a variety of cell types using the Hepes/MgSO$_4$ method and stained with Hoechst 33258 and Chromomycin A3: (A) fibroblast cell line HSF-7; (B) lymphoblast cell line GM131; (C) PHA-stimulated peripheral blood lymphocyte culture of a normal donor; (D) short-term culture of chronic myeloid leukemic blasts showing the products of the reciprocal 9;22 translocation; (E) amniocyte culture with a reciprocal 18;21 translocation; (F) erythroblast leukemic cell line HEL with multiple structural and numerical chromosome abnormalities. Numbers in part (A) indicate the chromosome(s) responsible for the peaks in the flow karyotype. [Data from van den Engh *et al.* (*77*) and B. Trask and G. van den Engh, unpublished data.]

lymphoblast cultures. It must also be possible to obtain flow karyotypes for cells from tissues or cultures containing few mitotic cells. Chromosomes can be isolated from the low number of cells found in small (25 cm^2) amniocytic cell cultures using the Hepes/MgSO$_4$ method (31,76). Flow karyotypes have been reported for PHA-stimulated peripheral blood lymphocyte cultures using the hexylene glycol, citrate/ethanol, polyamine, Hepes/MgSO$_4$ and PI method with psoralen stabilization (36,39,50,60,76,83,84). In our laboratory, 1 ml of blood is sufficient for one or two 5 ml PHA-stimulated cultures. One culture yields 3–4 ml of chromosome suspension using the Hepes/MgSO$_4$ procedure; only a fraction of this is required to obtain a high-resolution flow karyotype.

Leukemic blasts from peripheral blood or bone marrow biopsies of patients with chronic myeloid leukemia can be stimulated to divide in the presence of PHA-stimulated lymphocyte conditioned medium. Despite mitotic indices in these cultures as low as 4% and high levels of small fluorescent debris, the Philadelphia chromosome (t(9;22)) characteristic for CML was visible in the flow karyotype of 16 out of 17 cultures isolated with the Hepes/MgSO$_4$ procedure (77; B. Trask, G. van den Engh and P. Lansdorp, unpublished observations). The karyotype of the leukemic blasts can be compared to the karyotype of lymphoblast cultures of normal cells from the same patient. Arkesteijn and coworkers (3) have obtained high-resolution karyotypes of leukemic blasts by selecting blasts from patient samples before short-term culture and using a variation of the Tris/MgCl$_2$ isolation procedure.

Chromosome isolation is not restricted to cell lines of human origin. The procedures in Table I have been used variously to obtain flow karyotypes of Chinese hamster, muntjac, mouse, swine, rat and somatic cell hybrids (5–7,14,27,34,58,59,62,65,73,75,78). The hexylene glycol method has also been used to isolate chromosomes from an experimental ascites tumor in the rat: tumor cells were cultured *in vitro* in the presence of colcemid before chromosome isolation (40).

Several groups have succeeded in isolating chromosomes from primary cell populations, without culturing *in vitro*. To date, however, the quality of the flow karyotype and the information that can be derived from these cells is less than is possible from cultured cells. Chromosomes have been isolated after *in vivo* vincristine treatment from trypsinized rat rhabdomyosarcoma tumor tissue (42) or from the bone marrow of rats with BNML (brown norway myeloid leukemia) (58,73). In the latter case, the suspension of cells are enriched for mitotic cells using a density gradient. Mitotic cells are less dense than the majority of interphase cells. Cell suspensions with mitotic cell frequencies as high as 50% can be obtained. The flow karyotypes obtained, though crude, are similar to the karyotypes of *in vitro* cell lines of these tumors.

Flow karyotyping of plant cells is hampered by an additional barrier to chromosome isolation: the cell wall. Two groups have described methods for chromosome isolation from plant cells (19,24,25). These procedures call for enzymatic digestion of the cell wall in a sugar-containing solution to prepare naked protoplasts. Cells are then lysed either in a hypotonic buffer that is similar in content to the buffers used to isolate mammalian chromosomes (spermine, EDTA, dithiothreitol, KCl, NaCl, Triton X-100 and sucrose) (24) or in modified culture medium with added mannitol and 1% Triton X-100 (19). Flow karyotypes have been produced for two plant species, *Petunia hybrida* and *Haplopappus gracilis*, which have few chromosomes with considerable size differences.

B. MITOTIC CELL COLLECTION

a. Cell Harvest

After cell cultures are enriched for mitotic cells by the addition of a spindle poison (see

Chapter 3), cells are harvested. Mitotic cells can be selectively harvested from fibroblast monolayer cultures by shaking them loose from the flasks. In our laboratory, roughly 1.5×10^4 cells can be obtained per cm^2 of hamster or somatic cell hybrid monolayer cultures after a 4 h colcemid block, and 4×10^3 cells can be collected per cm^2 from human fibroblast cultures after a 10 h colcemid block. Mitotic indices range from 70 to 85%.

As noted above, sufficient mitotic cells can be obtained for flow karyotype analysis from as little as 1 ml of a lymphoblast or standard PHA-stimulated lymphocyte culture. Approximately 10–30% of the cells in these cultures are mitotic after 10–12 h incubation in the presence of colcemid. Interphase cells are collected and pelleted along with these mitotic cells. Nuclei are then isolated from the interphase cells and stained with the isolated chromosomes. The nuclei do not interfere with flow cytometric analysis, but it is common practice in some laboratories to let some of the interphase nuclei settle out of the chromosome suspension for several hours before analysis (38,51).

b. Number of Cells

The number of cells required to give optimal chromosome suspensions varies with the application, the cells being used and the isolation protocol. For example, concentrated chromosome suspensions (4×10^7/ml) make it possible to achieve high sorting rates possible on the Lawrence Livermore National Laboratory (LLNL) high-speed sorter (30,66). On the other hand, few mitotic cells may be available from some samples, such as amniocytes and leukemic cell cultures. We routinely prepare suspensions at $1–4 \times 10^6$ chromosomes/ml for high-resolution flow karyotyping at low flow rates.

The Hepes/MgSO$_4$ method has proved remarkably insensitive to large variations in starting cell number. Cell concentrations from as high as 10^8 to as few as several thousand per ml of isolation buffer result in high-resolution flow karyotypes. This feature makes the method useful for experiments ranging from high-resolution cytogenetics to preparative sorting. Carrano and coworkers (14) determined that optimal flow karyotype resolution was obtained using the hexylene glycol method with a cell concentration of 5×10^6/ml. Studies describing the effect of cell concentration on flow karyotype resolution have not been reported for the other methods. However, most reports indicate that high cell concentrations (from 10^6 to 10^8 cells per ml isolation solution) are routinely used (17,27,45,60,67).

c. Prewashing

Washing collected cells before they are swollen is called for in the hexylene glycol, citrate/ethanol, citric acid procedures and in some versions of the polyamine procedure. The Hepes/MgSO$_4$, Tris/MgCl$_2$, and PI methods have been simplified for increased yield and allow direct addition of the final buffer to the pellet of mitotic cells.

C. ISOLATION

a. Cell Swelling

One of the most important steps in any chromosome isolation procedure is the swelling of the cells in a hypotonic buffer. In this step, the cells increase in volume and the chromosomes move apart from one another within the cells. With proper swelling, the chromosomes can be easily dispersed when the cell is disrupted. If the swelling is insufficient or if the cell membrane breaks prematurely, the chromosomes will remain entwined. No amount of syringing or physical disruption will separate them without damage. The length of time and the temperature at which swelling takes place varies with the isolation procedure (see Table I). The optimal KCl concentration to minimize chromosome clumping and debris has been investigated for the Hepes/MgSO$_4$ buffer and the Tris/MgCl$_2$ buffer (9,75). In our laboratory, we have found that a buffer containing $50\,\mathrm{m}M$ KCl results in reproducible and adequate swelling of any cell type at room temperature, if care is taken to remove all remnants of cell culture fluid from the cell pellet before the hypotonic buffer is added. Other groups adjust the conditions of cell swelling to the number and type of cells being used. Cell swelling can be monitored microscopically after adding PI or Hoechst 33258 (HO) to a small aliquot of the chromosome suspension.

In several methods (hexylene glycol, polyamine and citric acid), cells are centrifuged after the swelling step and are then resuspended in buffer containing the chromosome-stabilizing agent. The propidium iodide, citrate/ethanol, Tris/MgCl$_2$ and Hepes/MgSO$_4$ methods call for the addition of the stabilizing agent at the swelling step. This has advantages in simplicity and reduced cell loss. Furthermore, any chromosomes released from prematurely disrupted cells during the swelling step will remain intact and will not contribute to the debris continuum in the flow karyotype.

b. Stabilization

The stabilization of chromosomes in suspension is crucial to any chromosome isolation procedure. Figure 2 shows an example of the flow karyotype that results if a stabilizing agent (in this case Mg^{2+}) is omitted from the isolation buffer.

Early chromosome isolation procedures utilized low pH and divalent cations to stabilize chromosome structure for morphological and biochemical studies [reviewed by Hanson (37)]. Low pH isolation methods have been used for flow cytometric analyses of chromosomes in several laboratories (17,68,69,70). A problem with these methods is the lability of chromosomal proteins and damage to nucleosomal structure observed at low pH (22,72). Therefore, efforts have been made to isolate chromosomes in solutions at neutral pH (57).

Various agents have been used to prevent the swelling and degradation of chromosomes

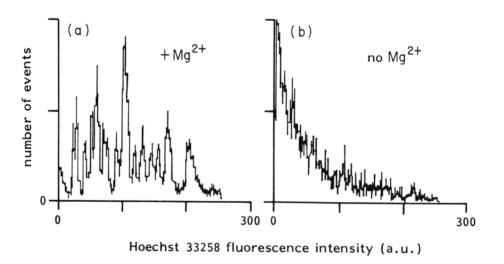

Fig. 2. Fluorescence distribution of human chromosomes isolated using the Hepes/MgSO$_4$ method demonstrating the necessity of the chromosome-stabilizing agent, Mg^{2+}, in the isolation buffer. The chromosomes were prepared from normal human fibroblasts (LLNL 811) with (A) or without (B) the addition of $10\,\mathrm{m}M$ MgSO$_4$. The chromosomes were stained with Hoechst 33258 and measured in a flow cytometer. [Data from Trask (73).]

in suspension at neutral pH prior to flow karyotyping. Chromosomes isolated with any of the procedures listed in Table I can be stored for at least several weeks without affecting resolution in conventional 64 × 64-channel flow karyotypes.

The stabilizing agents include:

(1) *Divalent cations, such as magnesium and calcium ions.* These ions stabilize the DNA helix, as evidenced by an increase in the melting temperature of DNA in their presence (*21,23,29*), especially at low ionic strengths (*71,85*). These ions also keep chromatin condensed, contributing to the stability of isolated chromosomes (*1,12,63*). The Tris/ $MgCl_2$ and Hepes/$MgSO_4$ methods rely on divalent cations to maintain chromosome integrity in suspension (*9,65,75,76*). Nucleases that may be present in the chromosome suspensions, however, are active in the presence of divalent cations. Therefore, these procedures are of limited use if high molecular weight DNA must be maintained during long-term storage of chromosome suspensions.

(2) *Cationic polyamines, such as spermine and spermidine.* These compounds have been shown to raise the thermal stability of DNA and to prevent mechanical damage to isolated DNA (*71*). The polyamine procedure, developed by Blumenthal *et al.* (*10*) and adapted for flow cytometry by Sillar and Young (*67*) and Lalande *et al.* (*45*), incorporates both spermine and spermidine to stabilize isolated chromosomes. Use of polyamines as stabilizing agents allows the addition of chelators, such as EDTA and EGTA, to the isolation buffer. Nuclease activity is inhibited and high molecular weight DNA can be extracted from isolated chromosomes even after long-term storage. However, chromomycin, an important stain for bivariate flow karyotyping, requires Mg^{2+} ions to bind to DNA. Thus, excess Mg^{2+} must be added with chromomycin to chromosomes isolated using the polyamine procedure to saturate the chelators and achieve equilibrium staining.

(3) *Hexylene glycol.* A chromosome isolation procedure using this agent to stabilize

chromosomes has been used for flow karyotyping with the addition of divalent cations (*14,32,34*). An isolation solution containing hexylene glycol and Ca^{2+} ions and buffered at pH 10 has been reported by Wray *et al.* (*82*).

(4) *Intercalating agents, such as propidium iodide (PI), ethidium bromide (EB), or 4,5',8-trimethylpsoralen.* Intercalating agents are known to stabilize the DNA helix (*54*). Isolation procedures employing PI as the stabilizing agent have been developed for flow karyotyping (PI, modified PI and citrate/ethanol). Intercalation of PI causes an unwinding of the DNA (*79*), resulting in "swollen" chromosomes. This feature can be used to advantage in slit-scan analysis, where long chromosomes with defined centromeres are desired (*55,56*). Procedures using PI as both the stabilizing agent and stain have the obvious disadvantage that one is restricted to quantifying the total DNA content of isolated chromosomes. The nonfluorescent intercalating agent, trimethylpsoralen, can also be used to stabilize chromosomes in place of PI. In its presence, chromosomes can be stained with HO for flow cytometric measurements (*84*).

c. Other Additives

Isolation buffers may contain other ingredients whose addition is intended to promote the isolation of individual and intact chromosomes. These ingredients include:

(1) *Chelators.* As discussed above, nuclease activity is inhibited and high molecular weight DNA can be obtained if divalent cations are chelated by EDTA, EGTA or sodium citrate.

(2) *Spindle poisons.* Colchicine or vincristine may be added to the isolation buffer to inhibit the spontaneous reformation of mitotic spindle apparatus during the isolation steps.

(3) *Buffers.* Isolation solutions may be buffered with Tris or Hepes to reproducibly maintain the pH of the solution from sample to sample.

(4) *Reducing agents*. Dithiothreitol or β-mercaptoethanol can be added to the isolation buffer to reduce disulfide bridges that may hold proteins and cell matrix together, thus promoting the isolation of single chromosomes.

(5) *Ribonuclease*. RNase is added to isolation solutions when chromosomes are to be stained with propidium iodide. This reduces the debris continuum in the flow karyotype produced by double-stranded RNA, to which PI can bind.

(6) *Trypsin*. Trypsin and RNase treatment of chromosomes isolated with the hexylene glycol procedure has been reported to decrease chromosome clumping, but to increase the amount of small debris (*14*).

(7) *Ethanol*.

(8) *Sucrose*.

d. Disruption of Cells

The detergents Triton X-100 and digitonin are commonly used to solubilize the plasma membrane of swollen cells for chromosome isolation (PI, modified PI, Tris/MgCl$_2$, Hepes/MgSO$_4$, polyamine procedures). Detergents are not used in the citric acid method, citrate/ethanol or hexylene glycol procedures.

After detergent treatment, cells are physically disrupted. If cells have been swollen properly, physical disruption need not be vigorous. Overshearing can result in damaged chromosomes. Cell disruption can be monitored under a fluorescence microscope after staining an aliquot of the chromosome suspension with PI or HO. Fibroblast cells are usually disrupted by passing the suspension several times through a needle on a syringe. Lymphoblast and lymphocyte cells can be easily disrupted by rapid vortexing for approximately 10 sec. The use of a homogenizer and sonicator has also been reported (*68*).

e. Staining

The DNA fluorochromes most commonly used for flow cytometric analysis of chromosomes are compatible with many of the isolation buffers. PI, HO, chromomycin A3 (CA3), DIPI and DAPI can be used with the Hepes/MgSO$_4$

or Tris/MgCl$_2$ procedures without method changes (*3,27,44,61,65,75,76,77*). Chromosomes isolated using the polyamine buffer can be stained with ethidium bromide (EB) (*67*) or PI, although degradation of the flow karyotype with this stain has been reported (*45*). Chromosomes isolated using the polyamine buffer can be stained with DIPI, DAPI, HO and CA3. As noted above, the stock solution of CA3 stain must contain sufficient MgCl$_2$ to allow homogeneous staining of chromosomes in the presence of the chelators in the polyamine buffer (*46*). Karyotype resolution can be improved for chromosomes isolated using both the Hepes/MgSO$_4$ and the polyamine procedures, if 10 mM sodium citrate and 25 mM sodium sulfite are added to the chromosome suspensions at least 10 min before flow analysis (*77*). Chromosomes isolated using the hexylene glycol method can be stained with any of the DNA stains, if MgCl$_2$ is added to the buffer (*40,47,48*). Chromosomes isolated using the citric acid method have been successfully stained with ethidium bromide or with a combination of EB and mithramycin. They can also be stained with HO and CA3, if they are first transferred to a neutral buffer containing MgCl$_2$ (*16*). PI serves both as the stabilizer and the stain in the PI methods, limiting these methods to univariate measurements of chromosome DNA content. However, the replacement of PI with the nonfluorescent intercalator, psoralen, has been shown to allow staining with Hoechst dyes (*84*).

III. CHARACTERISTICS OF ISOLATED CHROMOSOMES

A. YIELD

The number of single chromosomes obtained at the end of an isolation procedure from a given number of mitotic cells can be an important parameter in choosing an isolation method for chromosome sorting. Yield, for example, can be expected to be higher in methods with few steps than in methods in which cells are lost during recentrifugation. Unfortunately, chromosome yield is not routinely reported in the literature for many methods. In our laboratory, the efficiency of chromosome recovery using the polyamine and Hepes/$MgSO_4$ methods averages 30 and 45%, respectively (73,78; B. Trask and C. Lozes, unpublished observations).

B. IDENTIFICATION OF SORTED CHROMOSOMES

Sorting and identifying the chromosomes responsible for peaks in the flow karyotype can be an aid in the identification and cytogenetic study of chromosomes producing the peaks in a flow karyotype or a means to check the purity of sorted chromosomes (8,13,14). For these purposes, chromosomes are collected, fixed to slides and banded. The ease with which this can be done is determined by the chromosome isolation procedure. Chromosomes isolated using the hexylene glycol method can be successfully banded using quinacrine staining (13,14,34,84). Approximately 15% of sorted chromosomes can be identified after banding (14). Chromosomes prepared using the Tris/$MgCl_2$ method can be banded with quinacrine after sorting, if they are diluted during sorting in a sheath fluid containing hexylene glycol (27). Chromosomes isolated using the polyamine buffer can be banded using quinacrine staining (49). Bands can also be produced in

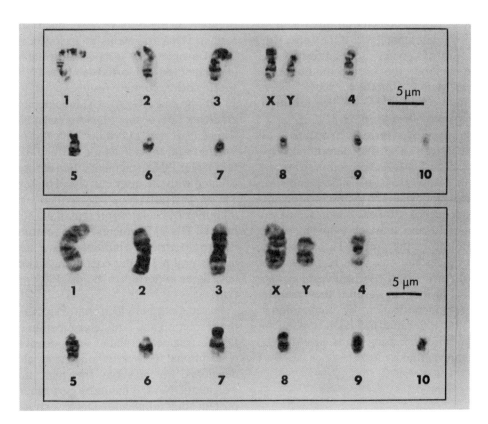

Fig. 3. Chinese hamster chromosomes isolated using the Hepes/$MgSO_4$ procedure, fixed with dimethylsuberimidate, attached to slides and stained with Giemsa to produce a banding pattern. The karyograms were compiled from photographs taken of two separate metaphases on two different slides. [Data from Trask (73).]

chromosomes isolated using the Hepes/MgSO$_4$ method. Such chromosomes are fixed with a protein cross-linker, attached to slides and Giemsa-banded (73) (Fig. 3). The majority of chromosomes isolated using the PI method can be banded, if they are fixed and stained with DAPI and actinomycin D (11). These results indicate that the structure in chromosomes required for the production of bands are preserved through most chromosome isolation procedures.

Individual chromosome types can also be sorted onto nitrocellulose filters for identification or for gene mapping. Hybridization of chromosome-specific DNA sequence probes to these filters can be used to identify the chromosomes responsible for the peaks in the fluorescence distribution (8,15,53). Bernheim et al. (8) report that contaminating chromosomes can be detected in this way at a frequency of greater than 10%, if they are all of one type. Chromosomes isolated using the polyamine method, the citrate method and the Hepes/MgSO$_4$ method have been demonstrated to be suitable for hybridization after sorting onto filters (8,15,53,73).

The purity of sorted fractions of chromosomes can also be determined by fluorescence in situ hybridization. Sorted chromosomes are fixed to slides and are hybridized to biotinylated chromosome-specific repetitive sequence probes. Fluoresceinated avidin is used to label those chromosomes that have hybridized to the probe. The frequency of hybridized chromosomes can be taken as a measure of sorting purity. Chromosomes can be identified in this way, even if they are too distorted for identification by conventional banding. This procedure has been applied to chromosomes isolated using the Hepes/MgSO$_4$ and polyamine methods (30,78).

C. Molecular Weight

The size of the DNA that can be extracted from sorted chromosomes may be a limiting factor if sorted chromosomes are to be used for the production of large-insert libraries in cosmids or yeast artificial chromosomes or if they are to be analyzed by pulsed field gel electrophoresis after digestion with rare-cutting restriction enzymes. The manner in which chromosomes are prepared has been shown to have an effect on the size of DNA fragments that can be isolated from them (10). Conventional agarose gel electrophoresis can be used to assay the molecular weight of DNA extracted from mitotic chromosomes. Figure 4 shows a comparison of the molecular weight of DNA from chromosomes isolated using the Hepes/MgSO$_4$, polyamine, hexylene glycol and citric acid procedures (73). In all cases the DNA fragments are larger than 50 kbp and are suitable for cloning in small-insert recombinant libraries. Chromosomes prepared with the psoralen method have also been shown to contain DNA fragments at least 50 kbp in size (84). Chromosomes isolated using the hexylene glycol method and sorted using the LLNL high-speed sorter contain fragments at least 15–150 kbp in size (33,66). Using alkaline sucrose gradients to assess DNA fragment size, Blumenthal et al. (10) found that chromosomes isolated with the polyamine buffer yielded 100-fold larger DNA fragments than chromosomes isolated with the hexylene glycol buffer (300 vs 3 kbp). This difference was ascribed to nuclease inhibition by chelators present in the polyamine buffer. DNA fragments from chromosomes isolated in citric acid buffer and later suspended in a neutral pH buffer containing 5 mM MgCl$_2$ have been reported to be 60–100 kbp in size (16).

Pulsed field gel electrophoresis now allows the resolution and size determination of larger DNA fragments than was possible with conventional gels. Methods have also been developed to preserve the high molecular weight of DNA fragments by performing gentle protein digestion and DNA extraction on chromosomes and nuclei trapped in agarose gels. Using these techniques (Fig. 5), it can be demonstrated that DNA fragments in excess

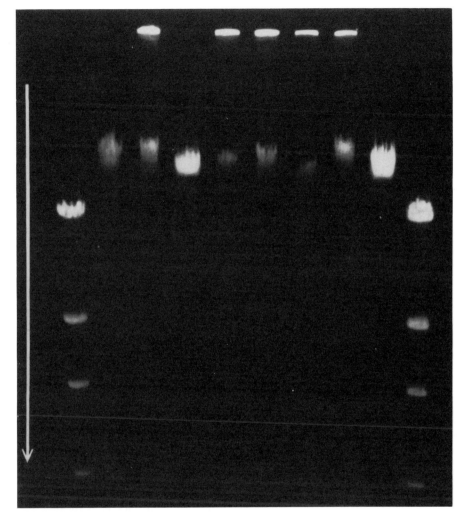

Fig. 4. A 0.3% agarose gel after electrophoresis showing the molecular weights of DNA extracted from chromosomes isolated from Chinese hamster fibroblasts according to several procedures. Lanes 1 and 10, molecular weight standards: (Hind III digest of lambda bacteriophage (23.5, 9.7, 6.6, 4.5 kbp bands shown)); lane 2, high molecular weight mouse DNA; lanes 4 and 9, intact lambda bacteriophage (50 kbp); lane 8, DNA isolated from cells trypsinized from cultures and not exposed to any of the chromosome isolation buffers; lane 3, chromosomes isolated using the Hepes/MgSO$_4$ procedure; lane 5, polyamine procedure; lane 6, citric acid procedure; lane 7, hexylene glycol procedure. [Data from Trask (73).]

of 500 kbp can be isolated from sorted chromosomes prepared using the Hepes/MgSO$_4$ procedure if chromosomes are used within several days after isolation. Chromosomes prepared using the polyamine procedure are significantly more resistant to nuclease digestion, and high molecular weight DNA (in excess of 700 kbp) can be reproducibly isolated from sorted chromosomes (Lynn Clark, unpublished results).

**Days
after
isolation**

λ 0 1 2 5 6 λ

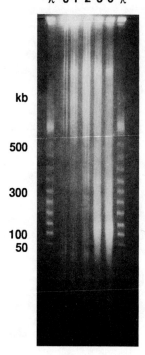

kb

500

300

100
50

Fig. 5. Field inversion electrophoresis gel demonstrating the molecular weight of DNA extracted from Hepes/MgSO$_4$-isolated chromosomes. DNA from Hepes/MgSO$_4$-isolated chromosomes is >500 kbp in size up to 2 days after isolation; by day 5, degradation has occurred resulting in fragments 100–500 kbp in size. Chromosomes were isolated from GM131, a human lymphoblast cell line. Nuclei were removed by centrifugation, and the chromosome-containing supernate was stored at 4°C. Samples were removed at various time points, and chromosomes were resuspended in 0.5% Hank's balanced salt solution and 0.5% low-melting point agarose. Agarose blocks suitable for electrophoresis were prepared and were incubated in ESP (0.5 M EDTA, 1% Lauroyl sarcosine, 2 mg/ml Proteinase K) at 50°C for 2 days. Samples were electrophoresed to resolve large DNA fragments using a field-inversion gel electrophoresis apparatus (260 V; 3:1 switching ratio; cycle time: 3 sec at $t = 0$, 60 sec at $t = 24$ h) in a 1% agarose gel with 0.5% TBE buffer containing 0.5 μg/ml ethidium bromide. A lambda ladder serves as the molecular weight standard (increments of 50 kb). (Figure from J. Fuscoe and B. Trask, unpublished results.)

D. PROTEIN CONTENT AND NUCLEOSOME STRUCTURE

Relatively little is known about the proteins retained in isolated chromosomes. Immuno-fluorescent labeling histone 2B in chromosomes isolated using the Hepes/MgSO$_4$ method has been performed (74). HPLC analysis has shown that the relative concentrations of histone proteins extracted from sorted chromosomes isolated using the Hepes/MgSO$_4$ method is similar to that predicted from the nucleosome structure (74).

The pattern of isolated proteins on polyacrylamide gels has been used to show that proteins are better preserved in chromosomes isolated in the presence of Mg^{2+} than in the presence of polyamines (2). Micrococcal nuclease digestion has been used to demonstrate that nucleosome structure is better preserved in chromosomes isolated in polyamine buffer than in hexylene glycol buffer (72). Green and Fantes (35) have demonstrated that the centromeres of chromosomes isolated and stained in a buffer based on the Tris/MgCl$_2$ procedure can be fluorescently labeled with an anti-kinetochore antibody.

IV. SUMMARY AND RECOMMENDATIONS

Several chromosome isolation procedures are being used with success for the analysis and sorting of chromosomes by flow cytometry. It may be unreasonable to expect that one protocol will be distilled from these procedures and be universally applicable. The specific aims and applications in each laboratory will continue to dictate which chromosome isolation procedure is employed.

For bivariate flow karyotyping and the detection of chromosome abnormalities at the detection limit of conventional banding techniques, we find the Hepes/MgSO$_4$ method to be the most useful. It is simple and reproducible, and high-resolution flow karyotypes can be obtained easily from a variety of cell types, from few cells and for a variety of chromosome stains. Other laboratories have found the polyamine procedure useful for the detection of chromosome abnormalities. For species with simple karyotypes and large differences in size among chromosomes, such as the Chinese hamster, karyotype resolution obtained with a single stain may be sufficient for many experiments. In this case, the simplicity of the PI methods may be attractive.

The presence of EDTA in the polyamine buffer prevents nuclease-mediated degradation of chromosomal DNA, making it ideal for chromosome sorting for the production of large-insert recombinant DNA libraries or for pulsed field gel electrophoresis analysis. Mg^{2+} in the Hepes/$MgSO_4$ or Tris/$MgCl_2$ buffers limits one to the production of small-insert recombinant libraries from sorted chromosomes. The citric acid procedure has the advantage that it can be coupled effectively with bulk methods for chromosome fractionation. Concentrated samples can be prepared using any of these procedures for high-speed chromosome sorting. It is possible to band chromosomes isolated with these procedures after sorting to check sorting purity. Fluorescence *in situ* hybridization using chromosome-specific repetitive DNA sequence probes, however, is a more convenient method for chromosome identification and can be applied to the compact chromosomes resulting from many isolation procedures.

Centromere detection using slit-scan flow cytometry requires stretched, enlarged chromosomes, and PI-based isolation procedures have proven useful for this application. The Hepes/$MgSO_4$ and Tris/$MgCl_2$ buffers are compatible with immunofluorescent-labeling reagents. Chromosomes isolated using these procedures can be probed for specific proteins or other structures (histones, kinetochores) using flow cytometry.

Much effort has been placed on methods to stabilize chromosomes to maintain chromosome integrity while allowing DNA denaturation and hybridization to fluorescently tagged DNA sequence probes. Fluorescence hybridization has been demonstrated on chromosomes isolated using a modified hexylene glycol procedure (*28*) and on chromosomes isolated using the Hepes/$MgSO_4$ procedure followed by dimethylsuberimidate fixation (B. Trask, unpublished observations). However, flow cytometric measurements of both DNA content and hybridized probes have met with limited success to date. Much work still needs to be done to optimize chromosome isolation buffers and stabilizing agents before fluorescence hybridization can be used to improve identification of chromosomes or detection of chromosome abnormalities by flow cytometry.

Flow analysis of plant chromosomes still lags behind mammalian flow karyotyping, largely because few groups have focused their attention on the specific problems of plant chromosome isolation. Further improvements in protoplast lysis, chromosome stabilization and removal of autofluorescent compounds will improve the flow karyotypes that can be obtained from plant protoplasts. This will permit karyotype comparisons to be made among plant species and hybrids. It will also make possible the construction of chromosome-enriched recombinant DNA libraries for the study of plant genomes at a molecular level.

ACKNOWLEDGEMENTS

Work performed under the auspices of the U.S. Department of Energy by the Lawrence Livermore National Laboratory under contact number W-7405-ENG-48 with support from USPHS grant 17665.

REFERENCES

1. Aaronson, R., and Woo, E. (1981). Organization in the cell nucleus: divalent cations modulate the distribution of condensed and diffuse chromatin. *J. Cell Biol.* **90**, 181–186.
2. Adolph, K. W. (1980). Isolation and structural organization of human mitotic chromosomes. *Chromosoma* **76**, 23–33.
3. Arkesteijn, G., Martens, A., and Hagenbeek, A. (1987). Dual laser beam flow karyotyping: clinical

studies in chronic myelocytic leukemia patients. *Cytometry, Suppl.* **1**, 90.

4. Aten, J. A., Kooi, M. W., Bijman, J. B., Kipp, J. B., and Barendsen, G. W. (1984). Flow-cytometric analysis of chromosome damage after irradiation: relation to chromosome aberrations and cell survival. *In* "Biological Dosimetry" (W. G. Eisert and M. L. Mendelsohn, eds), pp. 51–59. Springer-Verlag, Berlin.

5. Baron, B., Metezeau, P., Kelly, F., Bernheim, A., Berger, R., Guenet, J., and Goldberg, M. (1984). Flow cytometry isolation and improved visualization of sorted mouse chromosomes; purification of chromosomes X and ISO-1 from cell lines with Robertsonian translocations. *Exp. Cell Res.* **152**, 220–230.

6. Baron, B., Métézeau, P., Hatat, D., Roberts, C., Goldberg, M., and Bishop, C. (1986). Cloning of DNA libraries from mouse Y chromosomes purified by flow cytometry. *Somat. Cell Molec. Genet.* **12**, 289–295.

7. Bartholdi, M. F., Ray, F. A., Cram, S., and Kraemer, P. M. (1984). Flow karyology of serially cultured Chinese hamster cell lineages. *Cytometry* **5**, 534–538.

8. Bernheim, A., Metezeau, P., Guellaen, G., Fellous, M., Goldberg, M. E., and Berger, R. (1983). Direct hybridization of sorted human chromosomes: Localization of the Y chromosome on the flow karyotype. *Proc. Natl Acad. Sci. USA* **80**, 7571–7575.

9. Bijman, J. T. (1983). Optimization of mammalian chromosome suspension preparations employed in a flow cytometric analysis. *Cytometry* **3**, 354–358.

10. Blumenthal, A., Dieden, J., Kapp, L., and Sedat, J. (1979). Rapid isolation of metaphase chromosomes containing high molecular weight DNA. *J. Cell Biol.* **81**, 255–259.

11. Buys, C., Koerts, T., and Aten, J. (1982). Well-identifiable human chromosomes isolated from mitotic fibroblasts by a new method. *Hum. Genet.* **61**, 157–159.

12. Cantor, K., and Hearst, J. (1970). The structure of metaphase chromosomes. I. Electromagnetic titration, magnesium ion binding and circular dichroism. *J. Mol. Biol.* **49**, 213–229.

13. Carrano, A. V., Gray, J. W., Langlois, R. G., Burkhart-Schultz, K. J., Van Dilla, M. A. (1979). Measurement and purification of human chromosomes by flow cytometry and sorting. *Proc. Natl Acad. Sci. USA* **76**, 1382–1384.

14. Carrano, A. V., Van Dilla, M. A., and Gray, J. W. (1979). Flow cytogenetics: a new approach to chromosome analysis. *In* "Flow Cytometry and Sorting" (M. R. Melamed, P. F. Mullaney and M. L. Mendelsohn, eds), pp. 421–451. Wiley, New York.

15. Collard, J., de Boer, P., Janssen, J., Schijven, J., and de Jong, B. (1985). Gene mapping by chromosome spot hybridization. *Cytometry* **6**, 179–185.

16. Collard, J. G., Phillippus, E., Tulp, A., Lebo, R. V.,

Gray, J. W. (1984). Separation and analysis of human chromosomes by combined velocity sedimentation and flow sorting applying single- and dual-laser flow cytometry. *Cytometry* **5**, 9–19.

17. Collard, J. G., Tulp, A., Hollander, J. H., Bauer, F. W., and Boezeman, J. (1980). Separation of large quantities of chromosomes by velocity sedimentation at unit gravity. *Exp. Cell Res.* **126**, 191–197.

18. Collard, J. G., Tulp, A., Stegeman, J., and Boezeman, J. (1981). Separation of human metaphase chromosomes at neutral pH by velocity sedimentation at twenty times gravity. *Exp. Cell Res.* **133**, 341–346.

19. Conia, J., Bergouniox, C., Perennes, C., Muller, P., Brown, S., and Gadal, P. (1987). Flow cytometric analysis and sorting of plant chromosomes from *Petunia hybrida* protoplasts. *Cytometry* **8**, 500–508.

20. Cram, L. S., Bartholdi, M. F., Ray, F. A., Travis, G. L., and Kraemer, P. M. (1983). Spontaneous neoplastic change of Chinese hamster cells in culture: II. Multistep progression of karyotype. *Cancer Res.* **43**, 4828–4837.

21. Darzynkiewicz, Z., Traganos, F., Arlin, Z., Sharpless, T., and Melamed, M. (1976). Cytofluorometric studies of conformation of nucleic acids *in situ*. II. Denaturation of deoxyribonucleic acid. *J. Histochem. Cytochem.* **24**, 49–58.

22. Darzynkiewicz, Z., Traganos, F., Kapuscinski, J., Staiano-Coico, L., and Melamed, M. (1984). Accessibility of DNA *in situ* to various fluorochromes: relationship to chromatin changes during erythroid differentiation of Friend leukemia cells. *Cytometry* **5**, 355–363.

23. Darzynkiewicz, Z., Traganos, F., Sharpless, T., and Melamed, M. (1975). Thermal denaturation of DNA *in situ* as studied by acridine orange staining and automated cytofluorometry. *Exp. Cell Res.* **90**, 411–428.

24. de Laat, A., and Blass, J. (1984). Flow cytometric characterization and sorting of plant chromosomes. *Theor. Appl. Genet.* **67**, 463–467.

25. de Laat, A., and Schel, J. (1986). The integrity of metaphase chromosomes of *Haplopappus gracilis* (Nutt.) Gray isolated by flow cytometry. *Plant Sci.* **47**, 145–151.

26. de Taisne, C., Gegonne, A., Stehelin, D., Bernheim, A., and Berger, R. (1984). Chromosomal localization of the human proto-oncogene c-*ets*. *Nature* **310**, 581–583.

27. Disteche, C. M., Kunkel, L. M., Lojewski, A., Orkin, S. H., Eisenhard, M., Sahar, E., Travis, B., and Latt, S. A. (1982). Isolation of mouse X-chromosome specific DNA from an X-enriched lambda phage library derived from the flow sorted chromosomes. *Cytometry* **2**, 282–286.

28. Dudin, G., Cremer, T., Schardin, M., Hausmann, M., Bier, F., and Cremer, C. (1987). A method for nucleic acid hybridization to isolated chromosomes in suspension. *Human Genet.* **76**, 290–292.

29. Eichhorn, G. L. (1962). Metal ions as stabilizers or destabilizers of the deoxyribonucleic acid structure. *Nature* **194**, 474–475.

30. Gray, J., Dean, P., Fuscoe, J., Peters, D., Trask, B., van den Engh, G., and Van Dilla, M. (1987). High-speed chromosome sorting. *Science* **238**, 323–329.

31. Gray, J., Trask, B., van den Engh, G., Silva, A., Lozes, C., Grell, S., Schonberg, S., Yu, L.-C., and Golbus, M. (1988). Application of flow karyotyping in prenatal detection of chromosome aberrations. *Am. J. Hum. Genet.* **42**, 49–59.

32. Gray, J. W., Carrano, A. V., Steinmetz, L. L., Van Dilla, M. A., Moore III, D. H., Mayall, B. H., and Mendelsohn, M. L. (1975). Chromosome measurement and sorting by flow systems. *Proc. Natl Acad. Sci. USA* **72**, 1231–1234.

33. Gray, J. W., Cremer, C., Peters, D., Marsh, L., and Lord, D. (1981). Very high speed sorting: a preliminary report. *Cytometry* **2**, 101 (Abstract).

34. Gray, J. W., Langlois, R. G., Carrano, A. V., Burkhart-Schultz, K., and Van Dilla, M. A. (1979). High resolution chromosome analysis: one and two parameter flow cytometry. *Chromosoma* **73**, 9–27.

35. Green, D., and Fantes, J. (1987). Detection of dicentric chromosomes by flow analysis. *Cytometry, Suppl.* **1**, 13.

36. Green, D., Fantes, J., Buckton, K., Elder, J., Malloy, P., Carothers, A., and Evans, H. (1984). Karyotyping and identification of human chromosome polymorphisms by single fluorochrome flow cytometry. *Hum. Genet.* **66**, 143–146.

37. Hanson, C. V. (1975). Techniques in the isolation and fractionation of eukaryotic chromosomes. *In* "New Techniques in Biophysics and Cell Biology" (B. Smith and R. Pain, eds), pp. 43–83. Wiley, London.

38. Harris, P., Boyd, E., and Ferguson-Smith, M. (1985). Optimising human chromosome separation for the production of chromosome-specific DNA libraries by flow sorting. *Hum. Genet.* **70**, 59–65.

39. Harris, P., Boyd, E., Young, B., and Ferguson-Smith, M. (1986). Determination of the DNA content of human chromosomes by flow cytometry. *Cytogenet. Cell Genet.* **41**, 14–21.

40. Hutter, K., and Stöhr, M. (1985). Detection and separation of the submetacentric marker chromosome of the WALKER (W-256) carcinoma using flow cytometry and sorting. *Histochem.* **82**, 469–475.

41. Kanda, M., Schreck, R., Alt, F., Bruns, G., Baltimore, D., and Latt, S. A. (1983). Isolation of amplified DNA sequences from IMR-32 human neuroblastoma cells: facilitation by fluorescence-activated flow sorting of metaphase chromosomes. *Proc. Natl Acad. Sci. USA* **80**, 4069–4073.

42. Kooi, M. W., Aten, J. A., Stap, J., Kipp, J. B. A., Barendsen, G. W. (1984). Preparation of chromosome suspensions from cells of a solid experimental tumor for measurement by flow cytometry. *Cytometry* **5**, 547–549.

43. Krumlauf, R., Jeanpierre, M., and Young, B. D. (1982). Construction and characterization of genomic libraries of specific human chromosomes. *Proc. Natl Acad. Sci. USA* **79**, 2971–2975.

44. Kunkel, L. M., Tantravahi, U., Eisenhard, M., and Latt, S. A. (1982). Regional localization on the human X of DNA segments cloned from flow sorted chromosomes. *Nucl. Acids Res.* **10**, 1557–1578.

45. Lalande, M., Kunkel, L. M., Flint, A., and Latt, S. A. (1984). Development and use of metaphase chromosome flow sorting methodology to obtain recombinant phage libraries enriched for parts of the X chromosome. *Cytometry* **5**, 101–107.

46. Lalande, M., Schreck, R., Hoffman, R., and Latt, S. A. (1985). Identification of inverted duplicated # 15 chromosomes using bivariate flow karyotype analysis. *Cytometry* **6**, 1–6.

47. Langlois, R. G., Carrano, A. V., Gray, J. W., and Van Dilla, M. A. (1980). Cytochemical studies of metaphase chromosomes by flow cytometry. *Chromosoma* **77**, 229–251.

48. Langlois, R. G., Yu, L.-C., Gray, J. W., and Carrano, A. V. (1982). Quantitative karyotyping of human chromosomes by dual beam flow cytometry. *Proc. Natl Acad. Sci. USA* **79**, 7876–7880.

49. Lebo, R., Cheung, M.-C., Bruce, B., Riccardi, V., Kao, F.-T., and Kan, Y. (1985a). Mapping parathyroid hormone, beta-globin, insulin and LDH-A genes within the human chromosome 11 short arm by spot blotting sorted chromosomes. *Hum. Genet.* **69**, 316–320.

50. Lebo, R., Golbus, M., and Cheung, M.-C. (1986). Detecting abnormal human chromosome constitutions by dual flow cytogenetics. *Am. J. Med. Genet.* **25**, 519–529.

51. Lebo, R., Tolan, D., Bruce, B., Cheung, M.-C., and Kan, Y. (1985b). Spot–blot analysis of sorted chromosomes assigns a fructose intolerance disease locus to chromosome 9. *Cytometry* **6**, 478–483.

52. Lebo, R. V., Gorin, F., Fletterick, R. J., Kao, F.-T., Cheung, M.-C., Bruce, B. D., and Kan, Y. W. (1984). High resolution chromosome sorting and DNA spot–blot analysis assign McArdle's syndrome to chromosome 11. *Science* **225**, 57–59.

53. Lebo, R. V., Kan, Y. W., Cheung, M.-C., Carrano, A. V., Yu, L.-C., Chang, J. C., Cordell, B., and Goodman, H. M. (1982). Assigning the polymorphic human insulin gene to the short arm of chromosome 11 by chromosome sorting. *Hum. Genet.* **60**, 10–15.

54. LePecq, J. B., and Paoletti, C. (1967). A fluorescent complex between ethidium bromide and nucleic acids. Physical-chemical characterization. *J. Mol. Biol.* **37**, 87–106.

55. Lucas, J., and Gray, J. (1987). Centromeric index versus DNA content flow karyotypes of human chromosomes measured by means of slit-scan flow cytometry. *Cytometry* **8**, 273–279.

56. Lucas, J. N., Gray, J. W., Peters, D. C., and Van

Dilla, M. A. (1983). Centromeric index measurement by slit-scan flow cytometry. *Cytometry* **4**, 109–116.

57. Maio, J. J., and Schildkraut, C. L. (1967). Isolated mammalian metaphase chromosomes. I. General characteristics of nucleic acids and proteins. *J. Mol. Biol.* **24**, 29–30.

58. Martens, A., Arkesteijn, G., Hagemeijer, A., and Hagenbeek, A. (1987). Dual laser beam flow cytometry of chromosomes in experimental leukemia. *Cytometry, Suppl.* **1**, 90.

59. Matsson, P., Anneren, G., and Gustavsson, I. (1986). Flow cytometric karyotyping of mammals, using blood lymphocytes: detection and analysis of chromosomal abnormalities. *Hereditas* **104**, 49–54.

60. Matsson, P., and Rydberg, B. (1981). Analysis of chromosomes from human peripheral lymphocytes by flow cytometry. *Cytometry* **1**, 369–372.

61. Meyne, J., Bartholdi, M. F., Travis, G., and Cram, L. S. (1984). Counterstaining human chromosomes for flow karyology. *Cytometry* **5**, 580–583.

62. Nüsse, M., Egner, H., and Krämer, M. (1987). Flow karyotyping of mouse tumor cell lines. *Cytometry, Suppl.* **1**, 90.

63. Olins, D. E., and Olins, A. L. (1972). Physical studies of isolated eucaryotic nuclei. *J. Cell Biol.* **53**, 715–736.

64. Otto, F. J., and Oldiges, H. (1980). Flow cytogenetic studies in chromosomes and whole cells for the detection of clastogenic effects. *Cytometry* **1**, 13–17.

65. Otto, F. J., Oldiges, H., Göhde, W., Barlogie, B., and Schumann, J. (1980). Flow cytogenetics of uncloned and cloned Chinese hamster cells. *Cytogenet. Cell Genet.* **27**, 52–56.

66. Peters, D., Branscomb, E., Dean, P., Merrill, T., Pinkel, D., Van Dilla, M., and Gray, J. W. (1985). The Livermore high speed sorter (HISS): design features, operational characteristics, and biological utility. *Cytometry* **6**, 290–301.

67. Sillar, R., and Young, B. D. (1981). A new method for the preparation of metaphase chromosomes for flow analysis. *J. Histochem. Cytochem.* **29P**, 74–78.

68. Stöhr, M., Huttler, K. J., Frank, M., and Görttler, K. (1982). A reliable preparation of mono-dispersed chromosome suspensions for flow cytometry. *Histochem.* **74**, 57–61.

69. Stubblefield, E., Cram, S., and Deaven, L. (1975). Flow microfluorometric analysis of isolated Chinese Hamster chromosomes. *Exp. Cell Res.* **94**, 464–468.

70. Stubblefield, E., and Wray, W. (1978). Isolation of specific human metaphase chromosomes. *Biochem. Biophys. Res. Comm.* **83**, 1404–1414.

71. Tabor, H. (1962). The protective effect of spermine and other polyamines against heat denaturation of deoxyribonucleic acid. *Biochem.* **1**, 496–501.

72. Tien Kuo, M. (1982). Comparison of chromosomal structures isolated under different conditions. *Exp. Cell Res.* **138**, 221–229.

73. Trask, B. (1985). Studies of chromosomes and nuclei using flow cytometry. Ph.D. Thesis, Leiden University, Leiden, The Netherlands.

74. Trask, B., van den Engh, G., Gray, J., Vanderlaan, M., and Turner, B. (1984). Immunofluorescent detection of histone 2B on metaphase chromosomes using flow cytometry. *Chromosoma* **90**, 290–302.

75. van den Engh, G., Trask, B., Cram, S., and Bartholdi, M. (1984). Preparation of chromosome suspensions for flow cytometry. *Cytometry* **5**, 108–117.

76. van den Engh, G., Trask, B., Gray, J., Langlois, R., Yu, L.-C. (1985). Preparation and bivariate analysis of suspensions of human chromosomes. *Cytometry* **6**, 92–100.

77. van den Engh, G., Trask, B., Lansdorp, P., and Gray, J. (1988). Improved resolution of flow cytometric measurements of Hoechst- and chromomycin-A3-stained human chromosomes after addition of citrate and sulfite. *Cytometry* **9**, 266–270.

78. Van Dilla, M., Deaven, L., Albright, K., Allen, N., Aubuchon, M., Bartholdi, M., Browne, N., Campbell, E., Carrano, A., Clark, L., Cram, L., Fuscoe, J., Gray, J., Hildebrand, C., Jackson, P., Jett, J., Longmire, J., Lozes, C., Luedemann, M., Martin, J., McNinch, J., Meincke, L., Mendelsohn, M., Meyne, J., Moyzis, R., Munk, A., Perlman, J., Peters, D., Silva, A., and Trask, B. (1986). Human chromosome-specific DNA libraries: construction and availability. *Biotechnology* **4**, 537–552.

79. Waring, M. (1970). Variation of the supercoils in closed circular DNA by binding of antibiotics and drugs: Evidence for molecular models involving intercalation. *J. Mol. Biol.* **54**, 247–279.

80. Wirchubsky, Z., Perlmann, C., Lindsten, J., and Klein, G. (1983). Flow karyotype analysis and fluorescence-activated sorting of Burkitt-lymphoma-associated translocation chromosomes. *Int. J. Cancer* **32**, 147–153.

81. Wray, W., and Stubblefield, E. (1970). A new method for the rapid isolation of chromosomes, mitotic apparatus, or nuclei from mammalian fibroblasts at near neutral pH. *Exp. Cell Res.* **59**, 469–478.

82. Wray, W., Stubblefield, E., and Humphrey, R. (1972). Mammalian metaphase chromosomes with high molecular weight DNA isolated at pH 10.5. *Nature New Biol.* **238**, 237–238.

83. Young, B. D., Ferguson-Smith, M. A., Sillar, R., and Boyd, E. (1981). High-resolution analysis of human peripheral lymphocyte chromosomes by flow cytometry. *Proc. Natl Acad. Sci. USA* **78**, 7727–7731.

84. Yu, L.-C., Aten, J., Gray, J. W., and Carrano, A. V. (1981). Human chromosome isolation from short-term lymphocyte cultures for flow cytometry. *Nature* **293**, 154–155.

85. Zimmer, C., Reinert, K., Luck, G., Wahnert, U., Lober, G., and Thrum, H. (1979). Interaction of the oligopeptide antibiotics, netropsin, and distamycin A, with nucleic acids. *J. Mol. Biol.* **58**, 329–348.

DNA Stains as Cytochemical Probes for Chromosomes

RICHARD G. LANGLOIS

Biomedical Sciences Division
Lawrence Livermore National Laboratory
Livermore, CA 94550, USA

I. INTRODUCTION

The development of cytogenetic methods for the analysis of metaphase chromosomes has been dependent on the utilization of DNA-specific staining methods for visualizing chromosomes and for highlighting specific chromosomal features. Early studies used stains for total DNA content to identify individual chromosome types based on their size and the position of their centromeres. These staining methods allow human chromosomes to be visually resolved into seven groups (A through G), and many individual chromosome types can be identified by using quantitative image analysis methods to measure DNA content and centromeric index (*84*). In 1969, Cassperson and coworkers developed a staining method (Quinacrine banding) which produces transverse bands along the length of metaphase chromosomes (*19*). Since each human chromosome type has a distinctive banding pattern, all human chromosome types can be visually identified in banded metaphase spreads

(18). Banding methods have been successfully utilized to identify individual chromosome types in chromosome preparations from a variety of eukaryotic species. Transverse banding patterns on chromosomes have now been produced by a variety of newer banding methods including G-, R- and C-banding, as well as specialized staining methods for nucleolar-organizing regions and kinetochores (86,116). While the detailed mechanisms responsible for chromosome banding have yet to be fully elucidated, the differential staining of band *vs* interband regions clearly indicates that chromosomes are organized in regions which differ in cytochemical properties. The consistency of banding pattern within a species, and the similarity of banding patterns among closely related species (140), suggest that bands may reflect fundamental organizational units of metaphase chromosomes, with each unit corresponding to about 10^7 base pairs of DNA.

The introduction of flow cytometric methods for the analysis of stained suspensions of isolated metaphase chromosomes in 1975 (34,35), demonstrated that this approach could provide high-precision measurements of the fluorescence intensity of individual chromosome types. The observed high resolution of flow cytogenetic techniques suggested that this approach would be useful for quantitating stain interactions with individual chromosomes or chromosome segments, as well as for quantitative karyotype analysis and flow sorting. These applications, however, depend on staining methods that are compatible with the chemical environment in isolated chromosome suspensions and that provide sufficient information to allow most chromosome types to be separately resolved.

The purpose of this chapter is to introduce some of the criteria for selecting stains for flow cytogenetics, as well as to review some of the cytochemical characteristics of chromosomes that can be probed by appropriate staining methods. This chapter will focus on DNA-specific stains. Other methods, such as immunological labels or fluorescent poly-nucleotides, will be mentioned in other chapters. Initially, the physicochemical characteristics of chromosome suspensions which affect dye binding, and the basic specificities and spectral properties of the dyes will be discussed. Specific single- and multiple-dye staining methods which have proved useful for flow analysis of chromosomes are described. Finally, the relationship between flow cytometric measurements of stain fluorescence and biochemical variations among chromosome regions which may be important in chromosome banding are considered.

II. PRINCIPLES OF CHROMOSOME STAINING

A. CHROMOSOMAL CHARACTERISTICS AFFECTING DYES

Model system studies of the interactions of fluorescent dyes with DNA and chromatin in solution have demonstrated the potential of many dyes to be used as probes of the composition and structure of chromosomes. Fluorescent dyes are particularly sensitive to variations in composition or structure because such variations can affect both dye binding (e.g. binding affinity or number of binding sites per nucleotide) as well as the fluorescence efficiency and spectra of the bound dye molecules. While the fluorescence of stained chromosomes reflects the combination of many structural and compositional factors, it is often possible to determine the relative importance of different factors by the use of multiple stains with known differences in specificity, or by comparisons of chromosomal fluorescence before and after treatments having known effects on chromosomal proteins or structure.

The following is a brief listing of some of the characteristics which can be studied with dyes. For stains with no base specificity, the fluorescence intensity is generally proportional to the total DNA content of the chromosome (34,35,40). With base-specific stains, fluorescence intensity can be determined by the

DNA content, average base composition and by the frequency of specific base sequences (15,58,74,125). Other DNA features which can affect dyes include the presence of modified bases (70), single- vs double-stranded regions (26), supercoiled DNA (97,131) and altered DNA conformations, e.g. Z-DNA (102,124,130). Chromatin features that have been reported to affect dye binding or fluorescence include protein content (11) and degree of chromatin condensation (25,27).

B. CHEMISTRY OF CHROMOSOME SUSPENSIONS

A variety of methods have been developed for preparing suspensions of isolated metaphase chromosomes for flow analysis (see Chapter 4). While an isolation method is often chosen to provide chromosomes with specific characteristics (e.g. intact protein composition, high molecular weight DNA or long-term stability of the suspension) or to be compatible with specific cell types (e.g. fibroblasts, tumor cells or amniocytes), different isolation buffers can also have an important effect on chromosome staining. A number of cellular components and buffer components can effect fluorescence staining, and these will be discussed below using specific dyes as examples of these effects.

Chromosome suspensions usually are prepared by mechanical disruption of mitotic cells, so the final chromosome suspension also contains other cellular components, including membrane fragments, cytoplasmic organelles, and cytoplasmic proteins and RNA. While the presence of these other cellular components causes minimal problems with dyes having a high specificity for DNA (e.g. Hoechst 33258), substantial staining of nonchromosomal components can be observed with less-specific stains. Ethidium bromide, for example, binds with comparable affinity to both DNA and RNA, but background fluorescence from an RNA-bound stain can usually be reduced by RNase treatment of the chromosome suspension (121). Quinacrine, in contrast, shows substantial nonspecific binding with many other cellular components, and so this stain is poorly suited for flow cytogenetics.

Salt concentration, or ionic strength, effects dye–chromosome interactions in several ways. Ethidium bromide, like many other dyes, has two binding modes to DNA, a primary binding mode (intercalation between base pairs leading to high fluorescence efficiency) and a secondary mode (low fluorescence electrostatic interaction with the phosphate backbone of DNA) (75). High ionic strength inhibits secondary binding, but it can also reduce the binding affinity for intercalation (75). Salt concentrations of greater than 1 M can also indirectly affect staining by dissociating chromosomal proteins and increasing the accessibility of chromosomal DNA (9,94). Divalent cations (e.g. calcium and magnesium), as well as polyvalent cations (e.g. spermine or spermidine) also can have a significant effect on chromosome structure and degree of condensation, which may affect DNA accessibility to dye binding. For a few stains, however, divalent ion concentrations can dramatically effect dye binding. Chromomycin A3, for example, requires magnesium ions for binding (49,93). In general, the salt concentrations of different chromosome isolation buffers have been selected to optimize chromosome stability, so some modifications may be necessary for optimum staining with specific dyes.

Buffer pH varies considerably among chromosome isolation methods. The ionization state of dye molecules frequently varies with the solution pH, and changes in dye charge can profoundly affect both the binding characteristics as well as the fluorescence properties of dyes. Both the binding specificity and the emission wavelength of Hoechst 33258 change dramatically at acidic pH values (47). The DNA binding characteristics and fluorescence spectra of ethidium bromide, in contrast, is largely independent of pH (75,77). Two other buffer components, organic solvents and detergents, can affect the quality of stained chromosome suspensions indirectly by affecting the fluorescence of unbound dye. Since many fluorescent dyes exhibit enhanced fluorescence and spectral shifts in hydrophobic environments or nonpolar solvents (95), the presence of alcohols or glycols having lower polarity than water can lead to enhanced fluorescence from unbound dye. Similarly, the use of detergents at concentrations high enough to form micelles with hydrophobic cellular components can also provide a high-fluorescence environment for unbound dye.

C. BINDING KINETICS AND EQUILIBRIA

The intensity and uniformity of staining of chromosomes is dependant on the dye and chromosome concentrations, the affinity constants of the dye for both primary and secondary binding sites, and the degree of equilibration between the dye and chromosomes. In general, the resolution that can be obtained in flow analysis of metaphase chromosomes is determined by both the absolute fluorescence intensity of individual chromosomes and the uniformity of staining among chromosomes of the same type. The smallest human chromosome types contain only about 5×10^7 bp of DNA, so that even with saturating dye concentrations, peak widths are limited by photon statistics in most flow cytometers (36). Variations in condensation among chromosomes of the same type can also lead to subtle variations in binding affinity or kinetics, but these effects can be minimized by equilibrating the chromosome suspension with near-saturating concentrations of dye.

Chromosome suspensions derived from 10^7 mitotic cells/ml will have a DNA nucleotide concentration of about 4×10^{-4} M (58). Since the number of primary binding sites per nucleotide typically ranges from 0.2 to 0.01 (10,47,65,77,93), the concentration of accessible dye binding sites will range from about 10^{-4} to 10^{-5} M for commonly used dyes. Since dye dissociation constants (K_d) for primary binding sites typically range from 10^{-5}

to 10^{-7} M (10,65,77,93), near-stoichiometric dye binding can be expected when the dye and binding site concentrations are equal, and both are higher than the K_d. When only a small number of mitotic cells are available for chromosome analysis (e.g. amniotic cells or chorionic villus cells), a large molar excess of dye may be required to saturate the primary dye binding sites. Variations of dye concentration around the saturation value can alter the relative peak positions of individual chromosome types with stains having base-composition specificity (58) because of subtle variations in affinity for different base sequences (see discussion of base specificity below).

Excess dye concentrations can degrade the quality of the flow karyotype by forcing dye binding to secondary binding sites as well as nonspecific uptake of dye by nonchromosomal components (138). The presence of dye in secondary binding sites can reduce the fluorescence efficiency of dye molecules bound to primary binding sites (2,12,70). In some cases, dilution of the final stained chromosome suspension can improve flow resolution, because lowering the free dye concentration can reduce binding to low-affinity secondary and nonspecific binding sites.

The kinetics of dye association and dissociation also affect the quality of the flow karyotype, with the best resolution being obtained when the dye is at equilibrium with the chromosomes. Dye association times vary widely among stains. Hoechst 33258 binding, for example, is nearly complete after 1 or 2 min with typical chromosome concentrations, while chromomycin binding requires 30–60 min under the same conditions (7,122). Complete equilibration, however, can take much longer, as it has been shown that while most binding of propidium iodide to nuclei occurs in less than 5 min, peak coefficients of variation continue to improve up to 30 min after staining (122). While the dissociation kinetics of these dyes have not been studied in detail, it is likely that separation of the stained chromosomes from the free dye may lead to time-dependent changes in chromosome intensity and peak resolution.

One final point about binding equilibria is that flow analysis of stained chromosome suspensions provides a powerful tool for quantitating subtle cytochemical differences among chromosome types. Historically, solution studies of the partition of dyes between DNA or chromatin samples differing in base composition or sequence have utilized the technique of equilibrium dialysis (91). This approach can be quite tedious and imprecise because many days are often required to reach equilibrium in multiple sample chambers separated by dialysis membranes, and because many dyes are nonspecifically adsorbed on the membranes or chamber walls. With flow analysis, in contrast, suspensions of unfixed chromosomes having near-native structure and protein composition can be prepared (134–136) and equilibrated with dye. The relative fluorescence of individual chromosome types can then be determined with high precision as a function of dye concentration or solvent conditions allowing subtle variations in binding affinity among chromosome types to be measured (58,60). This approach has been particularly useful in understanding the interactions of multiple stains with chromosomes because the relative specificities of different stains can be directly compared in a single multiparameter analysis (58,60).

III. STAIN CHARACTERISTICS

Given the large number of DNA-specific stains, and the extensive literature on the spectroscopic, biophysical and biochemical properties of these stains, the following section will focus on only those stains which have shown some utility as cytological stains of cells or chromosomes. Specific values of binding and spectroscopic parameters are only estimates of the true values, because in most cases these parameters vary for different types of DNA or different solvent conditions. Comprehensive reviews of nucleic-acid-specific stains provide

additional details on these, and other, stains (63,67).

A. PHENANTHRIDIUM STAINS

The phenanthridium stains include the classical intercalators ethidium bromide (EB) and propidium iodide (PI). EB interacts with nucleic acids by two modes of binding. The primary mode involves intercalation with both double-stranded DNA and RNA (13,75,77). This primary mode has a dissociation constant (K_d) of about 10^{-6} M, and saturates at a dye to phosphate ratio (D/P) of about 0.2 (75). Intercalative binding induces a shift in the dye absorption spectrum from 480 to 520 nm, and an enhancement of about ten-fold in quantum efficiency (75,77). The emission maxima of bound EB is about 615 nm, and its quantum efficiency is about 0.19 (2). The binding of EB appears to be unaffected by DNA base composition or sequence (75,77), but the number of accessible binding sites is affected by the presence of chromosomal proteins (2,71). At high stain concentration, EB will also interact with nucleic acids by a secondary, electrostatic, mode of binding which can reduce the fluorescence efficiency from EB in primary intercalation sites (2,12). Secondary binding can be inhibited by staining in high ionic strength buffers. The binding and fluorescence of EB is relatively insensitive to pH over a pH range of 4–10 (75,77). PI appears to have the same biological specificities and modes of binding as EB (45), but its absorption and emission spectra are shifted about 20 nm to the red compared with EB.

B. BISBENZIMIDAZOLE AND RELATED STAINS

The bisbenzimidazole dye Hoechst 33258 (HO) binds strongly with double-stranded DNA with minimal interaction with RNA or single-stranded DNA (55,70). The primary binding mode of HO is not intercalation, but external binding along one of the grooves of

DNA ($10,61$). The primary binding mode has a K_d of about 10^{-6} M, and a saturation D/P of about 0.03 (10). HO shows a strong AT specificity in binding, and footprinting studies suggest that the strongest binding sites involve runs of four AT base pairs (82). HO will also bind with reduced affinity to less AT-rich sites so that the effective base specificity of this stain is reduced at high dye–base pair ratios. The free dye has an excitation maximum of 340 nm and an emission maximum of 500 nm. When HO binds to DNA the excitation shifts to 365 nm, the emission shifts to 470 nm, and the fluorescence quantum yield increases about 50-fold to a value of 0.6 ($10,55,70$). The binding specificity and spectral properties of HO are very pH sensitive. At pH values between 7 and 9, HO shows high DNA specificity and strong blue fluorescence at about 470 nm, while at pH values of less than 7, HO shows considerable nonspecific staining of other cellular components and the fluorescence emission shifts to green ($44,47$). The quantum efficiency of HO is reduced when it is bound to DNA containing the modified base bromo-deoxyuridine, and this sensitivity has been utilized to differentiate early- from late-replicating regions on chromosomes (61), and to visualize sister chromatid exchanges (64). Like other stains, HO shows a weak secondary mode of binding to DNA which appears to primarily involve electrostatic interactions (70). HO bound to secondary sites has a low quantum efficiency and these molecules may quench the fluorescence of HO molecules bound to primary binding sites (70). Secondary binding can be selectively reduced by increasing the solution ionic strength.

There are several other dyes with similar AT specificity and spectral properties to HO. Hoechst 33342 has been used for vital staining of living cells because of its increased membrane permeability (3), but this derivative appears to have no particular advantages over HO for chromosome staining. DAPI (4',6-diamidino-2-phenylindole) ($52,78$) and DIPI [4',6-bis(2'-imidazolinyl-4H,5H)-2-phenyl-indole] ($109,110$) both have similar spectral properties to HO and both have an AT specificity in binding. In detail, however, DAPI and DIPI appear to differ from HO in the DNA sequences which form strong binding sites. DAPI and DIPI show bright fluorescence in C-band regions of human chromosomes (in contrast to a moderate staining with HO) (109), and this difference has been exploited to shift the peaks of specific human chromosome types (e.g. chromosomes 1, 9, 16 and Y) in flow karyotypes (72).

C. CHROMOMYCINONE STAINS

The chromomycinone dye, chromomycin A3 (CA3), is a nonintercalator which binds with high specificity to double-stranded DNA and with negligible affinity for RNA ($46,49,131$). CA3 binding has a K_d of about 10^{-5} M and a saturation D/P of about 0.1 (93). Footprinting studies have shown that CA3 binds most strongly to runs of three GC base pairs with lower affinity binding to less GC-rich regions ($7,126$). Two related dyes, mithramycin and olivomycin, also show similar sequence specificities ($14,42,126$), and all three stains yield near-identical flow distributions with human chromosomes (R. G. Langlois, unpublished observation; Fig. 2). CA3 has an absolute requirement for magnesium ions for binding suggesting that at least one Mg^{2+} ion is associated with each bound CA3 molecule ($49,93$). Bound CA3 has maximal excitation at 425 nm (efficient excitation can be obtained using the 458 nm argon ion laser line), and a broad fluorescence emission with a maximum of about 580 nm. The quantum efficiency of CA3 increases about five-fold on binding, but the bound-dye quantum yield is only about 0.05 so efficient fluorescence collection optics are required to obtain optimal resolution of human chromosomes stained with CA3 (47).

D. OTHER FLUORESCENT STAINS

There are a number of other fluorescent DNA stains which have been useful for staining cell nuclei or metaphase chromosomes on slides, although they have not been applied to flow analysis of suspensions of isolated metaphase chromosomes. Quinicrine (Q) is an intercalating acridine dye which has been extensively used for chromosome banding of metaphase spreads (18,116). The binding of Q appears to be independent of base composition, but the quantum efficiency is much higher for Q bound to AT-rich DNA compared with GC-rich DNA (4,28,65). The primary binding mode of Q has a K_d of about 10^{-7} M, a saturation D/P of about 0.2 and a mean quantum efficiency of about 0.08 for calf thymus DNA (65). The use of Q in flow cytogenetics is complicated by the relatively high level of nonspecific binding to nonchromosomal cellular components and the low average quantum efficiency for DNA-bound Q (65). Thus, the free dye and the nonspecifically stained cellular debris can have higher fluorescence efficiency than the chromosomally bound dye (21,29). A second acridine dye, acridine orange, has been extensively used to quantitate single- vs double-stranded DNA in cells (26), but problems of nonspecific staining and a complex pattern of primary and secondary binding modes (50,51) has limited its usefulness for isolated chromosomes.

The anthracycline dyes, daunomycin and adriamycin, are similar to Q in that their binding is unaffected by base composition, but their fluorescence efficiency is higher for AT-rich regions vs GC-rich regions in DNA (48,120). In this case, however, the free dye has substantially higher quantum efficiency than dye bound to either AT- or GC-rich DNA (48,120). Finally, the actinomycin derivative, 7-amino actinomycin (7AA), is an intercalating dye which fluoresces in the red and has high specificity for DNA (87,142). The actinomycin dyes have a strong binding preference for GC-rich regions on DNA (57). The primary limitation of this dye is its very low quantum efficiency (32), but improvements in flow cytometer sensitivity may allow its other desirable features to be exploited in flow cytogenetics.

E. NONFLUORESCENT DNA COUNTERSTAINS

A number of nonfluorescent, DNA-specific counterstains have been described which can be combined with fluorescent stains to enhance or alter the specificity of the fluorescent stain. This approach has been applied to the analysis of metaphase spreads on slides to enhance normal banding patterns and to produce novel banding patterns (69,79,108,112,113), and these approaches are now beginning to be adapted to flow analysis (56,85,88). Counterstaining methods are based on two general principles (69,113). First, counterstains can block the binding of fluorescent dye molecules to specific sequences or chromosome regions. Secondly, if the counterstain absorption spectrum overlaps with the emission spectrum of the fluorescent dye, then energy transfer can quench the emission of fluorescent dye molecules that are bound in close proximity to counterstain molecules.

Examples of counterstains utilizing the first mechanism are distamycin A, netropsin and echinomycin. All three have no absorption bands at visible wavelengths (69), so contrast enhancement must be due to binding competition. Distamycin A and netropsin preferentially bind to AT-rich DNA (128), while echinomycin has some binding preference for GC-rich DNA (127). The nonfluorescent stains actinomycin D and methyl green provide two examples of counterstains which do have absorption bands at visible wavelengths (66), and energy-transfer quenching appears to be the dominant mechanism for contrast enhancement, although binding competition may also be involved (66,104). Recent footprinting studies on most of these nonfluorescent stains have provided detailed information on the specific DNA sequences which form strong binding sites for each stain (98,127,128). These studies also confirm that while most fluorescent and nonfluorescent base-specific stains fall into two broad categories (AT vs GC preference), each stain has a unique spectrum of DNA sequences which form preferred binding sites.

IV. CHARACTERISTICS OF SPECIFIC STAINING METHODS

A. SINGLE-STAIN METHODS

The initial development of flow cytogenetics utilized univariate analysis of isolated metaphase chromosomes stained with the DNA-specific stain, EB (35). While most human chromosome types could not be separately resolved in these early studies, improvements in both chromosome isolation methods and flow instrumentation have greatly improved the utility of univariate flow analysis for quantitative karyotyping. Univariate methods continue to be used for karyotype analysis because of the relative simplicity of the staining and instrumentation required for this type of analysis.

The intercalating stains EB and PI are particularly well suited for univariate flow karyotyping. Both stains are efficiently excited by the strong 488 nm line from an argon ion laser. These stains exhibit minimal base specificity in binding, so that the relative peak position of each chromosome type is primarily determined by its DNA content (34,35,40), and is relatively insensitive to variations in stain and chromosome concentrations (17,58). High-resolution analyses of karyotype instability in Chinese hamster cells have been reported using high-intensity illumination of PI-stained chromosomes after RNase pretreatment of the chromosome suspensions (5,6). Most human chromosome types have been separately resolved in univariate analyses of EB-stained chromosomes prepared by the polyamine isolation method (40,137). This approach has allowed normal human polymorphisms to be quantitated and their inheritance pattern determined (41). The primary limitation of this approach is that flow sorting is often required for unambiguous assignment of the chromosome types responsible for specific peaks in the flow distribution because of normal polymorphic variability among individuals (40,137).

HO also has been extensively used for univariate flow karyotypes. While HO has the disadvantage of requiring ultraviolet excitation at 360 nm, excellent resolution of individual chromosome types is possible because of the high fluorescence efficiency and high DNA specificity of HO (39,138). The AT binding specificity of HO can also be an advantage in shifting the peaks of specific chromosome types of interest away from interference from other peaks (39). Variations in stain concentration can also be used to alter relative peak positions (58) because the effective base specificity of this stain varies with the level of saturation of the DNA-binding sites.

Stain combinations utilizing EB (or PI) with CA3 (or mithramycin) are well suited for univariate flow analysis of chromosomes with flow cytometers that have mercury arc light sources (96). CA3 has high DNA specificity and is efficiently excited by the emission line at 435 nm from the mercury lamp (101), but its fluorescent efficiency is low. EB has a higher quantum efficiency than CA3, but it is poorly excited by a mercury lamp and is less DNA specific. Energy transfer between these two stains (47,59) provides the efficient DNA-specific excitation of CA3, with the high emission efficiency of EB.

B. PAIRS OF BASE-SPECIFIC STAINS

Bivariate analysis of chromosomes stained with pairs of base-specific fluorescent dyes allows individual chromosome types to be separately resolved due to differences in base composition as well as differences in DNA content. While a number of dye pairs have been successfully applied to chromosome analysis, the dye pair HO plus CA3 has been extensively studied (58–60,122) and illustrates principles which are also applicable to other dye pairs.

HO and CA3 have complementary base specificity (AT- vs GC-binding preference, respectively) and also differ in spectral properties, so the two stains can be selectively excited by two excitation lasers at 351 plus 363 nm and 458 nm, respectively. It has been

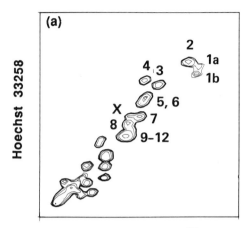
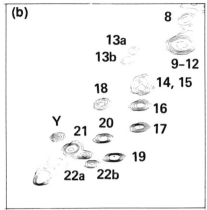

Chromomycin A3

Fig. 1. Bivariate flow karyotype of human chromosomes derived from a fibroblast cell strain (LLL#761). The chromosomes were isolated by the hexylene glycol procedure, stained with Hoechst 33258 and chromomycin A3, and analyzed as described previously (*60*). (a) The whole karyotype. (b) An expanded view of the smaller chromosome types.

shown that these two dyes do not compete for binding sites, but do interact spectrally by Forster energy transfer (*59,115,122*), so that part of the excitation energy absorbed by HO is transferred to CA3 and is emitted as CA3 fluorescence. The HO and CA3 contents can be measured separately by using two spatially separated excitation beams to differentiate fluorescence arising from HO excitation from fluorescence arising from CA3 excitation (*36,60*). This stain pair is compatible with a number of different chromosome isolation methods (*20,56,60,123*), if the chromosome suspension has a neutral to alkaline pH for DNA-specific HO binding, and magnesium ions are present for CA3 binding.

The bivariate flow karyotype of human chromosomes stained with HO and CA3 in Fig. 1 shows that most human chromosome types, and even individual homologs of some chromosome types, can be separately resolved by this staining method. While the peak position of some chromosome types varies among individuals because of normal polymorphisms, such variations do not interfere with identification of the chromosome types responsible for each peak (*60*). This is in contrast to univariate flow karyotypes where polymorphisms can lead to ambiguity in peak assignments (*40,137*). Similar bivariate flow karyotypes have been obtained using other pairs of base-specific stains including DAPI plus CA3, and DIPI plus CA3 (*72*). Fig. 2 shows flow karyotypes of chromosomes from one donor stained with three different stain pairs. While the overall flow karyotypes are very similar for different stain pairs, peaks corresponding to a few polymorphic chromosome types can show large shifts in individuals having large polymorphisms. Thus, subtle differences in specificity between dyes with the same base preference has been used to resolve specific chromosome types, e.g. chromosome 9, for flow sorting in chromosome suspensions from individuals with appropriately staining polymorphisms (*72*).

C. MULTIPLE STAINS AND CHEMICAL PERTURBATIONS

Nonfluorescent DNA counterstains have been used to highlight specific chromosomal regions in metaphase spreads

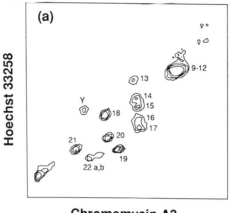

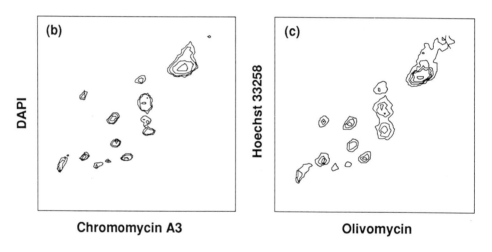

Fig. 2. Bivariate karyotypes of the smaller chromosome types from the human fibroblast strain LLL#811 stained with different pairs of fluorescent stains. Chromosome isolation and analysis conditions as in Fig. 1. The chromosomes were stained with (a) Hoechst 33258 plus chromomycin, (b) DAPI plus chromomycin and (c) Hoechst plus olivomycin.

(69,79,108,112,113). If the fluorescent dye and the counterstain have similar, but not identical, specificity then the counterstain will block the binding or quench the fluorescence of the dye in regions where both molecules can bind (79,113). Thus, the chromosomal fluorescence will be reduced in all regions except regions containing sequences which bind the fluorescent dye and not the counterstain. Figure 3 shows the effect of the nonfluorescent, AT-specific counterstain distamycin A on a bivariate flow karyotype of human chromosomes stained with HO and CA3. In this individual, chromosome 16 moves up into the peak corresponding to chromosomes 14 and 15, suggesting that there is less competition between HO and distamycin on this chromosome compared with other chromosome types. Other counterstaining methods which have been applied to flow cytogenetics

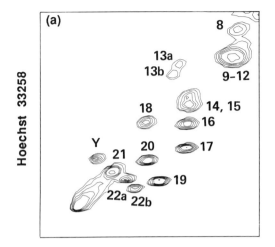

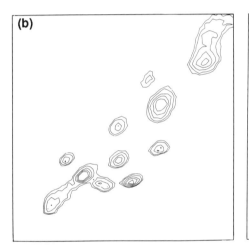 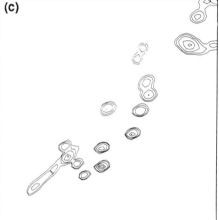

Chromomycin A3

Fig. 3. Bivariate flow karyotypes of the smaller chromosome types from human fibroblast strain LLL#761 stained with Hoechst 33258 and chromomycin as in Fig. 1. (a) No additional treatments. (b) Chromosomes counterstained with the nonfluorescent stain distamycin A. (c) Chromosomes treated with heparin to extract histone proteins before staining.

include netropsin plus HO plus CA3 (*56*), netropsin plus DAPI plus CA3 (*88*) and distamycin plus DAPI plus CA3 (*85*).

Chemical modification of chromosomal DNA or other chromosome components has been used to study differences among chromosome types in the timing of DNA replication and differences in DNA accessibility to dye due to chromosomal proteins. The replication timing of specific chromosomes or chromosome regions has been studied by pulse labeling with the base analog bromodeoxyuridine (BrdU). HO fluorescence is quenched when it is bound to DNA containing BrdU (*70*), but this analog appears to have minimal effect on CA3 fluorescence

(*117*). HO staining of metaphase spreads after pulse labeling with BrdU has been used to map the time of replication of specific chromosome regions (*61*). Double staining of chromosome suspensions with HO plus CA3 after BrdU incorporation has been used to separately resolve and sort late-replicating chromosomes (*23,24*).

The high precision of flow cytogenetic methods provides a quantitative tool for studying the effects of chromosomal proteins and structure on DNA accessibility to dyes in metaphase chromosomes. While there is clear evidence that chromosomal proteins limit the accessibility of DNA to dyes (*11*), it is less clear if the fraction of accessible DNA varies among chromosome types or if the accessible DNA differs in base composition or sequence from the total DNA. The report that metaphase chromosomes are kept intact by a scaffold of nonhistone proteins after most histones are removed by polyanions (*1,99*) raised the possibility that flow methods could be used to characterize protein-depleted chromosomes. Histone extraction of chromosomes with the polyanion, heparin, dramatically changes their structure and staining characteristics. After heparin treatment, chromosomes decondense and increase about four-fold in size, while their staining intensity with HO or CA3 increases by 50–100%. The bivariate flow karyotype of histone-depleted chromosomes in Fig. 3 shows similar peak positions for individual chromosome types before and after heparin treatment. This suggests that the staining characteristics of the DNA which is accessible in intact chromosomes is representative of the total chromosomal DNA, and that staining differences among chromosome types primarily reflects differences in DNA composition or sequence rather than differences in condensation or protein content.

V.   MECHANISMS OF CHROMOSOME STAINING AND BANDING

Model system studies using purified DNA or chromatin have shown that dye binding or fluorescence can be affected by many features, including DNA and protein composition and chromatin structure. Flow measurements provide the opportunity to make quantitative estimates of the relative importance of these factors in determining the fluorescence of intact chromosomes. An understanding of the significant biochemical factors which determine staining differences among chromosomes should simplify the design of new staining methods and the interpretation of peak shifts resulting from polymorphisms or rearrangements. Flow measurements may also contribute to our understanding of the large-scale organization of chromosomes at the level of bands (10^7 bp) or chromosomes (10^8 bp). It is becoming clear that band and inter-band regions differ not only in staining characteristics, but in time of replication (*62,83,105*), content of transcribed sequences (*139*), base composition (*54,118*), content of middle repetitive and highly repetitive sequences (*38,118*) and concentration of restriction sites for specific restriction enzymes (*8,30*).

Given the known base composition preference of HO, CA3 and other chromosomal stains, one hypothesis that can be directly tested is that staining differences among chromosome types is primarily determined by differences in average base composition. In this case there should be a strict inverse relationship between the relative AT-specific fluorescence and the relative GC-specific fluorescence for each chromosome type (e.g. a chromosome with high HO fluorescence per unit DNA should have a proportionately low CA fluorescence per unit DNA). One would also expect that chromosomes with high relative HO fluorescence would also have high relative fluorescence with other AT-specific dyes. If the staining differences between chromosome types are determined by differences in protein composition or chromosome structure, however, no correlation would be expected with stain base preference.

It has been shown previously that there is an inverse relationship between HO and CA3

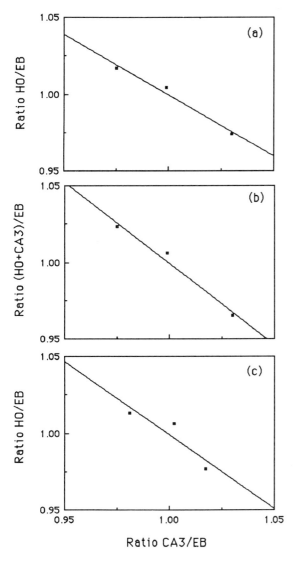

Fig. 4. Staining characteristics of the three chromosome types of the Indian muntjac [*Muntiacus muntjac (16,53)*]. Univariate analyses were used to determine the staining intensity of each chromosome type for each staining condition. Intensities with ethidium bromide (EB) were used as a measure of the DNA content of each chromosome to determine the relative intensity per unit DNA for the other stains. (a) The relative Hoechst 33258 (HO) intensity *vs* the relative chromomycin A3 (CA3) intensity showing an inverse relationship between these stains. (b) The relative intensity of the stain combination HO plus CA3 after HO excitation *vs* CA3 showing an inverse relationship with a steeper slope due to energy transfer between HO and CA3 (58). (c) The same as in part (a) except that histones were extracted with heparin before staining with HO or CA3 or EB. Note the inverse relationship between HO and CA3 is unaffected by histone removal.

fluorescence per unit DNA content for most Chinese hamster chromosome types when EB fluorescence is used as a measure of DNA content (58). Figure 4 shows a similar inverse relationship for the three chromosome types of the Indian muntjac. With human chromosomes, some curvature is seen in the inverse relationship between HO and CA3 (Fig. 5), but this nonlinearity is consistent with the calculated curve based on a previously described model for the frequency of clusters of AT and GC base pairs (58). The argument for average base composition being the primary determinant of staining differences among chromosome types is also supported by the similarity of the peak positions in bivariate flow distributions utilizing different pairs of base-specific stains (Fig. 2; 56,72,85), and by the observation that protein extraction and chromosome decondensation with heparin has minimal effect on the relative intensities of different chromosome types (Figs 3 and 4).

Base composition differences also appear to be the major determinant of staining differences between band and interband regions within chromosomes. AT-specific stains including HO and DAPI produce quinacrine-like banding patterns, while GC-specific stains produce a banding pattern which is the reverse of Q-banding (73,106,107,111). A common origin for staining differences among chromosome types stained in suspension and staining differences between band and interband regions in metaphase spreads is also supported by the correlation between average quinacrine brightness and HO to CA3 ratio for individual chromosomes shown in Fig. 6.

Differences in average base composition cannot, however, explain the staining characteristics of some heterochromatic and polymorphic regions on chromosomes. The centromeric heterochromatin in mouse, for example, stains dimly with quinacrine but brightly with HO (37,43,103). Distamycin plus DAPI selectively enhances the brightness of human C-band regions and some acro-centric satellites over other DAPI bright bands (100,113,114), while quinacrine plus

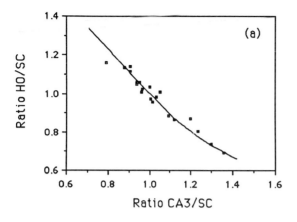

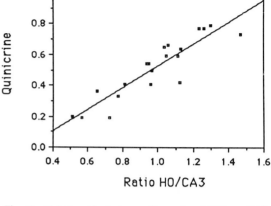

Fig. 6. Relationship between the ratio of HO to CA3 intensities measured in flow, and the average quinacrine brightness from metaphase spreads of human chromosomes. HO and CA3 values were obtained from double-stained chromosomes from strain LLL#761 analyzed as in Fig. 1. Average quinacrine intensities were obtained from literature values based on a different human donor (54). The linear correlation shown suggests that the ratio of HO to CA3 intensity provides a reasonable estimate of the average Q-band intensity of individual chromosome types.

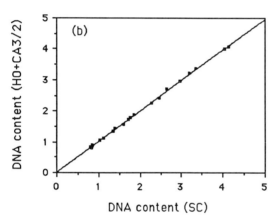

Fig. 5. Staining characteristics of human chromosomes derived from fibroblast strain LLL#761 stained and analyzed as in Fig. 1. DNA content values obtained from scanning cytophometry [SC (84)] on metaphase spreads from this cell strain were used to determine the relative HO and relative CA3 intensities per unit DNA shown in part (a). The curve represents predicted values using a previously described model for the frequency of clusters of AT and GC base pairs (58). Part (b) shows that the average of the HO and CA3 intensity of each chromosome type provides a measure of DNA content which is closely correlated with DNA contents from scanning cytophotometry.

actinomycin preferentially enhances the fluorescence of quinacrine bright satellites on human acrocentric chromosomes (68). The most likely explanation for these anomalies is the known association of different families of highly repetitive DNA in these regions, including satellite DNA (33), ribosomal DNAs

(129), and other families of chromosome-specific repetitive DNAs (22,88,133,141). While the average base composition of such a region may be AT rich, it will not stain brightly with a specific AT stain such as HO if the sequence of the repeat unit does not contain a strong HO binding site. Thus, the regions highlighted by distamycin plus DAPI may be AT rich and contain DAPI binding sites, but not contain the sequences required for distamycin binding.

In summary, the staining characteristics of individual chromosome types and chromosome regions appears to be primarily determined by DNA base composition and by the distribution of specific DNA sequences which form strong dye binding sites. The quantitative correlations between relative fluorescence and stain base preference for individual chromosome types determined by flow karyotyping suggests that the staining of most regions reflects average base composition. In regions containing repetitive DNA, however, staining may be dominated by the specific sequence of the repeat unit rather than the overall base composition of the region. Chromosome regions may also differ in protein composition or structure, but the heparin studies reported above suggest that such differences have a minor effect on the relative staining of individual chromosome types.

VI. FUTURE DIRECTIONS

Development of new staining methods utilizing dyes with new specificities, or increased specificity for specific DNA sequence classes should improve the ability of flow cytogenetics to both classify normal chromosome types and to differentiate normal polymorphic variability from abnormal chromosome rearrangements. While current methods allow most human chromosome types to be classified, chromosomes from other species important in biomedical research (e.g. mouse) are poorly resolved by current methods because they have many different chromosome types of similar size. The development of new cytochemical probes, which are described below, would be facilitated by improvements in two other areas of flow cytogenetics. The first is improvements in the sensitivity of flow cytometers so that dye systems having low fluorescence yield or low frequencies of binding sites on DNA could be used. Improvements in chromosome isolation methods which allow separation of chromosomes from other cellular debris would permit the evaluation of dyes that show promising specificity on metaphase spreads, but are currently unsuitable for flow because of nonspecific staining of other cellular components.

There are many currently known dyes which show promise for flow cytogenetics and should be evaluated. Quinacrine and many other acridine dyes show particular promise if cellular debris can be removed to reduce nonDNA background fluorescence. The same is true for daunomycin and other anthracycline dyes which show interesting specificities on metaphase spreads (48,80). An example of a low quantum efficiency dye which could be very useful is 7-amino actinomycin D, because of its strong GC binding preference and its long wavelength emission maximum (32,87,142).

There are also many possibilities for new combinations involving fluorescent dyes and nonfluorescent counterstains. The low fluorescence intensity of counterstained chromosomes currently limits this approach, but improved instrument sensitivity should allow the unique specificities of these methods to be exploited. Counterstains which have not been applied to flow, but are useful on metaphase spreads include methyl green and echinomycin (69). There are also many other nonfluorescent stains that are known to have base composition sensitivity which could be evaluated (89,90,92,132).

A third area which has not yet been evaluated in flow cytogenetics is the use of stain dimers (stains having two dye molecules chemically linked together). Several homodimers of one dye, as well as heterodimers involving different dyes, have been described (31,76,81). Such dimer stains have the potential of extremely high binding affinity, which would improve their DNA specificity. By choosing two different base-specific dyes for such dimers, very high sequence specificities should be possible.

While not within the scope of this chapter, improvements in suspension hybridization of fluorescently tagged oligonucleotides (119) could ultimately lead to dramatic improvements in chromosome and sequence-specific labeling.

ACKNOWLEDGEMENTS

This work was performed under the auspices of the US Department of Energy by the Lawrence Livermore National Laboratory under contract number W-7405-ENG-48.

REFERENCES

1. Adolph, K. W., Cheng, S. M., and Laemmli, U. K. (1977). Role of nonhistone proteins in metaphase chromosome structure. *Cell* **12**, 805–816.

2. Angerer, L. M., Georghiou, S., and Moudrianakis, E. N. (1974). Studies on the structure of deoxyribonucleoproteins. Spectroscopic characterization of the ethidium bromide binding sites. *Biochemistry* **13**, 1075–1082.

3. Arndt-Jovin, D. J., and Jovin, T. M. (1977). Analysis and sorting of living cells according to deoxyribonucleic acid content. *J. Histochem. Cytochem.* **25**, 585–589.

4. Arndt-Jovin, D. J., Latt, S. A., Striker, G., and Jovin, T. M. (1979). Fluorescence decay analysis in solution and in a microscope of DNA and chromosomes stained with quinicrine. *J. Histochem. Cytochem.* **27**, 87–95.

5. Bartholdi, M. F., Ray, F. A., Cram, L. S., and Kraemer, P. M. (1987). Karyotype instability of Chinese hamster cells during in vivo tumor progression. *Somat. Cell Mol. Genet.* **13**,1–10.

6. Bartholdi, M. F., Sinclair, D. C., and Cram, L. S. (1983). Chromosome analysis by high illumination flow cytometry. *Cytometry* **3**, 395–401.

7. Behr, W., Honikel, K., and Hartmann, G. (1969). Interaction of the RNA polymerase inhibitor chromomycin with DNA. *European J. Biochem.* **9**, 82–92.

8. Bianchi, M. S., Bianchi, N. O., Pantelias, G. E., and Wolff, S. (1985). The mechanism and pattern of banding induced by restriction endonucleases in human chromosomes. *Chromosoma* **91**, 131–136.

9. Bidney, D. L., and Reeck, G. R. (1978). Analysis of the effectiveness of sodium chloride in dissociating non-histone chromatin proteins of cultured hepatoma cells. *Biochem. Biophys. Acta* **521**, 753–761.

10. Bontemps, J., Houssier, C., and Fredericq, E. (1975). Physico-chemical study of the complexes of 33258 Hoechst with DNA and nucleohistone. *Nucleic Acids Res.* **2**, 971–984.

11. Brodie, S., Giron, J., and Latt, S. A. (1975). Estimation of accessibility of DNA in chromatin from fluorescence measurements of electronic excitation energy transfer. *Nature* **253**, 470–471.

12. Burns, V. W. F. (1969). Fluorescence decay time characteristics of the complex between ethidium bromide and nucleic acids. *Arch. Biochem. Biophys.* **133**, 420–424.

13. Burns, V. W. F. (1971). Fluorescence polarization characteristics of the complexes between ethidium bromide and rRNA, tRNA, and DNA. *Arch. Biochem. Biophys.* **145**, 248–254.

14. Burns, V. W. F. (1977). Studies with a fluorescent vital probe for DNA in mammalian cells. *Exptl Cell Res.* **107**, 459–462.

15. Capriglione, T., Olmo, E., Odierna, G., Improta, B., and Morescalchi, A. (1987). Cytofluorometric DNA base determination in vertebrate species with different genome sizes. *Bas. Appl. Histochem.* **31**, 119–126.

16. Carrano, A. V., Gray, J. W., Moore, D. H., Minkler, J. L., Mayall, B. H., Van Dilla, M. A., and Mendelsohn, M. L. (1976). Purification of the chromosomes of the Indian muntjac by flow sorting. *J. Histochem. Cytochem.* **24**, 348–354.

17. Carrano, A. V., Van Dilla, M. A., and Gray, J. W. (1979). Flow cytogenetics: a new approach to chromosome analysis. *In* "Flow Cytometry and Sorting" (M. Melamed, M. Mendelsohn, and P. Mullaney, eds), pp. 263–284. Wiley, New York.

18. Caspersson, T., Lomakka, G., and Zech, L. (1971). The 24 fluorescence patterns of the human metaphase chromosomes — distinguishing characters and variability. *Hereditas* **67**, 89–102.

19. Caspersson, T., Zech, L., Modest, E. J., Foley, G. E., Wagh, U., and Simmonsson, E. (1969). Chemical differentiation with fluorescent alkylating agents in Vicia Faba metaphase chromosomes. *Exptl Cell Res.* **58**, 128–140.

20. Collard, J. G., Philippus, E., Tulp, A., Lebo, R. V., and Gray, J. W. (1984). Separation and analysis of human chromosomes by combined velocity sedimentation and flow sorting applying single- and dual-laser flow cytometry. *Cytometry* **5**, 9–19.

21. Comings, D. E., Kovacs, B. W., Avelino, E., and Harris, D. C. (1975). Mechanisms of chromosome banding. V. Quinicrine banding. *Chromosoma* **50**, 111–145.

22. Cooke, H. J., Schmidtke, J., and Gosden, J. R. (1982). Characterisation of a human Y chromosome repeated sequence and related sequences in higher primates. *Chromosoma* **87**, 491–502.

23. Cremer, C., and Gray, J. W. (1983). Application of the BrdU/thymidine method to flow cytogenetics: differential quenching/enhancement of Hoechst 33258 fluorescence of late-replicating chromosomes. *Somatic Cell Genet.* **8**, 319–327.

24. Cremer, C., and Gray, J. W. (1983). Replication kinetics of Chinese hamster chromosomes as revealed by bivariate flow karyotyping. *Cytometry* **3**, 282–286.

25. Darzynkiewicz, Z., Gledhill, B. L., and Ringertz, N. R. (1969). Changes in deoxyribonucleoprotein during spermiogenesis in the bull. *Exptl Cell Res.* **58**, 435–438.

26. Darzynkeiwicz, Z., Traganos, F., Arlin, Z. A., Sharpless, T., and Melamed, M. R. (1976). Cytofluorometric studies on conformation of nucleic acids *in situ*. II. Denaturation of deoxyribonucleic acid. *J. Histochem. Cytochem.* **24**, 49–58.

27. Darzynkiewicz, Z., Taganos, F., Kapuscinski, J., Staiano-Coico, L., and Melamed, M. R. (1984). Accessibility of DNA *in situ* to various fluorochromes: relationship to chromatin changes during erythroid differentiation of friend leukemia cells. *Cytometry* **5**, 355–363.

28. Disteche, C., Bontemps, J., Houssier, C., Frederic, J., and Fredericq, E. (1980). Quantitative analysis of

fluorescence profiles of chromosomes. Influence of DNA base composition on banding. *Exptl Cell Res.* **125**, 251–264.

29. Duportail, G., Mauss, Y., and Chambron, J. (1977). Quantum yields and fluorescence lifetimes of acridine derivatives interacting with DNA. *Biopolymers* **16**, 1397–1413.

30. Ferrucci, L., Romano, E., and Stefano, G. F. (1987). The Alu1-induced bands in great apes and man: implication for heterochromatin characterization and satellite DNA distribution. *Cytogenet. Cell Genet.* **44**, 53–57.

31. Gaugain, B., Barbet, J., Capelle, N., Roques, B. P., and Le Pecq, J.-B. (1978). DNA bifunctional intercalators. 2. Fluorescence properties and DNA binding interaction of an ethidium homodimer and an acridine ethidium heterodimer. *Biochem.* **17**, 5078–5088.

32. Gill, J. E., Jotz, M. M., Young, S. G., Modest, E. J., and Sengupta, S. K. (1975). 7-amino-actinomycin D as a cytochemical probe. I. Spectral properties. *J. Histochem. Cytochem.* **23**, 793–799.

33. Gosden, J. R., Mitchell, A. R., Buckland, R. A., Clayton, R. P., and Evans, H. J. (1975). The location of four human satellite DNA's on human chromosomes. *Exptl Cell Res.* **92**, 148–158.

34. Gray, J. W., Carrano, A. V., Moore II, D. H., Steinmetz, L. L., Minkler, J., Mayall, B. H., Mendelsohn, M. L., and Van Dilla, M. A. (1975). High-speed quantitative karyotyping by flow microfluorometry. *Clin. Chem.* **21**, 1258–1262.

35. Gray, J. W., Carrano, A. V., Steinmetz, L. L., Van Dilla, M. A., Moore II, D. H., Mayall, B. H., and Mendelsohn, M. L. (1975). Chromosome measurement and sorting by flow systems. *Proc. Natl Acad. Sci. USA* **72**, 1231–1234.

36. Gray, J. W., Langlois, R. G., Carrano, A. V., Burkhart-Schulte, K., and Van Dilla, M. A. (1979). High resolution chromosome analysis: one and two parameter flow cytometry. *Chromosoma* **73**, 9–27.

37. Gropp, A., Hilwig, I., and Seth, P. K. (1973). Fluorescence chromosome banding patterns produced by a benzimidazole derivative. *In* "Nobel Symposia, Medicine and Natural Sciences, Chromosome Identification" (T. Caspersson, and L. Zech, eds), Vol. 23, pp. 300–306. Academic Press, New York.

38. Hanania, N., Caneva, R., Tapiero, H., and Harel, J. (1975). Distribution of repetitious DNA in randomly growing and synchronized Chinese hamster cells. *Exptl Cell Res.* **90**, 79–86.

39. Harris, P., Boyd, E., and Ferguson-Smith, M. A. (1985). Optimising human chromosome separation for the production of chromosome-specific DNA libraries by flow sorting. *Hum. Genet.* **70**, 59–65.

40. Harris, P., Boyd, E., Young, B. D., and Ferguson-Smith, M. A. (1986). Determination of the DNA content of human chromosomes by flow cytometry. *Cytogenet. Cell Genet.* **41**, 14–21.

41. Harris, P., Cooke, A., Boyd, E., Young, B. D., and Ferguson-Smith, M. A. (1987). The potential of family flow karyotyping for the detection of chromosome abnormalities. *Hum. Genet.* **76**, 129–133.

42. Hill, B. T., and Whatley, S. (1975). A simple, rapid microassay for DNA. *FEBS Letters* **56**, 20–23.

43. Hilwig, I., and Gropp, A. (1972). Staining of constitutive heterochromatin in mammalian chromosomes with a new fluorochrome. *Exptl Cell Res.* **75**, 122–126.

44. Hilwig, I., and Gropp, A. (1975). pH-dependent fluorescence of DNA and RNA in cytologic staining with "33258 Hoechst". *Exptl Cell Res.* **91**, 457–460.

45. Hudson, B., Upholt, W. B., Devinny, J., and Vinograd, J. (1969). The use of an ethidium analogue in the dye-buoyant density procedure for the isolation of closed circular DNA: the variation of the superhelix density of mitochondrial DNA. *Proc. Natl Acad. Sci. USA* **62**, 813–820.

46. Jensen, R. H. (1977). Chromomycin A3 as a fluorescent probe for flow cytometry of human gynecological samples. *J. Histochem. Cytochem.* **25**, 573–579.

47. Jensen, R. H., Langlois, R. G., and Mayall, B. H. (1977). Strategies for choosing a deoxyribonucleic acid stain for flow cytometry of metaphase chromosomes. *J. Histochem. Cytochem.* **25**, 954–964.

48. Johnston, F. P., Jorgenson, K. F., Linn, C. C., and van de Sande, J. H. (1978). Interaction of anthracyclines with DNA and chromosomes. *Chromosoma* **68**, 115–129.

49. Kamiyama, M. (1968). Mechanism of action of chromomycin A3. III. On the binding of chromomycin A3 with DNA and physicochemical properties of the complex. *J. Biochem.* **63**, 566–572.

50. Kapuscinski, J., and Darzynkiewicz, Z. (1983). Increased accessibility of bases in DNA upon binding of acridine orange. *Nucleic Acids Res.* **11**, 7555–7568.

51. Kapuscinski, J., Darzynkiewicz, Z., and Melamed, M. R. (1982). Luminescence of the solid complexes of acridine orange with RNA. *Cytometry* **2**, 201–211.

52. Kapuscinski, J., and Skoczylas, B. (1978). Fluorescent complexes of DNA with DAPI 4',6-diamidine-2-phenyl indole. 2HCl or DCI 4',6-dicarboxyamide-2-phenyl indole. *Nucleic Acids Res.* **5**, 3775–3799.

53. Kato, H., Tsuchiya, K., and Yosida, T. H. (1974). Constitutive heterochromatin of Indian muntjac chromosomes revealed by DNAse treatment and a C-banding technique. *Can. J. Genet. Cytol.* **16**, 273–280.

54. Korenberg, J. R., and Engels, W. R. (1978). Base ratio, DNA content, and quinicrine-brightness of human chromosomes. *Proc. Natl Acad. Sci. USA* **75**, 3382–3386.

55. Labarca, C., and Paigen, K. (1980). A simple, rapid, and sensitive DNA assay procedure. *Anal. Biochem.* **102**, 344–352.

56. Lalande, M., Schreck, R. R., Hoffman, R., and Latt, S. A. (1985). Identification of inverted duplicated # 15 chromosomes using bivariate flow cytometric analysis. *Cytometry* **6**, 1–6.

57. Lane, M. J., Dabrowiak, J. C., and Vournakis, J. N. (1983). Sequence specificity of actinomycin D and netropsin binding to pBR322 DNA analyzed by protection from DNase I. *Proc. Natl Acad. Sci. USA* **80**, 3260–3264.

58. Langlois, R. G., Carrano, A. V., Gray, J. W., and Van Dilla, M. A. (1980). Cytochemical studies of metaphase chromosomes by flow cytometry. *Chromosoma* **77**, 229–251.

59. Langlois, R. G., and Jensen, R. H. (1979). Interactions between pairs of DNA-specific fluorescent stains bound to mammalian cells. *J. Histochem. Cytochem.* **27**, 72–79. Erratum p. 1559.

60. Langlois, R. G., Yu, L.-C., Gray, J. W., and Carrano, A. V. (1982). Quantitative karyotyping of human chromosomes by dual beam flow cytometry. *Proc. Natl Acad. Sci. USA* **79**, 7876–7880.

61. Latt, S. A. (1973). Microfluorometric detection of deoxyribonucleic acid replication in human metaphase chromosomes. *Proc. Natl Acad. Sci. USA* **70**, 3395–3399.

62. Latt, S. A. (1977). Fluorescent probes of chromosome structure and replication. *Can. J. Genet. Cytol.* **19**, 603–623.

63. Latt, S. A. (1979). Fluorescent probes of DNA microstructure and synthesis. *In* "Flow Cytometry and Sorting" (M. Melamed, N. Mendelsohn, and P. Mullaney, eds), pp. 263–284. Wiley, New York.

64. Latt, S. A., Allen, J. W., Rogers, W. E., and Juergens, L. A. (1977). *In vitro* and *in vivo* analysis of sister chromatid exchange formation. *In* "Handbook of Mutagenicity Test Procedures" (B. J. Kilbey, M. Legator, and W. Nichols, eds), pp. 275–291. Elsevier, New York.

65. Latt, S. A., Brodie, S., and Munroe, S. H. (1974). Optical studies of complexes of quinacrine with DNA and chromatin: implications for the fluorescence of cytological preparations. *Chromosoma* **49**, 17–40.

66. Latt, S. A., Juergens, L. A., Matthews, D. J., Gustashaw, K. M., and Sahar, E. (1980). Energy transfer-enhanced chromosome banding. An overview. *Cancer Genet. Cytogenet.* **1**, 187–196.

67. Latt, S. A., and Langlois, R. G. (1989). Fluorescent probes of DNA microstructure and DNA synthesis. *In* "Flow Cytometry and Sorting" (M. R. Melamed, T. Lindmo, and M. L. Mendelsohn, eds). Alan R. Liss, New York. In Press.

68. Latt, S. A., Sahar, E., and Eisenhard, M. E. (1979). Pairs of fluorescent dyes as probes of DNA and chromosomes. *J. Histochem. Cytochem.* **27**, 65–71.

69. Latt, S. A., Sahar, E., Eisenhard, M. E., and Juergens, L. A. (1980). Interactions between pairs of DNA-binding dyes: results and implications for chromosome analysis. *Cytometry* **1**, 2–12.

70. Latt, S. A., and Wohlleb, J. C. (1975). Optical studies of the interaction of 33258 Hoechst with DNA, chromatin, and metaphase chromosomes. *Chromosoma* **52**, 297–316.

71. Lawrence, J.-J., and Daune, M. (1976). Ethidium bromide as a probe of conformational heterogeneity of DNA in chromatin. The role of histone H1. *Biochemistry* **15**, 3301–3307.

72. Lebo, R. V., Gorin, F., Fletterick, R. F., Kao, F.-T., Cheung, M.-C., Bruce, B. D., and Kan, Y. W. (1984). High resolution chromosome sorting and DNA spot-blot analysis assign McArdle's syndrome to chromosome 11. *Science* **225**, 57–59.

73. Leemann, U., and Ruch, F. (1978). Selective excitation of mithramycin or DAPI fluorescence on double-stained cell nuclei and chromosomes. *Histochem.* **58**, 329–334.

74. Leemann, U., and Ruch, F. (1982). Cytofluorometric determination of DNA base content in plant nuclei and chromosomes by the fluorochromes DAPI and chromomycin A3. *Exptl Cell Res.* **140**, 275–282.

75. Le Pecq, J.-B. (1971). Use of ethidium bromide for separation and determination of nucleic acids of various conformational forms and measurement of their associated enzymes. *Meth. Biochem. Anal.* **20**, 41–86.

76. Le Pecq, J.-B., Le Bret, M., Barbet, J., and Roques, B. (1975). DNA polyintercalating drugs: DNA binding of diacridine derivatives. *Proc. Natl Acad. Sci. USA* **72**, 2915–2919.

77. Le Pecq, J.-B., and Paoletti, C. (1967). A fluorescent complex between ethidium bromide and nucleic acids. Physical-chemical characterization. *J. Mol. Biol.* **27**, 87–106.

78. Lin, C. C., Comings, D. E., and Alfi, O. S. (1977). Optical studies of the interaction of 4'-6-diamidino-2-phenylindole with DNA and metaphase chromosomes. *Chromosoma* **60**, 15–25.

79. Lin, C. C., Jorgenson, K. F., and Van de Sande, J. H. (1980). Specific fluorescent bands on chromosomes produced by acridine orange after prestaining with base specific non-fluorescent ligands. *Chromosoma* **79**, 271–286.

80. Lin, C. C., and Van de Sande, J. H. (1975). Differential fluorescent staining of human chromosomes with daunomycin and adriamycin—the D bands. *Science* **190**, 61–63.

81. Markovits, J., Roques, B. P., and Le Pecq, J. B. (1979). Ethidium dimer: a new reagent for the fluorometric determination of nucleic acids. *Anal. Biochem.* **94**, 259–264.

82. Martin, R. F., and Holmes, N. (1983). Use of an 125-

I-labeled DNA ligand to probe DNA structure. *Nature* **302**, 452–454.

83. Meer, B., Hameister, H., and Cerrillo, M. (1981). Early and late replication patterns of increased resolution in human lymphocyte chromosomes. *Chromosoma* **82**, 315–319.

84. Mendelsohn, M. L., Mayall, B. H., Bogart, E., Moore II, D. M., and Perry, B. H. (1973). DNA content and DNA-based centromeric index of the 24 human chromosomes. *Science* **179**, 1126–1129.

85. Meyne, J., Bartholdi, M. F., Travis, G., and Cram, L. S. (1984). Counterstaining human chromosomes for flow karyology. *Cytometry* **5**, 580–583.

86. Miller, O. J., Miller, D. A., and Warburton, D. (1973). Application of new staining techniques to the study of human chromosomes. *Prog. Med. Genet.* **9**, 1–47.

87. Modest, E. J., and Sengupta, S. K. (1974). 7-substituted actinomycin D (NSC-3053) analogs as fluorescent DNA-binding and experimental antitumor agents. *Cancer Chemother. Reports* **58**, 35–48.

88. Moyzis, R. K., Albright, K. L., Bartholdi, M. F., Cram, L. S., Deaven, L. L., Hildebrand, C. E., Joste, N. E., Longmire, J. L., Meyne, J., and Schwarzacher-Robinson, T. (1987). Human chromosome-specific repetitive DNA sequences: novel markers for genetic analysis. *Chromosoma* **95**, 375–386.

89. Muller, W., Bunemann, H., and Dattagupta, N. (1975). Interactions of heteroaromatic compounds with nucleic acids. 2. Influence of substituents on the base and sequence specificity of intercalating ligands. *Eur. J. Biochem.* **54**, 279–291.

90. Muller, W., and Crothers, D. M. (1975). Interactions of heteroaromatic compounds with nucleic acids. 1. The influence of heteroatoms and polarizability on the base specificity of intercalating ligands. *Eur. J. Biochem.* **54**, 267–277.

91. Muller, W., Crothers, D. M., and Waring, M. J. (1973). A non-intercalating proflavine derivative. *Eur. J. Biochem.* **39**, 223–234.

92. Muller, W., and Gautier, F. (1975). Interactions of heteroaromatic compounds with nucleic acids: AT-specific non-intercalating DNA ligands. *Eur. J. Biochem.* **54**, 385–394.

93. Nayak, R., Sirsi, M., and Podder, S. K. (1975). Mode of action of antitumor antibiotics. Spectrophotometric studies on the interaction of chromomycin A3 with DNA and chromatin of normal and neoplastic tissue. *Biochem. Biophys. Acta* **378**, 195–204.

94. Ohlenbusch, H. H., Olivera, B. M., Tuan, D., and Davidson, N. (1967). Selective dissociation of histones from calf thymus nucleoprotein. *J. Mol. Biol.* **25**, 299–315.

95. Olmsted III, J., and Kearns, D. R. (1977).

Mechanism of ethidium bromide fluorescence enhancement on binding to nucleic acids. *Biochemistry* **16**, 3647–3654.

96. Otto, F. J., and Oldiges, H. (1980). Flow cytogenetic studies in chromosomes and whole cells for the detection of clastogenic effects. *Cytometry* **1**, 13–17.

97. Paoletti, J., and Le Pecq, J.-B. (1971). Resonance energy transfer between ethidium bromide molecules bound to nucleic acids. Does intercalation wind or unwind the DNA helix? *J. Mol. Biol.* **59**, 43–62.

98. Patel, D. J., Kozlowski, S. A., Rice, J. A., Broka, C., and Itakura, K. (1981). Mutual interaction between adjacent dG.dC actinomycin binding sites and dA.dT netropsin binding sites on the self-complementary d(C-G-C-G-A-A-T-T-C-G-C-G) duplex in solution. *Proc. Natl Acad. Sci. USA* **78**, 7281–7284.

99. Paulson, J. R., and Laemmli, U. K. (1977). The structure of histone-depleted metaphase chromosomes. *Cell* **12**, 817–828.

100. Perez-Castillo, A., Martin-Lucas, M. A., and Abrisqueta, J. A. (1987). Evidence for lack of specificity of the DA/DAPI technique. *Cytogenet. Cell Genet.* **45**, 62.

101. Peters, D. C. (1979). A comparison of mercury arc lamp and laser illumination for flow cytometers. *J. Histochem. Cytochem.* **27**, 241–245.

102. Pohl, F., Jovin, T., Baehr, W., and Holbrook, J. J. (1972). Ethidium bromide as a cooperative effector of a DNA structure. *Proc. Natl Acad. Sci. USA* **69**, 3805–3809.

103. Raposa, T., and Natarajan, A. T. (1974). Fluorescence banding pattern of human and mouse chromosomes with a benzimidazol derivative (Hoechst 33258). *Humangenetik* **21**, 221–226.

104. Sahar, E., and Latt, S. A. (1978). Enhancement of banding patterns in human metaphase chromosomes by energy transfer. *Proc. Natl Acad. Sci. USA* **75**, 5650–5654.

105. Schmidt, M. (1980). Two phases of DNA replication in human cells. *Chromosoma* **76**, 101–110.

106. Schnedl, W. (1978). Structure and variability of human chromosomes analyzed by recent techniques. *Hum. Genet.* **41**, 1–9.

107. Schnedl, W., Breitenbach, M., Mikelsaar, A.-V., and Stranzinger, G. (1977). Mithramycin and DIPI: a pair of fluorochromes specific for GC- and AT-rich DNA respectively. *Hum. Genet.* **36**, 299–305.

108. Schnedl, W., Dann, O., and Schweizer, D. (1980). Effects of counterstaining with DNA binding drugs on fluorescent banding patterns of human and mammalian chromosomes. *Eur. J. Cell. Biol.* **20**, 290–296.

109. Schnedl, W., Mikelsaar, A.-V., Breitenbach, M., and Dann, O. (1977). DIPI and DAPI: fluorescence banding with only negligible fading. *Hum. Genet.* **36**, 167–172.

110. Schnedl, W., Roscher, U., Van der Ploeg, M., and Dann, O. (1977). Cytofluorometric analysis of nuclei and chromosomes by DIPI staining. *Cytobiologie* **15**, 357–362.

111. Schweizer, D. (1976). Reverse fluorescent chromosome banding with chromomycin and DAPI. *Chromosoma* **58**, 307–324.

112. Schweizer, D. (1980). Simultaneous fluorescent staining of R bands and specific heterochromatic regions (DA–DAPI bands) in human chromosomes. *Cytogenet. Cell Genet.* **27**, 190–193.

113. Schweizer, D. (1981). Counterstain-enhanced chromosome banding. *Hum. Genet.* **57**, 1–14.

114. Schweizer, D., Ambros, P., and Andrle, M. (1978). Modification of DAPI banding on human chromosomes by prestaining with a DNA-binding oligopeptide antibiotic, distamycin A. *Exptl Cell Res.* **111**, 327–332.

115. Stryer, L., and Haugland, R. (1967). Energy transfer: a spectroscopic ruler. *Proc. Natl Acad. Sci. USA* **58**, 719–726.

116. Sumner, A. T. (1982). The nature and mechanisms of chromosome banding. *Cancer Genet. Cytogenet.* **6**, 59–87.

117. Swartzendruber, D. E. (1977). Microfluorometric analysis of cellular DNA following incorporation of bromodeoxyuridine. *J. Cell. Physiol.* **90**, 445–454.

118. Tobia, A. M., Schildkraut, C. L., and Maio, J. J. (1970). Deoxyribonucleic acid replication in synchronized cultured mammalian cells. I. Time of synthesis of molecules of different average guanine + cytosine content. *J. Mol. Biol.* **54**, 499–515.

119. Trask, B., Van den Engh, G., Landegent, J., Jansen in de Wal, N., and Van der Ploeg, M. (1985). Detection of DNA sequences in nuclei in suspension by in situ hybridization and dual beam flow cytometry. *Science* **230**, 1401–1403.

120. Tsou, K. C., Yip, K. F., and Go, K. J. (1976). Fluorescence in human metaphase chromosomes produced by adriamycin with and without deoxyribonuclease treatment. *J. Histochem. Cytochem.* **24**, 752–756.

121. Van den Engh, G., Trask, B., Cram, S., and Bartholdi, M. (1984). Preparation of chromosome suspensions for flow cytometry. *Cytometry* **5**, 108–117.

122. Van den Engh, G. J., Trask, B. J., and Gray, J. W. (1986). The binding kinetics and interaction of DNA fluorochromes used in the analysis of nuclei and chromosomes by flow cytometry. *Histochem.* **84**, 501–508.

123. Van den Engh, G. J., Trask, B. J., Gray, J. W., Langlois, R. G., and Yu, L.-C. (1985). Preparation and bivariate analysis of suspensions of human chromosomes. *Cytometry* **6**, 92–100.

124. Van de Sande, J. H., and Jovin, T. M. (1982).

Z DNA, the left-handed helical form of poly [d(G-C)] in MgCl$_2$-ethanol, is biologically active. *EMBO J.* **1**, 115–120.

125. Van Dilla, M. A., Langlois, R. G., Pinkel, D., Yajko, D., and Hadley, W. K. (1983). Bacterial characterization by flow cytometry. *Science* **220**, 620–622.

126. Van Dyke, M. M., and Dervan, P. B. (1983). Chromomycin, mithramycin, and olivomycin binding sites on heterogeneous deoxyribonucleic acid. Footprinting with (methidiumpropyl–EDTA) iron (II). *Biochemistry* **22**, 2373–2377.

127. Van Dyke, M. M., and Dervan, P. B. (1984). Echinomycin binding sites on DNA. *Science* **225**, 1122–1127.

128. Van Dyke, M. M., Hertzberg, R. P., and Dervan, P. B. (1982). Map of distamycin, netropsin, and actinomycin binding sites on heterogeneous DNA:DNA cleavage-inhibition patterns with methidiumpropyl–EDTA.Fe(II). *Proc. Natl Acad. Sci. USA* **79**, 5470–5474.

129. Van Prooijen-Knegt, A. C., Van Hoek, J. F. M., Bauman, J. G. J., Van Duijn, P., Wool, I. G., and Van der Ploeg, M. (1982). *In situ* hybridization of DNA sequences in human metaphase chromosomes visualized by an indirect fluorescent immunocytochemical procedure. *Exptl Cell Res.* **141**, 397–407.

130. Viegas-Pequignot, E., Malfoy, B., Sabatier, L., and Dutrillaux, B. (1987). Different reactivity of Z-DNA antibodies with human chromosomes modified by actinomycin D and 5-bromodeoxyuridine. *Hum. Genet.* **75**, 114–119.

131. Waring, M. (1970). Variation of the supercoils in closed circular DNA by binding of antibiotics and drugs: evidence for molecular models involving intercalation. *J. Mol. Biol.* **54**, 247–279.

132. Wartell, R. M., Larson, J. E., and Wells, R. D. (1975). The compatibility of netropsin and actinomycin binding to natural deoxyribonucleic acid. *J. Biol. Chem.* **250**, 2698–2702.

133. Willard, H. F., and Waye, J. S. (1987). Hierarchical order in chromosome-specific human alpha satellite DNA. *Trends Genet.* **3**, 192–198.

134. Wray, W., and Stubblefield, E. (1970). A new method for the rapid isolation of chromosomes, mitotic apparatus, or nuclei from mammalian fibroblasts at near neutral pH. *Exptl Cell Res.* **59**, 469–478.

135. Wray, W., and Stubblefield, E. (1982). Techniques to isolate specific human metaphase chromosomes. *In* "Techniques in Somatic Cell Genetics" (J. W. Shay, ed.), pp. 349–373. Plenum Press, New York.

136. Wray, W., and Wray, V. P. (1980). Proteins from metaphase chromosomes treated with fluorochromes. *Cytometry* **4**, 18–20.

137. Young, B. D., Ferguson-Smith, M. A., Sillar, R.,

and Boyd, E. (1981). High-resolution analysis of human peripheral lymphocyte chromosomes by flow cytometry. *Proc. Natl Acad. Sci. USA* **78**, 7727–7731.

138. Yu, L.-C., Aten, J., Gray, J., and Carrano, A. V. (1981). Human chromosome isolation from short-term lymphocyte culture for flow cytometry. *Nature* **293**, 154–155.

139. Yunis, J. J., Kuo, M. T., and Saunders, G. F. (1977). Localization of sequences specifying messenger RNA to light-staining G-bands of human chromosomes. *Chromosoma* **61**, 335–344.

140. Yunis, J. J., Sawyer, J. R., and Dunham, K. (1980). The striking resemblance of high-resolution

G-banding chromosomes of man and chimpanzee. *Science* **208**, 1145–1148.

141. Yurov, Y. B., Mitkevich, S. P., Alexandrov, I. A. (1987). Application of cloned satellite DNA sequences to molecular-cytogenetic analysis of constitutive heterochromatin heteromorphisms in man. *Hum. Genet.* **76**, 157–164.

142. Zelenin, A. V., Poletaev, A. I., Stepanova, N. G., Barsky, V. E., Kolesnikov, V. A., Nitikin, S. M., Zhuze, A. L., and Gnutchev, N. V. (1984). 7-amino-actinomycin D as a specific fluorophore for DNA content analysis by laser flow cytometry. *Cytometry* **5**, 348–354.

Methods for Estimating Components of Multipeaked Flow Histograms

DAN H. MOORE II

Biomedical Sciences Division
Lawrence Livermore National Laboratory
Livermore, CA 94500, USA

I. INTRODUCTION

The focus of this chapter is on statistical methods for resolving multipeaked flow histograms (flow karyotypes) into separate components. Typically, it can be assumed that the components are Gaussian (normal) in shape and that a complicating factor is the existence of a background continuum which underlies the components of interest. A typical univariate flow histogram is shown in Fig. 1. The goal of the statistical analysis is to determine the location (means), spread (standard deviations or, equivalently, the coefficients of variation) and the area (number of data points) for each component of the histogram.

There are several properties of these multipeaked flow histograms that are unique to flow cytogenetics and make it difficult to apply "standard" techniques for decomposition reported in the statistical literature. First

FLOW CYTOGENETICS
ISBN 0-12-296110-2

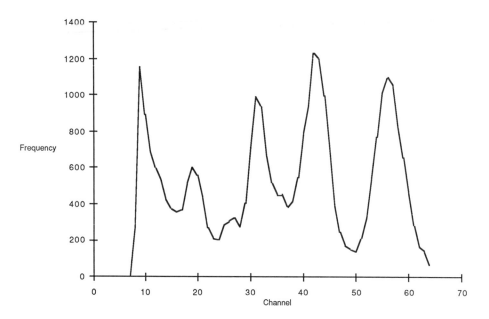

Fig. 1. Univariate flow histogram for chromosomes 9–22 from a normal human karyotype. The chromosomes were stained with Hoechst 33258 and chromomysin 193. The histogram shows the distributions of Hoechst 33258 fluorescence following ultraviolet excitation.

sample sizes for flow data are enormous; typically 100 000–300 000 data points are collected in flow karyotypes. This averages out to 5000–15 000 observations per component. Most methods reported in the statistical literature are concerned with samples in the range of tens to hundreds. This means that methods judged inefficient for small samples may prove efficient with flow data and *vice versa*.

Most, if not all, flow data are "contaminated" by a background component whose shape is not easily characterized. Methods which rely on exact parametric representations of the data may perform poorly or not at all with flow data.

Finally, the objective of the analysis of flow data may not be primarily to obtain estimates for component parameters but, rather, to make a comparison with a "standard" (control) in order to measure degree of abnormality. If this is the case, then precise estimates may be of less importance than a reliable test for comparing multipeaked distributions.

I begin with a mathematical description of the problem followed by a brief history of methods from the statistical literature. These methods are then applied to example flow histograms of human chromosomes. Next, I describe methods currently in use for analyzing flow data in the form of univariate and bivariate histograms. Finally, I describe adjuvant procedures which aid in the display, computing efficiency or interpretation of the data.

II. MATHEMATICAL DESCRIPTION OF THE PROBLEM

The problem may be formulated succinctly in the univariate case as that of estimating the parameters (A_i, M_i, S_i) in the mixture

$$H(x) = \Sigma A_i F(x; M_i, S_i) + B(x)$$

where $H(x)$ is the histogram frequency at channel x, A_i are the (unknown) areas for the components, $i = 1, ..., k$; $F(x; M_i, S_i)$ is the theoretical fitting function with parameters M_i and S_i and $B(x)$ is an unknown background function. In the univariate case $F(x; M_i, S_i)$ is usually assumed to be a Gaussian (normal) distribution with mean M_i and standard deviation S_i. In this case

$$F(x; M_i, S_i) = \int_{x-1/2}^{x+1/2} \frac{1}{\sqrt{2\pi} S_i}$$
$$\times \exp\left[-\tfrac{1}{2}(t - M_i)/S_i\right]^2 dt.$$

The relationship between the coefficient of variation (CV) and the parameters M_i and S_i is given by

$$CV_i = \frac{S_i}{M_i},$$

so that it is possible to express initial estimates and results in terms of this more familar (to biologists) parameter. In the bivariate case the fitting function becomes

$$f(x_1, x_2) = \frac{1}{2\pi S_1 S_2 \sqrt{1 - R^2}}$$
$$\times \exp\{-1/2(1 - R^2)[((x_1 - M_1)/S_1)^2$$
$$- 2R((x_1 - M_1)/S_1) \times ((x_2 - M_2)/S_2)$$
$$+ ((x_2 - M_2)/S_2)^2]\}$$

where M_i are the bivariate means, S_i are the bivariate standard deviations and R is the correlation between the variables x_1 and x_2.

There is no absolute requirement that the fitting function be Gaussian, but the literature is very thin on non-Gaussian functions and the underlying biology supports a Gaussian distribution for the data. If the variation in measured stain content of each object is believed to arise from many small contributors, such as variation in stain uptake, measurement sensitivity as well as underlying biologic variabilities, the central limit theorem suggests that the result will be a normal distribution. Thus, until it is demonstrated that the distributional shapes are not Gaussian and a tractable method is developed for non-Gaussian components, it is reasonable to assume that the $F(x; M_i, S_i)$ are Gaussians. Of course, there are situations where the distributional shape will not be Gaussian; for example if there is a change in gain in the middle of a sample run. Unfortunately, at present there are no reliable methods for handling the analysis of such unintentional variations in data gathering.

In the univariate case the shape of the background function $B(x)$ is usually assumed to be exponential, i.e.

$$B(x) = C e^{-Dx}$$

or to follow a power law, known as a Pareto distribution in the statistical literature, i.e.

$$B(x) = C x^D$$

These functions describe a continuum that starts high and decreases rapidly with x. They model a situation where the continuum is made up of randomly broken chromosome fragments. Problems may arise with these functions when applied to data which is clumpy since they cannot model lumpy data. There is a dearth of information in the literature on background functions for the bivariate or multivariate case. As mentioned earlier, the great majority of the statistical literature deals only with the case of no background, i.e. $B(x) = 0$. In the next section I review the published univariate methods·

III. UNIVARIATE METHODS

A. METHOD OF MOMENTS

The earliest published method is based on equating sample and theoretical moments (23). This method can be used to separate two Gaussian components by solving a ninth-degree equation based on setting the first five sample moments equal to theoretical moments for the mixture. The sample moments can be computed from the following equations:

$$m_1 = \Sigma H(x) x$$
$$m_i = \Sigma H(x) (x - m_1)^2 \text{ for } i = 2, ..., 5$$

The theoretical moments for a mixture of two normals (Gaussians) are

$$\mu_1 = \alpha\,\theta_1 + (1-\alpha)\,\theta_2$$

$$\mu_2 = \alpha(\sigma_1^2 + (1-\alpha)^2\,\delta^2) + (1-\alpha)\,(\sigma_2^2 + \alpha^2\,\delta^2)$$

$$\mu_3 = \alpha(1-\alpha)\,(\theta_1 - \theta_2) \times (3\sigma_1^2 + (1-\alpha)^2\,\delta^2) + \alpha(1-\alpha) \times (\theta_2 - \theta_1)\,(3\sigma_2^2 + \alpha^2\,\delta^2)$$

$$\mu_4 = \alpha[3\sigma_1^4 + 6(1-\alpha)^2\,\delta^2\sigma_1^2 + (1-\alpha)^4\,\delta^4] + (1-\alpha) \times [3\sigma_2^4 + 6\alpha^2\,\delta^2\,\sigma_2^2 + \alpha^4\,\delta^4]$$

$$\mu_5 = \alpha(1-\alpha)\,(\theta_1 - \theta_2)\,[15\sigma_1^4 + 10(1-\alpha)^2\,\delta^2\sigma_1^2 + (1-\alpha)^4\,\delta^4] + \alpha(1-\alpha)\,(\theta_2 - \theta_1) \times [15\sigma_2^4 + 10\alpha^2\,\delta^2\,\sigma_2^2 + \alpha^4\,\delta^4],$$

where θ_i and σ_i^2 are the (true) mean and variance for the ith component and $\delta = \theta_2 - \theta_1$. The idea is to set the m_i equal to the μ_i and solve for the five unknown parameters $\{\theta_1, \theta_2, \sigma_1^2, \sigma_2^2, \alpha\}$. After much algebra and rearranging of terms the task boils down to solving the ninth-degree equation

$$a_9\,v^9 + a_8\,v^8 + a_7\,v^7 + a_6\,v^6 + a_5\,v^5 + a_4\,v^4 + a_3\,v^3 + a_2\,v^2 + a_1\,v^1 + a_0 = 0,$$

where

$$a_9 = 24$$
$$a_8 = 0$$
$$a_7 = 84k_4$$
$$a_6 = 36m_3^2$$
$$a_5 = 90k_4^2 + 72k_5\,m_3$$
$$a_4 = 444k_4\,m_3^2$$
$$a_3 = 288m_3^2 - 108m_3\,k_4\,k_5 + 27k_4^2$$
$$a_2 = -(63m_3^2\,k_4^2 + 72m_3^2\,k_5)$$
$$a_1 = -96m_3^4\,k_4$$
$$a_0 = -24m_3^6.$$

k_4 and k_5 are the fourth- and fifth-order sample cumulants and are related to the sample moments by

$$k_4 = m_4 - 3m_2^2$$
$$k_5 = m_5 - 10m_2m_3.$$

This method was applied to sample data from a chromosome flow histogram shown in Fig. 2. In this sample there are two peaks riding on a high background continuum. I applied the method of moments to the raw data as well as to the same data with a linear background component removed. The calculations for the fit to data with background removed are shown in Table I. The resulting fits are shown in Fig. 3. In Fig. 3(a) it is evident that the method of moments cannot provide an acceptable fit due to the unaccounted for background component. The fit appears to be reasonably good in Fig. 3(b), although it fails to meet the maximum frequencies at the center of each peak. A measure of the adequacy of the fit is given by the chi-square statistic which is formed by calculating

$$X^2 = \Sigma\,\frac{[\text{observed frequency}\,(x) - \text{fit}\,(x)]^2}{\text{fit}\,(x)}.$$

An assessment of the fit is provided by comparing this sum with a table of the chi-square distribution. The degrees of freedom are equal to the number of channels minus the number of parameters estimated. In our example there are 18 channels, and five parameters were used in the fit so the degrees of freedom are 13. The upper 95% point for a chi-square distribution with 13 degrees of freedom is 22.36 and the upper 99% point is 27.69. Thus, with a lack-of-fit chi-square of 26.9 (Table I, last column) our fit is somewhat unsatisfactory from a statistical point of view. (The interpretation is that if we are sampling data from a mixture of two normals with the parameters as specified, only rarely, i.e. somewhere between 1 and 5% of the time, would we realize a sample as poorly fitting as ours.) However, the subtraction of a background component reduced the chi-square from 453.7 (fit to raw data) to 26.9 (Table I) and the fit appears reasonable.

As is evident from Table I, the method of moments is algebraically cumbersome. Also, according to a recent report, it is likely to fail

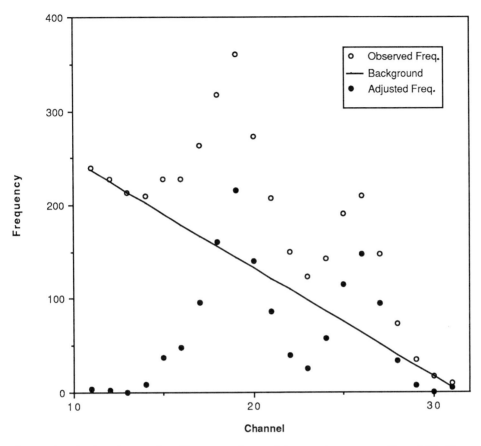

Fig. 2. Sample data from a flow histogram of normal human chromosomes 21 and 22. These chromosomes are confounded with a background (debris) continuum which has been modeled by the linear component shown as a solid line. The result of subtracting the background from the observed data is shown as the adjusted frequency (solid line).

(30). The authors of this report found that the moment method produced negative estimates for variances in as many as 25% of simulated samples from simple mixtures of two normal distributions. In carrying out the fits, I found negative variance estimates whenever the "wrong" root of the nonic equation was used. In the fit to the data with a background component removed, there is a negative root at -1.5 which led to an estimate of -51.72 for σ_1^2. It was necessary to find the next root (proceeding away from the origin on the negative axis) at -11.2 in order to realize admissible parameter estimates.

Estimation by the method of moments can be simplified by assuming that the variances are equal. In this case the solution of a cubic equation based on the first four sample moments is required. Cohen (5) shows that the equation to be solved is

$$2v^3 + k_4 v + m_3{}^2 = 0.$$

In this case there will be a single negative root, say v^*. This leads to $r^* = -m_3/v^*$. Estimation then proceeds as outlined for the general case, except that the estimate for the common variance is given by

$$\sigma^2 = v^* + m_2$$

When I applied this method to our example

Table I

METHOD OF MOMENTS FIT TO EXAMPLE DATA ADJUSTED FOR BACKGROUND

Channel	Observed frequency	Linear background	Adjusted frequency	Hxx	$Hx(x-m_1)^2$	$Hx(x-m_1)^3$	$Hx(x-m_1)^4$	$Hx(x-m_1)^5$	Fit Peak 1	Peak 2	Sum	Chi-square
13	213	212	1	0	0	0	0	0	2	0	2	0.4
14	210	201	9	112	423	−3076	22366	−162639	8	0	8	0.2
15	228	189	39	555	1455	−9127	57240	−358984	25	0	25	8.1
16	227	177	50	768	1334	−7032	37068	−195403	61	0	61	2.0
17	263	166	97	1632	1752	−7482	31960	−136521	115	0	115	2.9
18	317	154	163	2898	1723	−5637	18443	−60338	165	0	165	0.0
19	360	143	217	4104	1115	−2532	5751	−13064	182	0	182	6.9
20	273	131	142	2800	226	−288	366	−465	152	0	152	0.7
21	207	119	88	1806	6	−2	0	0	97	0	97	0.9
22	150	108	42	880	21	15	11	8	48	1	49	0.9
23	123	96	27	575	75	129	223	386	18	10	28	0.0
24	143	85	58	1368	424	1158	3159	8619	5	47	52	0.7
25	190	73	117	2875	1599	5961	22223	82859	1	114	116	0.0
26	210	61	149	3822	3287	15541	73484	347466	0	147	147	0.0
27	147	50	97	2565	3117	17858	102299	586016	0	99	99	0.1
28	73	38	35	924	1494	10052	67635	455082	0	35	35	0.0
29	35	27	8	203	418	3231	24973	193001	0	7	7	0.3
30	17	15	2	0	0	0	0	0	0	1	1	2.8
Sums	3386		1341	27887	18469	18770	467204	746023	879	462	1341	26.9

$m_1 = \Sigma Hxx/\Sigma Hx = 20.80$
$m_2 = \Sigma Hx(x-m_1)^2/\Sigma Hx = 13.77$
$m_3 = \Sigma Hx(x-m_1)^3/\Sigma Hx = 14.00$
$m_4 = \Sigma Hx(x-m_1)^4/\Sigma Hx = 348.40$
$m_5 = \Sigma Hx(x-m_1)^5/\Sigma Hx = 556.32$
$k_4 = m_4 - 3m_2^2 = -220.67$
$k_5 = m_5 - 10m_2 m_3 = -1371.44$

$a_9 = 24$
$a_8 = 0$
$a_7 = -20079$
$a_6 = 7379$
$a_5 = 3\,650\,040$
$a_4 = -59\,493\,454$
$a_3 = -891\,875\,560$
$a_2 = -431\,946\,162$
$a_1 = 964\,207\,360$
$a_0 = -206\,705\,747$

$\nu = $ (sol. 9th deg. eq.) $= -11.2$
$w = $ (see below) $= -24.5$
$r = w/\nu = 2.191$
$\delta_1 = 0.5[r - \sqrt{r^2 - 4\nu}] = -2.43$
$\delta_2 = 0.5[r + \sqrt{r^2 - 4\nu}] = 4.617$
$\beta = (2r - m_3/\nu)/3 = 1.886$
$\sigma_1^2 = \delta_1\beta + m_2 - \delta_1^2 = 3.621$ and 1.903
$\sigma_2^2 = \delta_2\beta + m_2 - \delta_2^2 = 1.478$ and 1.216
$\theta_1 = \delta_1 + m_1 = 18.84$
$\theta_2 = \delta_2 + m_1 = 25.89$
$\alpha = m_2/(m_2 - m_1) = 0.655$
$w = (-8m_3\nu^3 + 3k_5\nu^2 + 6m_3 k_4 \nu + 2m_3^3)/(2\nu^3 + 3k_4 \nu + 4m_3^2)$

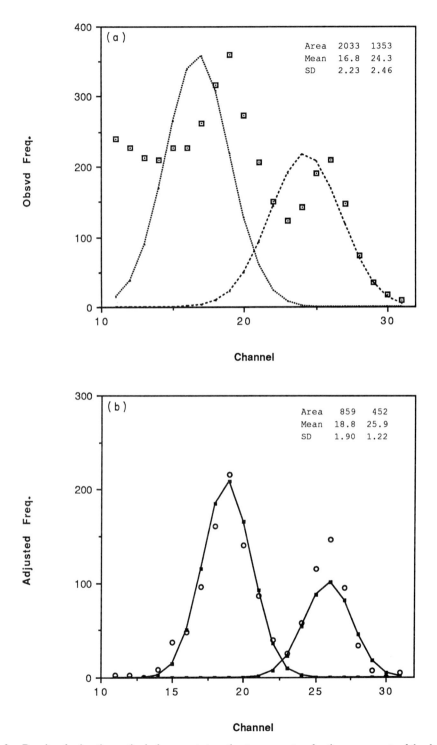

Fig. 3. Results of using the method of moments to estimate parameters for the components of the data shown in Fig. 2. (a) Result when there is no adjustment for background. (b) Result when linear background component is subtracted prior to fit.

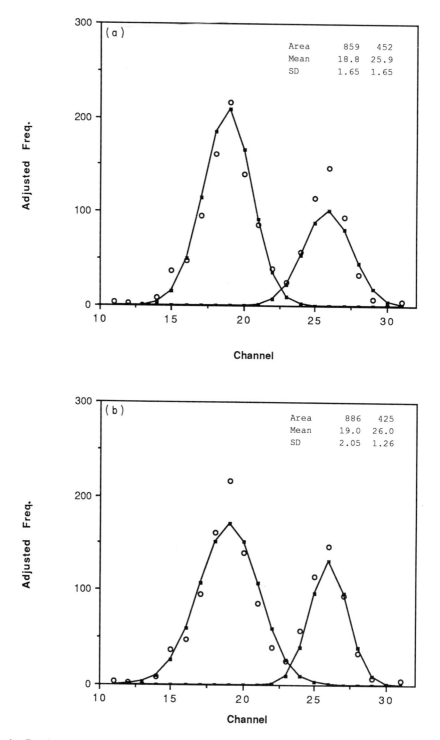

Fig. 4. Results of using the method of moments applied to adjusted data of Fig. 2. (a) Fit when components are assumed to have equal variances. (b) Fit when components are assumed to have known means (equal to channel with local maximum frequency).

data (with background subtracted), the fit shown in Fig. 4(a) was realized. Visually, the fit appears better on the first peak but worse on the second. The overall chi-square has increased to 98.8.

Estimation can be further simplified by assuming that the component means are known. In this case estimation of component areas is particularly easy. The area estimates are

$$A_i = N \frac{|\bar{x}_i - m_1|}{|\bar{x}_2 - \bar{x}_1|}$$

where N is the total frequency in both components, $\bar{x}_i$ is the (known) mean for the ith component and m_1 is the mean for the combined sample. The estimates for the variances are now given by

$$\sigma_1^2 = m_2 + 2\frac{(\bar{x}_1 - m_1)(\bar{x}_2 - m_1)}{3}$$
$$- \frac{(\bar{x}_1 - m_1)^2}{3} - \frac{m_3}{3(\bar{x}_2 - m_1)}$$

$$\sigma_2^2 = m_2 + 2\frac{(\bar{x}_1 - m_1)(\bar{x}_2 - m_1)}{3}$$
$$- \frac{(\bar{x}_2 - m_1)^2}{3} - \frac{m_3}{3(\bar{x}_1 - m_1)}.$$

The result of applying this method to the example data is shown in Fig. 4(b). In this case the fit is more similar to the unrestricted case [Fig. 3(b)] but the chi-square has increased to 45.9.

This method is interesting from a theoretical standpoint—much can be learned about the complexity of the decomposition problem by studying the moments of the mixture—but is not practical for flow data for two reasons: there are nearly always more than two components to the mixture and the presence of an unknown background will make the method unreliable.

B. GRAPHICAL METHODS

In 1967, Bhattacharya (3) published a method for resolving univariate Gaussian components based on plotting log [$h(x + 1)/$ $h(x)$], where $h(x)$ is the observed histogram frequency at channel x, against x. The component is estimated by adding ½ to the lines with negative slopes (by eye) to the resulting sequence of points. The mean for each component if estimated by adding ½ to the intersection of the line with the x axis. The variance is equated to $-1/\text{slope}$ and the area is found as the ratio of the sum of the observed frequencies over a range of channels near the mean divided by the area under a normal curve over the same range. An example of application of this method to (adjusted) flow karyotype analysis is shown in Table II. The result of the graphical method is shown in Fig. 5. The fit obtained by this method appears similar to that obtained by the method of moments [Fig. 3(b)] but chi-square is almost twice as large (51.9 vs 26.9). The method is approximate (especially when the straight lines are drawn by hand as was advocated in the original 1967 publication of the method) and does not work well with overlapped components. Furthermore, its performance in the presence of background is unreliable. Figure 6 shows the result of using the graphical method on the raw (unadjusted for background) data. The chi-square for the fit is 319. While somewhat better than the fit obtained by the method of moments, it is still unacceptable.

C. FOURIER METHODS

These methods seem to have originated with Doetsch (7,8). The idea is to subtract a constant (c) from the standard deviation (S_i) for each component without changing any of the other parameters. When this is done the ith component of the mixture becomes

$$F^*(x; M_i, S_i) = \frac{1}{\sqrt{2\pi}(S_i - c)}$$
$$\times \exp\left[-\tfrac{1}{2}((x - M_i)/(S_i - c))^2\right].$$

This has the effect of making each component less overlapped with its neighbours so that it is possible to get cleaner estimates for peak means and areas. Since neither the M_i nor the S_i are

Table II

GRAPHICAL FIT TO ADJUSTED DATA

Channel	Observed frequency	Linear background	Adjusted frequency	$\ln[h(x+1)/h(x)]$	Linear fit	Normal $Z(x)$	Comp. 1	Comp. 2	Fit	Chi-square
13	213	212	1	2.20			1	0	1	0.0
14	210	201	9	1.47			5	0	5	3.0
15	228	189	39	0.25			19	0	19	21.0
16	227	177	50	0.66			53	0	53	0.1
17	263	166	97	0.52	0.43	0.12685	108	0	108	1.1
18	317	154	163	0.29	0.12	0.19276	164	0	164	0.0
19	360	143	217	−0.42	−0.18	0.21771	185	0	185	5.6
20	273	131	142	−0.48	−0.49	0.18217	155	0	155	1.1
21	207	119	88	−0.74	−0.79	0.11327	96	1	97	0.9
22	150	108	42	−0.44			44	7	51	1.6
23	123	96	27	0.76	0.53	0.15186	15	29	44	6.6
24	143	85	58	0.70	0.04	0.25183	4	76	80	6.2
25	190	73	117	0.24	−0.45	0.26193	1	127	127	0.9
26	210	61	149	−0.43	−0.93	0.17071	0	132	132	2.2
27	147	50	97	−1.02	−1.42	0.06969	0	86	86	1.4
28	73	38	35	−1.48			0	35	35	0.0
29	35	27	8	−1.39			0	9	9	0.1
30	17	15	2				0	1	1	0.2
Sums	3386		1341				849	503	1352	51.9

b_1 (slope) = −0.306

a_1 (intcp) = 5.6344

b_2 (slope) = −0.487

a_2 (intcp) = 12.21

$\theta_1 = -a_1/b_1 =$ 18.91

$\theta_2 = -a_2/b_2 =$ 25.58

$\sigma_2^2 = -1/b_1 =$ 3.2666 and 1.80737

$\sigma_1^2 = -1/b_2 =$ 2.0544 and 1.43331

Area$_1 = \Sigma Hx/\Sigma Zx =$ 849

Area$_2 = \Sigma Hx/\Sigma Zx =$ 503

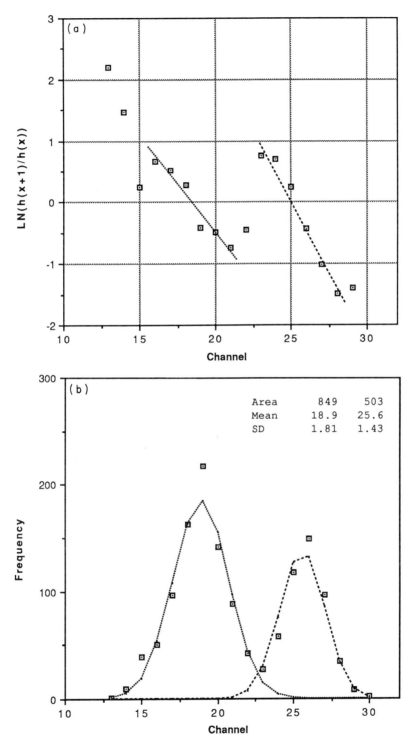

Fig. 5. Results of using graphical method to estimate parameters for the adjusted data shown in Fig. 2. (a) The log of the ratio of adjacent channels is plotted against channel to find location of peak means (one-half added to where straight lines cross the x-axis). The slopes are related to the variances by the equation $\sigma^2 = -1/\text{slope}$. (b) Resulting fit.

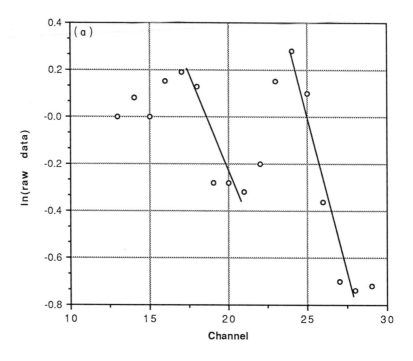

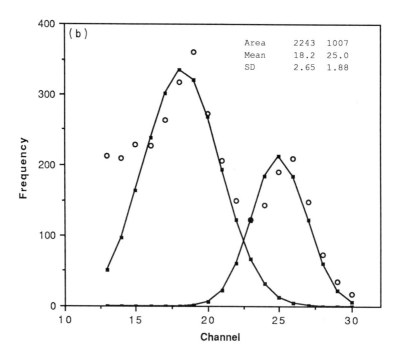

Fig. 6. Results of applying the graphical method to the raw (unadjusted) data of Fig. 2. (a) Log of ratio of adjacent channels. (b) Resultant fit.

known, reduction of the S_i by the constant c must be done indirectly. This is accomplished by a Fourier transform. First the histogram of observed frequencies is represented by the harmonic series

$$H(x) = \Sigma\, C_k \sin(kAx + B_k)$$

where $Ax = 2x\pi/T$ with T set equal to the range of x. Reduction of overlap is now accomplished by forming

$$H^*(x) = \Sigma \exp(k^2 \pi^2 c/T^2)\, C_k \sin(kAx + B_k),$$

which is the Fourier transform of the Gaussian mixture $\Sigma\, F(x; M_i, S_i)$. Means M_i, are estimated by the transformed peak modes, standard deviations by the constant c which makes each component, in turn, a single spike, and areas, A_i, by summing counts in the transformed and (theoretically) well-separated, peaks. Details of this procedure may be found in Gregor (12) and Medgyessy (18).

Figure 7 shows the result of applying the Fourier method to our sample chromosome flow data. Twenty terms were used to obtain the Fourier approximation in Fig. 7(a). The effect of increasing the variance-reducing parameter c is shown in Fig. 7(b–f). It is clear that the method fails since it introduces a plethora of spikes at values of c well below the "true" standard deviation for either of the peaks (which is somewhat greater than 1.0 according to the fits by other methods).

Everitt and Hand (10) claim that "there appears to have been no investigations of the sampling properties of this method or attempts to compare it with statistically more conventional techniques such as moments or maximum likelihood." However, they seem to be unaware of results by Essenwanger (9). In summarizing his results with simulated data he states that "although this procedure may be satisfactory from the theoretical point of view, it cannot be recommended in practical work because it would only lead to unnecessary computer expenses without adding real information." The basic problem seems to be the creation of false components due to random noise in the data (see example above). This

method is appealing for data contaminated with a background continuum, since in theory this would be eliminated by removing "large", (i.e. slowly varying) terms in the Fourier transformation, but it failed when applied to our sample flow data.

D. TRUNCATION METHOD

This method is based on the assumption that one side of the histogram for each peak can be considered to be uncontaminated by neighboring peaks. Starting at the component with the largest mean, a truncation point x^* is selected. The portion of the histogram from x^* to x_{max}, the maximum channel containing data, is assumed to be uncontaminated by other components, and therefore can be used to estimate the parameters for that component. Tables for the truncated Gaussian distribution by Hald (13) can be used to facilitate parameter estimation. Once the largest component's parameters have been estimated, the contribution of this component is subtracted from the mixture and the process is repeated stepwise until all components have been estimated. Essenwanger (9) proposes a final "balancing step" where the residual differences between the stepwise fit and the observed frequences are minimized by adjustments in the previously determined components.

Hasselblad (15) used this method to obtain initial values for maximum likelihood estimates (see below). Unfortunately, he does not report on the accuracy of these estimates. It is not clear how successful this method would be in the presence of a contaminating background continuum; clearly the estimates based on Hald's tables would be biased without any adjustment for background. A modification of the method which first subtracts an estimate of the background component, either by a polynomial or a decaying function such as the exponential, prior to application of the method to the remaining frequencies should work. The result of applying this modified procedure to the adjusted chromosome flow histogram data is shown in Fig. 8. The fitting was accomplished

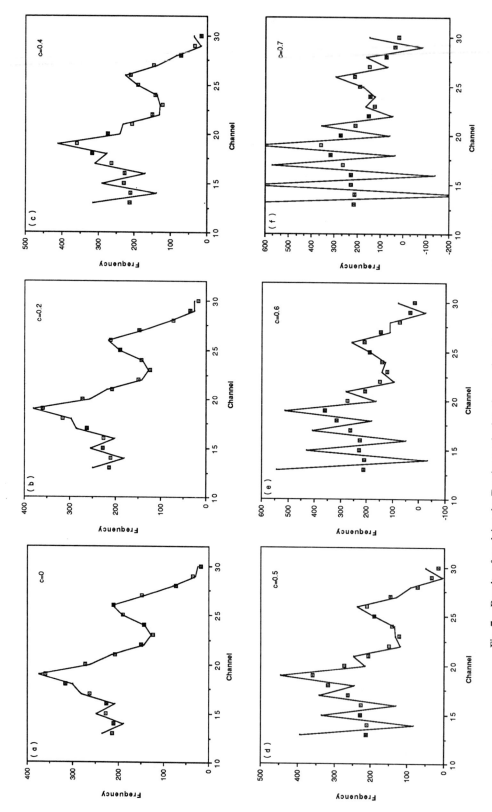

Fig. 7. Results of applying the Fourier method to the raw (unadjusted) data of Fig. 2. (a) Fit with 20 Fourier terms and constant $c = 0$. (See text for description of c parameter.) (b)–(f) Result of increasing c from 0.2 to 0.7. Note the "ringing" for $c \geq 0.4$. It was not possible to discern a value for c which transformed either peak into a spike (which was supposed to occur when c equals the standard deviation of the peak).

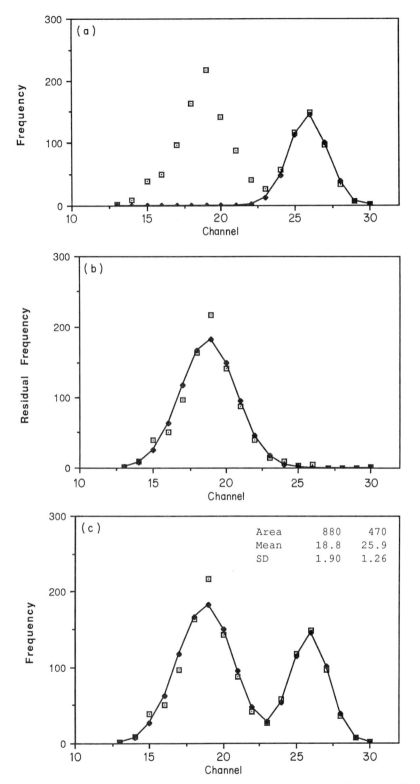

Fig. 8. Results of applying the truncation method to the adjusted data of Fig. 2. (a) Result of fitting a Gaussian to the right side of the right peak. (b) Result of subtracting the fit to the right peak and then fitting a Gaussian to the left peak. (c) The combined fit.

by least squares (see below for a description of this popular method). The parameter estimates are similar to those found by the method of moments (shown above). The chi-square for this fit is 24.5.

E. MAXIMUM LIKELIHOOD METHOD

Currently, this is the most widely studied and most frequently used method for separating mixtures of components. It has the advantage that the (asymptotic) properties of maximum likelihood estimation are well established in the statistical literature. The idea behind the method is to find the collection of parameter values $\{A_i, M_i, S_i\}$ which maximize the log of the likelihood function:

$$L = \Sigma_x h(x) \log [\Sigma_i A_i F(x; M_i, S_i)].$$

The maximum is found by equating the first partial derivatives to zero. This leads to the following series of equations:

$$\frac{dL}{dA_i} = \sum_{x=1}^{N} \frac{h(x)}{G(x)} [F(x; M_i, S_i) - F(x; M_k, S_k)]$$

$$\frac{dL}{dM_i} = \sum_{x=1}^{N} \frac{h(x)}{G(x)} A_i F(x; M_i, S_i) \frac{x - M_i}{S_i^2}$$

$$\frac{dL}{dS_i} = \sum_{x=1}^{N} \frac{h(x)}{G(x)} A_i F(x; M_i, S_i) \left[\frac{x - M_i}{S_i^3} - \frac{1}{S_i} \right],$$

where

$$G(x) = \Sigma A_i F(x; M_i, S_i).$$

Hasselblad (15) found an ingenious way to solve these equations iteratively. First he replaced the integrals $F(x; M_i, S_i)$ by their values at the midpoints of the intervals. Mathematically speaking, he used

$$f(x; M_i, S_i) = \frac{1}{\sqrt{2\pi} S_i}$$
$$\times \exp [-\tfrac{1}{2}(x - M_i/S_i)^2]$$

as an approximation for

$$F(x; M_i, S_i) = \int_{x-1/2}^{x+1/2} \frac{1}{\sqrt{2\pi} S_i}$$
$$\times \exp [-\tfrac{1}{2}((t - M_i)/S_i)^2] \, dt.$$

In a study of the accuracy of this approximation, Moore (21) found that the maximum difference between F and f is less than 0.01 whenever $S_i > 1.20$. Solving $dL/dM_i = 0$ for M_i and rearranging terms leads to

$$M_i = \frac{\displaystyle\sum_x \frac{h(x)}{G(x)} f(x; M_i, S_i) x}{\displaystyle\sum_x \frac{h(x)}{G(x)} f(x; M_i, S_i)}$$

Solving $dL/dS_i^2 = 0$ for S_i^2 and rearranging terms leads to

$$S_i^2 = \frac{\displaystyle\sum_x \frac{h(x)}{G(x)} f(x; M_i, S_i) (x - M_i)^2}{\displaystyle\sum_x \frac{h(x)}{G(x)} f(x; M_i, S_i)}$$

Finally, multiplying dL/dA_i by A_i and summing over i gives

$$A_i = \frac{\displaystyle\sum_x \frac{h(x)}{G(x)} f(x; M_i, S_i) A_i}{\displaystyle\sum_x h(x)}.$$

Thus there are $3k - 1$ presumably independent equations and $3k - 1$ unknowns. Solutions are found iteratively. At each step, new estimates for the parameters A_i, M_i and S_i are found by substituting the current estimates on the right sides of the equations. Table III shows the calculations necessary for the first step in applying this method to the sample data. Figure 9 shows the fit when the parameter estimates stabilized (relative change $< 0.1\%$) after five iterations.

Hasselblad found that this iterative method for solution always worked when the truncation method (described above) was used to obtain initial estimates. Redner and Walker (25), who studied the method from a theoretical viewpoint, essentially confirmed Hasselblad's claim. They point out that the method is a direct application of the so-called EM (E for "expectation" and M for "maximization") method which was formalized by Dempster et al. (6). Redner and Walker state that the method has the advantages of "reliable global

Table III

MAXIMUM LIKELIHOOD FIT TO ADJUSTED DATA. FIRST ITERATION STEP

Channel	Observed frequency	Linear background	Adjusted frequency	Initial estimates 671 19.00 2.00	Initial estimates 671 26.00 1.00	F	hf_1/F	$hf_1 x/F$	$hf_1 \Delta^2/F$	hf_2/F	$hf_2 x/F$	$hf_1 \Delta^2/F$	Log likelihood $h\log(F)$	Fit	Chi-square
13	213	212	1	0.00	0.00	0.00	2.00	26.00	72.00	0.00	0.00	0.00	−6.81	1	0.16
14	210	201	9	0.01	0.00	0.00	18.00	252.00	450.00	0.00	0.00	0.00	−48.87	6	1.66
15	228	189	39	0.03	0.00	0.01	78.00	1170.00	1248.00	0.00	0.00	0.00	−167.90	18	24.13
16	227	177	50	0.06	0.00	0.03	100.00	1600.00	900.00	0.00	0.00	0.00	−171.51	43	1.00
17	263	166	97	0.12	0.00	0.06	194.00	3298.00	776.00	0.00	0.00	0.00	−272.11	81	3.11
18	317	154	163	0.18	0.00	0.09	326.00	5868.00	326.00	0.00	0.00	0.00	−396.13	118	17.13
19	360	143	217	0.20	0.00	0.10	434.00	8246.00	0.00	0.00	0.00	0.00	−500.24	134	51.82
20	273	131	142	0.18	0.00	0.09	284.00	5680.00	284.00	0.00	0.00	0.00	−345.09	118	4.87
21	207	119	88	0.12	0.00	0.06	176.00	3695.95	703.99	0.00	0.05	0.05	−246.86	81	0.58
22	150	108	42	0.06	0.00	0.03	83.83	1844.19	754.44	0.17	3.81	2.77	−143.98	44	0.05
23	123	96	27	0.03	0.00	0.02	46.38	1066.85	742.16	7.62	175.15	68.54	−112.14	21	1.67
24	143	85	58	0.01	0.05	0.03	16.20	388.80	405.00	99.80	2395.20	399.20	−200.78	42	6.03
25	190	73	117	0.00	0.24	0.12	2.12	53.09	76.45	231.88	5796.91	231.88	−246.05	164	13.34
26	210	61	149	0.00	0.40	0.20	0.33	8.47	15.95	297.67	7739.50	0.00	−240.04	268	52.69
27	147	50	97	0.00	0.24	0.12	0.05	1.45	3.43	193.95	5236.55	193.95	−204.85	162	26.26
28	73	38	35	0.00	0.05	0.03	0.01	0.29	0.84	69.99	1959.71	279.96	−126.42	36	0.04
29	35	27	8	0.00	0.00	0.00	0.00	0.08	0.27	16.00	463.92	143.98	−48.90	3	8.51
30	17	15	2	0.00	0.00	0.00	0.00	0.05	0.19	4.00	119.95	63.97	−19.22	0	40.65
Sums	3386		1341	1.00	1.00	1.00	1760.93	33199.20	6758.73	921.07	23891	1384.29	−3497.88	1341	253.70

$M_1 = 18.85$
$S_1 = 1.96$
$A_1 = 880$

$M_2 = 25.94$
$S_2 = 1.23$
$A_2 = 461$

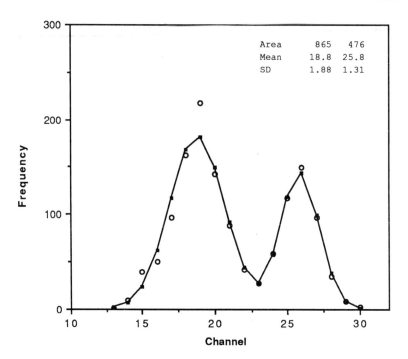

Fig. 9. Result of using maximum likelihood to estimate the component parameters of the adjusted data of Fig. 2.

convergence, low cost per iteration, economy of storage and ease of programming." However, it has the disadvantage of "hopelessly slow convergence in some seemingly innocuous applications" [Redner and Walker (25, p. 197)]. Peters and Walker (24) describe a method for speeding up the algorithm; however, in practice the improvement is not very great.

Redner and Walker also specify some general problems with the maximum likelihood method. One problem is that the decomposition may not be unique, particularly when the components are close together. Unfortunately, this problem occurs frequently with flow chromosome data. Another problem is that the local maximum, that converged to by the EM algorithm from a particular starting point, may not be a global maximum. Fortunately, this problem is encountered infrequently in practice.

Newton–Raphson iteration (which is based on a Taylor series approximation to the log

likelihood) provides a substantially faster method for finding maximum likelihood estimates. A disadvantage of this method is that at each iterative step a matrix of second partial derivatives of the log likelihood function must be calculated and then inverted (see the paper by Hasselblad (15) for details of this method). Hasselblad reported that the Newton–Raphson scheme did not always increase the likelihood, whereas the EM-based procedure always did. Redner and Walker (25) also report that although the EM procedure was slow to converge, over 95% of the change in the log likelihood from the initial to the final estimates was accomplished in the first five iterations. Therefore, if accuracy to many significant digits is not of primary importance the EM procedure will provide satisfactory estimates with a minimum of labor.

In a simulation study of the accuracy of the maximum likelihood method when there are two univariate components, Scholz (26) found that peak areas, A_i, could be estimated reasonably well as long as peak means were separated by at least $2S_i$. In samples of 2500 from each component of a 0.5/0.5 mixture, the areas could be estimated to within ±0.05 for 95% of the time.

F. LEAST SQUARES METHODS

These methods are by far the most widely used for flow karyotype analysis. The usual procedure requires finding values for the parameters A_i, M_i and S_i which minimize the sum of squared deviations:

$$SSQ = \Sigma [h(x) - \Sigma_i A_i F(x; M_i, S_i)]^2.$$

This can be accomplished by setting the partial derivatives of SSQ equal to zero. This results in a nonlinear system of equations which can be solved by the Marquardt method (described in many optimization texts, see Bevington (2) for example) or the EM algorithm described above. The system of equations which must be solved is

$$\frac{dSSQ}{dA_i} = \Sigma [h(x) - G(x)] [F(x; M_i, S_i)$$
$$- F(x; M_k, S_k)] = 0$$

$$\frac{dSSQ}{dM_i} = \Sigma [h(x) - G(x)] A_i F(x; M_i, S_i)$$
$$\times \frac{x - M_i}{S_i^2} = 0$$

$$\frac{dSSQ}{dS_i} = \Sigma [h(x) - G(x)] A_i F(x; M_i, S_i)$$
$$\times \left[\frac{x - M_i}{S_i^3} - \frac{1}{S_i} \right] = 0.$$

Note the similarity between these equations and those for the maximum likelihood method. In the least squares equations $[h(x) - G(x)]$ replaces $h(x)/G(x)$ in the maximum likelihood equations.

The popular least squares method can also be modified to include weights which, when set equal to the inverse of the observed histogram counts, $h(x)$, lead to what is known in the statistical literature as modified minimum Chi-square estimates. Charnes *et al.* (4) demonstrated the asymptotic equivalence of the weighted least squares and maximum likelihood methods, so that it makes little difference which method is used in flow cytometry, where the samples are always large.

I wrote a computer program to carry out least squares minimization. The results are shown in Table IV and the fit is depicted in Fig. 10. It is evident from both the table and figure that there is little difference between the maximum likelihood and least squares methods. The chi-squares for these fits are 24.1 and 23.7 respectively. These are the smallest values found among all of the methods described.

G. DERIVATIVE DOMAIN FITTING

This method was developed by Moore and Gray (20) specifically to deal with a background continuum of unknown shape. This method fits the derivatives of a mixture of Gaussian components to numerical differences between adjacent histogram channels by (weighted or unweighted) least squares. The function minimized is

$$SSR = \Sigma \{[h(x) - h(x + 1)]$$
$$- \Sigma_i A_i [f(x; M_i, S_i) - f(x + 1; M_i S_i)]\}^2.$$

The idea behind the method is that the derivative of the background continuum will vary slowly compared to the overlying peaks. In this situation, the derivative of the continuum is near zero so that its effect on the fit is negligible. This method can readily be installed with minor modification in an already functional least squares program.

Moore and Gray (20) report, on the basis of a large number of Monte Carlo trials, that component peak areas can be estimated to within 10% of their true values and that means and standard deviations can be estimated to within 0.3σ (where σ is the true standard deviation) of their true values when there is adequate separation between adjacent peaks (i.e. three or more σ).

Table IV

MAXIMUM LIKELIHOOD AND LEAST SQUARES FITS TO DATA

			Least squares	
Channel	Adjusted frequency	Maximum likelihood	Chi-square	Mod. chi-square
13	1	2	2	1
14	9	7	8	6
15	39	24	26	22
16	50	62	64	58
17	97	117	120	114
18	163	168	169	168
19	217	182	181	184
20	142	149	147	151
21	88	92	90	93
22	42	44	44	44
23	27	28	29	28
24	58	57	57	58
25	117	118	116	121
26	149	144	142	147
27	97	99	98	100
28	35	38	38	37
29	8	8	8	8
30	2	1	1	1
Sums	1341	1341	1341	1341
Component 1				
Area =		865	869	858
Mean =		18.78	18.74	18.82
Standard deviation =		1.88	1.90	1.85
Component 2				
Area =		476	472	483
Mean =		25.85	25.86	25.84
Standard deviation =		1.31	1.32	1.30
Fit statistics				
Log likelihood =		−3400.48	−3400.8	−3401.02
Chi-square =		24.10	23.71	26.84
Mod. chi-square		20.47	22.31	19.72

Figure 11 shows the result of applying this method to our sample chromosome flow data. The fit is not shown separately since it will go exactly through all the data points. This is due to the fact that the "residuals" are defined as the difference between the fit and the observed values and include both the contribution of the background continuum and the random error due to sampling. The criterion for judging the fit is in the derivative domain.

IV. COMPARISON OF UNIVARIATE METHODS

Table V summarizes the results of applying each of the methods described above to the sample chromosome flow histogram data. It is evident that there is agreement among all the methods which require estimation of a background component prior to fitting (these are all of the methods except derivative domain fitting). There is substantial disagreement

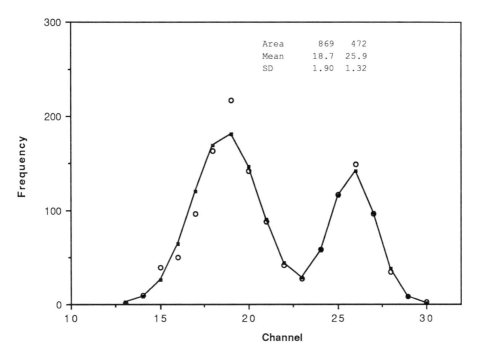

Area	869	472
Mean	18.7	25.9
SD	1.90	1.32

Fig. 10. Result of using least squares to estimate the component parameters of the adjusted data of Fig. 2.

between these methods and the derivative domain method, particularly in the area estimates. These data arise from a sample of human chromosomes 21 and 22 from a genotypically normal individual so that I expect to see equal numbers in each peak. This result was seen *only* for the derivative domain fit where the areas of the two peaks are approximately the same.

These results are typical for samples of the smallest human chromosomes where the signals from the chromosomes are confounded with those from the background continuum. In retrospect, the choice of a linear function to model the background continuum was probably a poor one although there is no way to tell from looking at the plots.

Given that there is no background continuum or that its shape can be accurately characterized, there is little to choose from among the methods. The method of moments requires considerable calculation and is not suitable for dealing with more than two overlapped peaks. The graphical method can handle more than two peaks but may not perform well when peaks are highly overlapped (i.e. less than two standard deviations apart). I could not get the Fourier method to work on our data. There is little to choose between maximum likelihood and least squares. The maximum likelihood iterating equations are easier to program for the computer but the rate of convergence may be slow. Finding least squares estimates requires matrix inversion, at least when the efficient Marquardt algorithm is used, but the convergence is rapid. Derivative domain fitting is simply a modification of least squares and no specification of a background function is required. Thus it appears to be the method of choice, particularly when there is a substantial slowly varying background component of unknown shape.

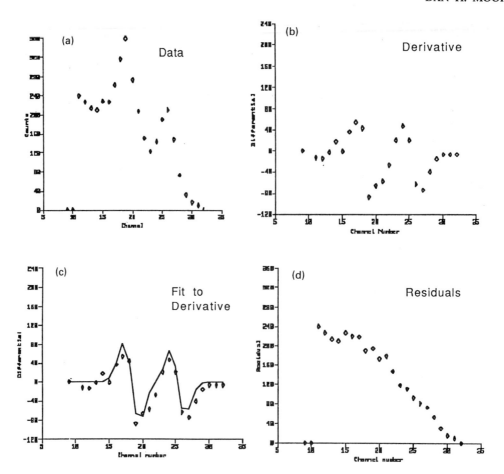

Fig. 11. Result of applying the derivative domain least squares method to the unadjusted data of Fig. 2. (a) The raw data. (b) The (numerical) derivative of the raw data. (c) Fit by least squares to the derivative. (d) Residuals remaining after the fit.

Table V

SUMMARY OF FIT PARAMETER ESTIMATES

Parameter	Method					
	Moments	Graphical	Truncation	ML	LS	DDLS
Area 1	859	849	880	865	869	406
Mean 1	18.84	18.91	18.80	18.78	18.74	18.80
Standard deviation 1	1.90	1.81	1.90	1.88	1.90	0.85
Area 2	452	503	470	476	472	351
Mean 2	25.89	25.58	25.90	25.85	25.86	25.83
Standard deviation 2	1.22	1.43	1.26	1.31	1.32	1.04
Chi-square	26.9	51.9	24.45	24.10	23.71	

ML, maximum likelihood. LS, least squares. DDLS, derivative domain least squares.

V. BIVARIATE METHODS

The bivariate methods are mainly extensions of the univariate methods to more variables. I have not seen extensions of the method of moments nor of the graphical methods of Bhattacharya (3). Tarter and Silvers (27) extended the Fourier method to two dimensions. Parameters are estimated by matching two-dimensional contour plots with the bivariate Fourier transform of the bivariate Gaussian. Since this method did not work on our unidimensional data (see above), I did not attempt to apply it to two-dimensional data.

Figure 12 shows data from a bivariate flow histogram of human chromosomes. I tested three methods on this data. The first I call simple boxing.

A. Simple Boxing Method

This method requires the user to select a polygonal area to be used to obtain estimates of the mean and volume of each peak. All counts within the polygon are summed to estimate the peak volume. The peak mean is estimated by the center of mass. For example, if the histogram height (number of counts) at location (x, y) is denoted by $h(x, y)$, the center of mass is that point (M_x, M_y), which minimizes $\Sigma\Sigma$ $(x - M_x)(y - M_y)h(x, y)$, where the summation is over points where both x and y are inside the polygon. The solution is easily seen to be given by

$$M_x = \frac{\Sigma\, xh(x, y)}{\Sigma\, h(x, y)}$$

and

$$M_y = \frac{\Sigma\, yh(x, y)}{\Sigma\, h(x, y)}.$$

This method is depicted graphically in Fig. 13. It is evident that the accuracy of this method is highly dependent on the ability of the user to exclude background points and to include chromosome data. Our results, derived from the hand of an experienced user, are surprisingly good (see Section VI below).

B. Least Squares Method

This method uses the same algorithm as the univariate least squares. However, this time six parameters (two means, two standard deviations, a correlation coefficient and

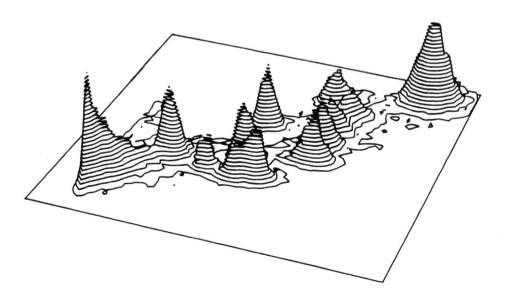

Fig. 12. Bivariate plot of data from a flow histogram of human chromosomes 9–22.

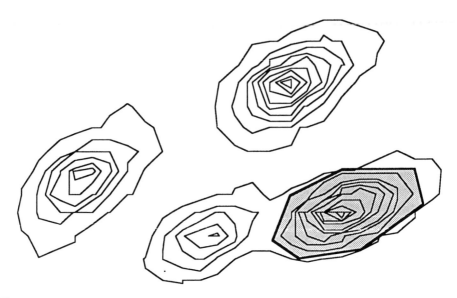

Fig. 13. Bivariate boxing requires selection of a polygon which encloses all chromosomes of a particular peak. The total frequency within the polygon is counted by computer to determine peak volume.

volume) must be estimated for each peak. The method is accurate as long as the peaks are well separated (i.e. two or more Mahalanobis' distance units apart) and the debris continuum is low. (Mahalanobis distance is a multivariate measure of distance which takes into account the scaling in terms of covariance matrices for the distribution.)

Figure 14 shows graphically the application of this method to a peak for chromosome 21 which was riding on top of a complex, slowly varying background continuum. After fitting by bivariate least squares, a plot of the residuals suggests that the peak volume was over-estimated. This is typical behaviour with least squares algorithms which do not include a separate component for the background continuum.

C. Derivative Domain Least Squares Method

This method uses the same algorithm as the bivariate least squares. However, the directional derivatives (actually the numerical differences in the diagonal direction) are used in

place of the histogram heights. The numerical derivative at point (x, y) is found from

$$\Delta h(x, y) = h(x + 1, y + 1) - h(x, y).$$

The fitting function is given by

$$\Delta F(x, y) = F(x + 1, y + 1; \mu, \Sigma) - F(x, y; \mu, \Sigma),$$

where F is the bivariate Gaussian mixture distribution with parameters μ for the vector of bivariate means and Σ for the covariance matrix.

Figure 15 shows the result of applying this method to the chromosome 21 peak of the previous figure. This time the volume is more accurately estimated, without having to specify a mathematical shape for the background continuum.

VI. APPLICATION OF BIVARIATE METHODS TO FLOW CHROMOSOME DATA

Figure 16 summarizes the results, in graphical form, of applying the three methods described

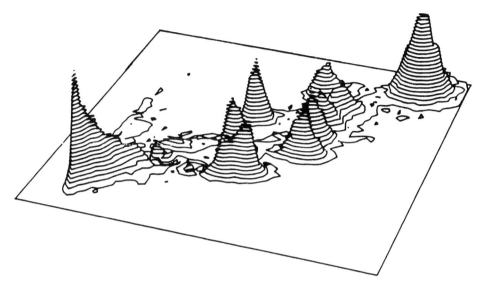

Fig. 14. Result of applying least squares to fit the chromosome 21 peak of the histogram shown in Fig. 12. Note that the volume has been overestimated by including most of the underlying background continuum.

above to our bivariate chromosome sample data. Comparisons among the estimators for the variances and covariances are not shown since the boxing method does not produce such estimates. The following conclusions summarize our experience:

(1) In the absence of background all three methods lead to similar estimates for relative numbers of chromosomes. It was surprising to find that simple polygonal boxing gave results consistent with the more computationally intensive methods. It is likely however, that the

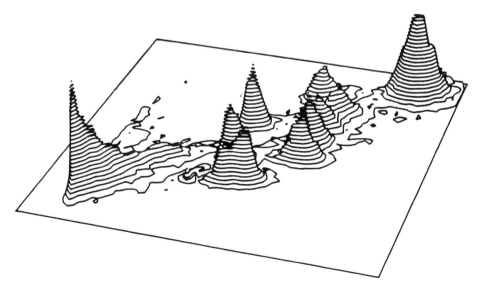

Fig. 15. Result of applying derivative domain least squares to fit the peak for chromosome 21 in the histogram shown in Fig. 12. Note that the residual suggests that this method has made a correct adjustment for the underlying background continuum.

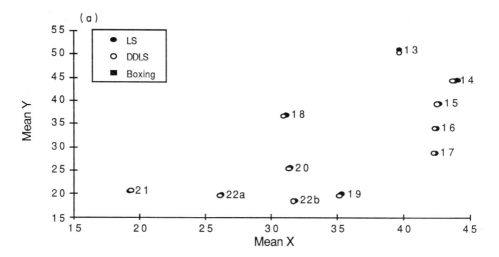

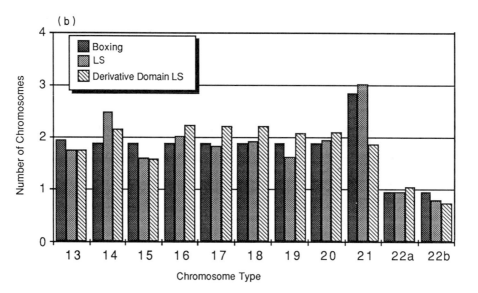

Fig. 16. Comparison of bivariate estimates for chromosomes 13–22 of Fig. 12. The three methods were least squares (LS), derivative domain least squares (DDLS) and polygonal boxing. (a) Estimates for peak means are virtually identical for all three methods. (b) Estimates for the relative peak volumes do not agree as well as those for the means. In particular, both least squares and boxing overestimate the contribution of the chromosome 21. Fit by derivative domain least squares was relatively unaffected by the presence of a background continuum.

boxing technique will fail in situations where the peaks are more heavily overlapped (e.g. for the larger human chromosomes).

(2) When a peak is riding on a substantial background continuum the only method which produced reliable component frequency

estimates was the derivative domain least squares. Figure 16(b) shows that both boxing and least squares fitting overestimated the chromosome 21 numbers. This could lead to erroneous diagnosis of trisomy 21.

(3) Peak means can be reliably estimated by

any of the methods. The results of Monte Carlo simulations (not shown in the figure) confirm that the means are the most stable of the estimates in the least squares and derivative domain least squares methods.

VII. CURRENT APPROACHES IN FLOW CYTOMETRY

There are three main problems encountered in chromosome flow cytometry data which require special attention:

(1) Each human flow karyotype histogram is composed of a mixture of 24 chromosome types (22 autosomes plus sex chromosomes X and Y). Fitting a Gaussian for each type would require estimation of 72 parameters (mean, standard deviation and area for each type). This is computationally intensive.

(2) Underlying each flow cytometry histogram is a background continuum which, in general, does not have a Gaussian shape. The shape and size of this continuum may vary from sample to sample.

(3) Many of the chromosome components are highly overlapped so that only a single peak is observed in the flow cytometry histogram for several clusters of chromosomes. This is particularly true for the chromosomes 9–12.

A survey of the recent literature in flow cytometry reveals a variety of methods for dealing with these problems.

Moore (*19*) proposed a template method for analyzing univariate data which reduced the number of parameters by assuming constant CV for all chromosomes and a common parameter for the areas (since the template assumes there are two of each chromosome type for males and two of each autosome and one of each sex chromosome for females). The background continuum was modeled by the exponential function

$$b(x) = ae^{-bx}$$

Residual analysis was used to measure the adequacy of the fit. It was suggested that in those regions where the fit was judged

inadequate, by the presence of large residuals, further analysis would be performed. This might include refitting with a relaxation of the equal homolog assumption and, therefore, require the addition of another Gaussian.

In treating the same type of univariate data Matsson (*17*) also assumed equal CVs and areas but modeled the background with the function

$$b(x) = p_1 + p_2(x - p_3)^{-p_4}$$

where the parameters p_1 to p_4 were estimated from the data. He showed that the relative positions for the component means from five different donors were reproducible.

Voet *et al.* (*29*) employed a variant of the truncation method to estimate parameters in bivariate flow data. They used the function

$$b(x, y) = ae^{-bx}e^{-cy}$$

to model the background continuum which was subtracted from all histogram data in the first step. Each Gaussian component was then estimated sequentially, by drawing a box around the peak whose parameters were to be estimated, and then using a general nonlinear least squares routine from the Harwell Subroutine Library to obtain parameter estimates. Chi-square deviations between fit and observed values were used to judge adequacy of the overall fit. This procedure appeared to work reasonably well on a mixed population of five different cell types. Its performance on chromosome data was not tested.

At the Los Alamos National Laboratory, Bartholdi *et al.* (*1*) used a polynomial (of unspecified degree) to model the continuum in analyzing univariate data in flow karyotype histograms. Unfortunately, no details of the fitting procedure nor of its general utility are given in this report.

More recently, Moore and Gray (*20*) have been working on a method which eliminates the need for specifying the shape of the background. In their method the derivatives, or numerical differences between adjacent histogram channels, are fitted, via weighted least squares, to the derivatives of the Gaussian mixture. The idea behind this method is that the background is varying slowly compared to the chromosome-containing peaks so that the derivative of the background components will be very small compared to those for the peaks. Results to date indicate that the method works quite well, producing reasonable estimates for parameters.

VIII.   FUTURE DIRECTIONS

Most of the methods employed today use standard nonlinear curve-fitting procedures to match a mixture of Gaussians plus, in most cases, a background function to the observed data. This method is not always reliable, especially when individual chromosomes are not well separated, as is the case particularly with univariate data. With the advent of dual-beam flow cytometry there is increased separation of most chromosome peaks but human chromosomes 9–12 are still grouped into a single peak and the background continuum must now be dealt with in two dimensions. In addition, the number of parameters for each component has increased from three in the univariate case (mean, standard deviation or CV and area) to six in the bivariate case (two means, two standard deviations or CVs, a cross-product or correlation and volume). Thus, reliance on least squares (or maximum likelihood) to obtain parameter estimates will require increased computer time. Two methods have been suggested in the literature for dealing with this problem.

Mann *et al.* (*16*) describe a method for reducing bivariate data to univariate, based on finding the first principal component (eigenvalue) of the composite data. The bivariate histogram is then projected onto this single component and the problem is reduced to the univariate one. Actually, Mann *et al.* apply the method twice to reduce three-dimensional flow data to univariate. This seemed to work well on their cell data, but its performance with chromosome data is untested. The concern here is that in reducing the data we will decrease the

separation between the components that was gained by going to more dimensions (variables).

More recently, Murphy (*22*) has proposed using the multivariate *K*-means clustering algorithm (*14*) on list mode data for obtaining estimates. Under this method, *K* seeds (means) are selected in a *p*-dimensional space. Sequentially each data point is "assigned" to the nearest mean and its value is updated by recomputing the mean from all assigned members. This procedure is continued until all points have been assigned. In the second stage, pairs of clusters which are deemed close enough, say less than *J* standard deviations apart, are pooled and their means and standard deviations recomputed.

This procedure has the virtue of being trivially easy to implement and will work in any number of dimensions. It is ideally suited for list-mode data, which can easily be run through the procedure a second time to ensure that the final "clusters" are stable. This appears to be an interesting method but it needs to be tested further. (In the report only simulated data from three well-separated clusters and samples of stained microspheres were tested.) Its performance on chromosome data has not been reported.

A general idea, which I have not seen suggested in the literature, is to use image-processing methods on flow data. In the bivariate case the number of events in the channel pair (*x*, *y*) could be represented on a gray scale. This would produce an image which could then be compared, via standard image processing tools, with other "standard" images. This idea comes from a report by Fantes *et al.* (*11*) on the detection of radiation damage to human chromosomes by measuring the increase in the noise level of the background. It may be possible to remove the background component by treating it as an unwanted distortion of an image, the image in this case being the pure histogram of chromosomes.

I believe that new methods will be required to resolve individual chromosome components when the dimensionality is higher than two. Least   squares   and   maximum   likelihood methods are just too computationally intensive to be practical in the truly multivariate case.

ACKNOWLEDGEMENTS

Work performed under the auspices of the U.S. Department of Energy by the Lawrence Livermore National Laboratory under contract number W-7405-ENG-48.

REFERENCES

1. Bartholdi, M. F., Roy, F. A., Cram, L. S., and Kraemer, P. M. (1984). Flow karyology of serially cultured Chinese hamster cell lineages. *Cytometry,* 5, 534–538.

2. Bevington, P. R. (1969) "Data Reduction and Error Analysis for the Physical Sciences." McGraw Hill, New York.

3. Bhattacharya, C. G. (1967). A simple method of resolution of a distribution into Gaussian components. *Biometrics* 23, 115–135.

4. Charnes, A., Frome, E. L., and Yu, P. L. (1976). The equivalence of generalized least squares and maximum likelihood estimates in the exponential family. *J. Amer. Statist. Ass.* 71, 169–71.

5. Cohen, A. C. (1967). Estimation in mixtures of two normal distributions. *Technometrics* 9, 15–28.

6. Dempster, A. P., Laird, N. M., and Rubin, D. B. (1977). Maximum likelihood from incomplete data via the EM algorithm. *J. R. Statist. Soc., Ser. B.* 39, 1–38.

7. Doetsch, G. (1928). Die elimination des doppler-effeks bei spektroskopischen feinstrukturen und exakte bestimming der componenten. *Zh. Phys.* 49, 705–730.

8. Doetsch, G. (1936). Zerlegung einer funktion in Gauß'sche fehlerkurven und zeitliche zuruck-verfolgung eines temperaturzustandes. *Math. Z.* 41, 283–318.

9. Essenwanger, O. (1976). Applied statistics in atmospheric science. Part A. Frequencies and curve fitting. *In* "Developments in Atmospheric Science, 4A", Elsevier, New York.

10. Everitt, B. S. and Hand, D. J. (1981). "Finite Mixture Distributions." Chapman & Hall, London.

11. Fantes, J. A., Green, D. K., Elder, J. K., Malloy, P., and Evans, H. J. (1983). Detecting radiation damage to human chromosomes by flow cytometry. *Mutat. Res.* 119, 161–168.

12. Gregor, J. (1969). An algorithm for the decomposition of a distribution into Gaussian components. *Biometrics* 25, 79–93.

13. Hald, A. (1949). Maximum likelihood estimation of the parameters of a normal distribution which is truncated at a known point. *Skand. Aktuarietidskr.* 32, 119–134.

14. Hartigan, J. A. (1975). "Clustering Algorithms." Wiley, New York.

15. Hasselblad, V. (1966). Estimation of parameters for a mixture of normal distributions. *Technometrics* 8, 431–44.

16. Mann, R. C., Popp, D. M., and Hand, Jr R.E. (1984). The use of projections for dimensionality reduction of flow cytometric data. *Cytometry* 5, 304–307.

17. Matsson, P. and Rydberg, B. (1981). Analysis of chromosomes from human peripheral lymphocytes by flow cytometry. *Cytometry* 1, 369–372.

18. Medgyessy, P. (1977). "Decomposition of Superpositions of Density Functions and Discrete Distributions." Wiley, New York.

19. Moore II, D. H. (1979). A template method for decomposing flow cytometry histograms of human chromosomes. *J. Histochem. Cytochem.* 27, 305–310.

20. Moore II, D. H. and Gray, J. W. (1989). Derivative domain least squares analysis: a new method for decomposition of a mixture of normal distributions in the presence of a contaminating background. *Cytometry* (submitted).

21. Moore II, D. H. (1975). Use of residuals in fitting mixtures of normal distributions. Lawrence Livermore National Laboratory UCRL-76507.

22. Murphy, R. F. (1985). Automated identification of subpopulations in flow cytometric list mode data using cluster analysis. *Cytometry* 6, 302–9.

23. Pearson, K. (1894). Contribution to the mathematical theory of evolution. *Phil. Trans. A.* 185, 71–110.

24. Peters, B. C. and Walker, H. F. (1978). An iterative procedure for obtaining maximum-likelihood estimates of the parameters for a mixture of normal distributions. *SIAM J. Appl. Math.,* 35, 362–378.

25. Redner, R. A. and Walker, H.F. (1984). Mixture densities, maximum likelihood and the EM algorithm. *SIAM Review* 26, 195–239.

26. Scholz, K. U. (1981). Mathematical evaluation of two parameter flow cytometric histograms. *Cytometry* 2, 159–164.

27. Tarter, M. and Silvers, A. (1975). Implementation and application of bivariate Gaussian mixture decomposition. *J. Am. Statist. Ass.,* 70, 47–55.

28. Titterington, D. M., Smith, A. F. M. and Makov, U. E. (1985). "Statistical Analysis of Finite Mixture Distributions." Wiley, Chichester.

29. Voet, L., Krefft, M., Mairhofer, H., and Williams, K. L. (1984). An assay for pattern formation in dictyostelium discoiderim using monoclonal antibodies, flow cytometry and subsequent data analysis. *Cytometry* 5, 26–33.

30. Woodward, W. A., Parr, W. C., Schucany, W. R., and Lindsey, H. (1984). A comparison of minimum distance and maximum likelihood estimation of a mixture proportion. *J. Am. Statist. Ass.* 79, 590–598.

Univariate Flow Karyotype Analysis

L. SCOTT CRAM, MARTY F. BARTHOLDI, F. ANDREW RAY,
MARY CASSIDY AND PAUL M. KRAEMER

Los Alamos National Laboratory, US Department of Energy
Los Alamos, NM 87545, USA

I.   INTRODUCTION

Flow karyotyping is defined as the use of flow cytometry for classification of chromosomes isolated from mitotic cells. For univariate classification, the technique refers to the analysis of a single variable (parameter) on each chromosome, normal or aberrant, that passes through the laser beam of a flow cytometer. Flow karyotyping encompasses the isolation of intact metaphase chromosomes, staining each chromosome with a fluorescent tag, and the rapid quantitative analysis of the stained chromosomes in a flow cytometer. Chromosome isolation consists of freeing individual chromosomes from mitotic cells and stabilizing their structure. Staining reactions are designed to label a chromosome constituent in a way that helps distinguish one chromosome type from another. The critical steps of chromosome isolation, stabilization, staining and analysis are described in Chapters 3–5. The ultimate goal is to resolve each chromosome type. In the case of chromosomes isolated from human cells, this would mean resolving

23 populations when using cells of female origin (22 autosomes, XX) and 24 populations in cells of male origin (22, XY) (*56*).

Flow karyotype analysis provides the research investigator with a powerful tool for the quantitative analysis of a large number of single chromosomes in a short period of time. The application of these techniques to fundamental biological questions is having a major impact on our understanding of the human genome (*18,52*), its organization (*19,37*) and the critical consequences of karyotype instability (*14,32*). Several independent variables can be measured for each chromosome as it passes through the interrogation (analysis) region at typical analysis rates of 1000–2000 chromosomes/second. The ability to make quantitative measurements on single chromosomes at such high throughput rates has been recognized as a profound capability by the cytogenetics community. This is due in part to the way in which flow karyotype analysis compliments extant cytogenetic capabilities. Cytogenetic analysis currently relies on the structural information obtained from a relatively small number of banded chromosomes. Flow karyotyping compliments this approach by providing quantitative, precise information about chromosome properties, such as DNA content, for several hundred thousand chromosomes. This chapter will focus on univariate analysis; in particular analysis of chromosomes stained with a single DNA-specific fluorochrome. Other chromosome properties, such as chromosome length (*15*), RNA content (*2*), total protein (*48*) and specific protein content (*50*) can also be measured, but these are currently less useful for cytogenetic studies and thus will not be included here.

As this chapter explains, the attractive advantages of univariate flow karyotype analysis are based on its quantitation, statistical precision and flexibility. Several examples will be reviewed that illustrate the power the technique has to (1) eliminate uncertainties in chromosome classification, (2) uncover and monitor subtle chromosomal polymorphisms and (3) to quantitate chromosomal DNA

content with unprecedented precision. The options, expectations and limitations of univariate flow karyotyping will also be explained in terms of how one can best utilize the technique and in what situations the technique, at its present stage of development, might not be the method of choice.

The ability to resolve all the chromosome types from any mammalian species depends upon differences in interchromosomal DNA content and instrumental resolution. These aspects of the measurement will be discussed in terms of how they affect the application of the technology. In general, flow karyotype analysis and banding analysis are complimentary tools. However, there are situations where only one of the techniques can be applied.

The utility of chromosomal DNA content as a single variable that would resolve a large number of chromosome types from as many species as possible was illustrated by the work of Mendelsohn and Mayall (*41*). This study demonstrated that the DNA content of individual chromosome types was largely invariant for chromosomes of the same type and thus could serve as a good chromosome descriptor. Chromosome length has also been shown to correlate well with DNA content. The uniformity of DNA content among normal euploid chromosomes of the same type has proven to be of paramount importance in the success of univariate flow karyotype analysis.

Several applications of univariate flow karyotype analysis are cited in this chapter to illustrate the following points: (1) Univariate flow karyotypes provide a rapid technique for determining and monitoring karyotype instability. (2) Variation in the frequency of chromosome types (due to the occurrence of a trisomy or monosomy for one chromosome type) can be determined with a good degree of accuracy but with less precision than changes in DNA content. (3) Chromosomal polymorphisms can be resolved and quantitated where it is not otherwise possible. (4) Chromosome rearrangements occurring with sufficient frequency to be considered "marker chromosomes" can be quickly and quantitatively assessed. This is particularly important in a heteroploid population where a very large variety of chromosome types are observed. (5) Errors associated with flow karyotype analysis are of an entirely different origin and nature than those associated with banding analysis. (6) Chromosome sorting is useful for the identification of chromosome types and has been extensively used for gene mapping, cloning and molecular characterization of normal and rearranged chromosomes. (See Chapters 14–16).

II. TECHNIQUES FOR HIGH-RESOLUTION UNIVARIATE FLOW KARYOTYPE ANALYSIS

The first reports describing flow karyotype analysis were published in 1975 (25,26,49). From this beginning, the ability to resolve different chromosome types has steadily increased and further improvements are expected. Protocols for chromosome isolation, staining and analysis (see Chapters 3–8) have developed to the point where univariate flow karyotypes for several species now allow all chromosome types to be resolved as separate peaks. These high-resolution flow karyotypes are produced using the specialized procedures discussed below. Using these procedures, the coefficient of variation (CV) of univariate distributions is usually better than either of the two single variables, taken alone, of a bivariate distribution.

A. Chromosome Isolation

High-quality flow karyotypes require the production of suspensions of intact, mono-disperse chromosomes in which the frequency of clumps and DNA-containing debris is minimal. Production of such suspensions begins with the establishment of cell cultures in which the cell number increases exponentially over several population doublings so that mitotic cells can be collected with high efficiency for chromosome isolation. In the case of monolayer

fibroblasts, contact inhibition should be minimal. Two sequential colcemid blocks are sometimes used to increase the mitotic fraction, reduce background debris and clumps, and to obtain cells from within a smaller portion of mitosis. Cells shaken off after the first block are discarded and those from the second block used for chromosome isolation. A mitotic index greater than 95% is achievable with fibroblasts. The mitotic cells collected in this fashion are then hypotonically swollen to facilitate chromosome separation within the mitotic nucleus. This step is critical for proper chromosome isolation. Ideally, the cells will retain membrane integrity during the swelling process so that the chromosomes separate as completely as possible. Cell swelling can be monitored by the fluorescent microscope by using a combination of the vital fluorescent dye Hoechst 33342 and propidium iodide. Under conditions favorable for obtaining excellent preparations, the chromosomes in a mitotic cell with an intact cell membrane will appear blue (Hoechst 33342 will pass through an intact cell membrane whereas propidium iodide will not). The loss of membrane integrity can be visualized by the change in chromosome fluorescence from blue to pink. Chromosomes can be observed to disentangle as the cells swell. Excellent preparations correlate with the percentage of cells that swell in a uniform fashion and in which the chromosomes disentangle prior to membrane lysis. The fraction of ideally swollen cells can exceed 90%. This technique is effective for monitoring cell swelling in any of the chromosome isolation protocols.

B. Chromosome Analysis Considerations

Several instrument and sample adjustments are needed to maximize the resolution of univariate flow karyotypes. It is especially important to increase the uniformity of chromosome illumination during flow cytometry. This can be accomplished by (1) matching the index of refraction of the sample and sheath streams by using the same buffer in each, (2) keeping the sample concentration as high as possible so that the chromosome analysis rate can be kept high while maintaining a small ($< 2\,\mu$m diameter) sample stream diameter and (3) spreading the laser beam so that variation in intensity across the sample stream is less than 1%. A sample concentration of about $1-5 \times 10^6$ chromosomes/ml is necessary to maintain a reasonable count rate (100–200 events/sec or less for high-resolution analysis) while constricting the sample stream diameter to about $2\,\mu$m. These conditions help ensure uniform illumination across the stream diameter and also act to hydrodynamically orient the chromosomes along the direction of flow.

High-resolution flow karyotyping also requires that sufficient photons be measured from each chromosome to allow a statistically precise estimate of its chromosomal dye content. The number of photons can be increased by maximizing the intensity of chromosome illumination, selecting spectral filters that efficiently pass the fluorescence emitted from each chromosome and selecting photodetectors that are maximally sensitive in the spectral region over which the chromosomes fluoresce and maximizing the efficiency of fluorescence collection.

As little as $15\,$mW at $488\,$nm can be used to generate recognizable univariate flow karyotypes using commercially available FCMs. However, increased illumination intensity improves resolution and the signal-to-noise ratio. Only limited quantitative information has been published on the factors, and their relative importance, that influence the resolution of flow karyotypes. Bartholdi *et al.* (*8*) have measured the response of fluorescence intensity to illumination intensity. At very high power density [$4\,$MW/cm^2 achieved by using high laser output ($2\,$W at $488\,$nm) focused on a $2\,\mu$m spot], one can achieve dye saturation. Saturating all the available dye molecules offers improved resolution (CV = 1.3%) and small trajectory variations at the sample stream/laser

beam intersection point are of less importance, providing the chromosome remains in a region of illumination where saturation will occur. With high illumination intensity, more options are available for improving resolution, such as spreading out the laser beam shape to achieve more uniform illumination.

High measurement precision requires detection of as much fluorescence as possible for each chromosome. A single-color glass barrier filter, one with a wavelength cut-off just adequate to block scattered excitation light, allows the largest number of photons to reach the photodetector. Beam splitters, dichroic filters, narrow band pass filters and interference filters should be removed for high-resolution univariate flow karyotype analysis. As much of the emitted fluoresence as possible should be incident to the photomultiplier.

Figure 1 shows examples of high-resolution flow karyotypes of Chinese hamster and human chromosomes where the above recommendations were employed (*9,14*). Note that all of the Chinese hamster chromosome types and most of the human chromosome types are resolved in these flow karyotypes. Statistical analysis of this type of data is straightforward because of the high resolution and low noise level.

III. DATA INTERPRETATION

A univariate flow karyotype measured for propidium iodide stained chromosomes contains quantitative information about the (1) frequency of occurrence and DNA content of each chromosome type in a population, (2) chromosome fragility and/or the frequency of randomly occurring aberrant chromosomes and (3) measurement resolution. The DNA content and frequency of occurrence of each chromosome type are proportional to the mean fluorescence intensity and area of the peak for that chromosome type. The CV of one or more peaks can be considered as an indication of the measurement resolution. The magnitude of the continuum underlying the peaks in a

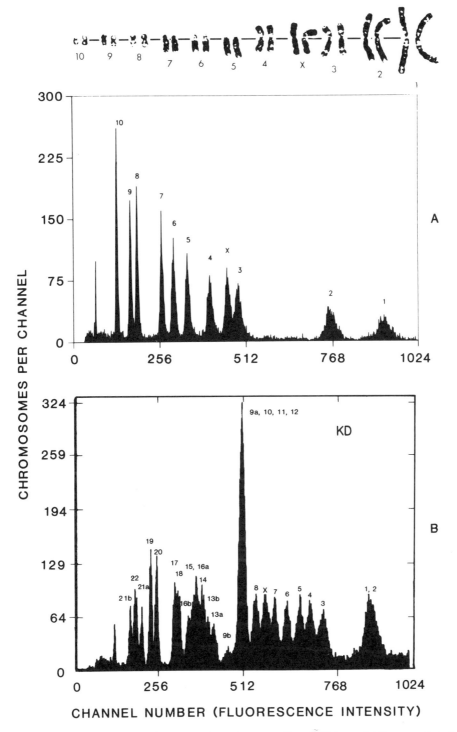

Fig. 1. Flow karyotypes of (A) low-passage Chinese hamster cells and (B) the euploid human cell strain, KD (30). Chromosomes were isolated in 75 m*M* KCl and stained with propidium iodide as described by Aten (3). Analysis was done using the Los Alamos National Laboratory high-resolution imaging sorter (15). Trypsin–Giemsa-banded Chinese hamster chromosomes are shown above their respective peaks. A lower amplifier gain was used to record the Chinese hamster flow karyotype.

flow karyotype is an estimate of the frequency of debris particles and random aberrant chromosomes in the population. Karyotypic changes that occur during cell culture, as a result of *de novo* genetic changes, or as a result of exposure to clastogenic agents, can be detected by analysis of flow karyotype features such as peak means, areas, CVs and continuum magnitude. Procedures for estimation of these features are discussed in Chapter 6.

Flow karyotypes do not contain information about the association of individual chromosomes from individual cells. After mitotic swelling and chromosome release, all associative information about which chromosomes came from which cell is lost. Such information can only be obtained from metaphase spreads.

In Fig. 1(A), each chromosome type of the female Chinese hamster is represented by a separate and clearly resolved peak. The human karyotype with twice as many chromosome types (2) does not resolve into single peaks for each chromosome [Fig. 1(B)]. In Fig. 1(A), the area of each peak (number of counts under the peak) is proportional to the number of chromosomes of that type. For chromosomes isolated from a euploid cell population, all autosomal peaks should have the same areas. The sex chromosomes from a male (XY) cell will produce peaks that are one-half the area of the autosomes. The relative peak positions (fluorescence intensity) are proportional to the chromosomal DNA contents in the examples illustrated (9).

Although most peaks in the flow karyotypes illustrated in Fig. 1 have the same area, they do not have the same height. The brighter peaks are broader (spread over more channels) and have fewer counts in the peak channel. This shape effect is due to the way in which the data is recorded and does not have any biological significance. (Peak standard deviations increase approximately linearly with peak mean.) In general, the peaks produced by chromosomes of one type are Gaussian in shape. However, the shape of a peak can be influenced by the presence of an aberrant chromosome type with a slightly different DNA content that is not resolved and is present at low frequency. In this situation, the peak may appear to be skewed.

Peak appearance can also be influenced by the type of amplifier used on the flow cytometer. Figure 2 compares the visual differences that result from the use of either a linear or logarithmic amplifier. This data set was acquired by splitting the signal from the preamplifier and sending the two identical signals to linear [Fig. 2(A)] and logarithmic (Fig. 2(B)] amplifiers. The information in both distributions is exactly the same. Data acquisition with a logarithmic amplifier offers advantages for visual interpretation. Chromosomes from a truly euploid cell (no aberrations and two copies of each homologue) will, when analyzed using a logarithmic amplifier, have all peaks of equal height and equal width. Chromosomes present in only one copy (chromosome X) and homologues with a difference in DNA content (chromosome 9) are readily detected as their peak heights are about half as high as chromosome types present in two copies per cell, as seen in Fig. 2(B). This occurs because peaks with constant CV in the linear domain have constant width in the logarithmic domain. The peaks representing chromosomes 1 and 2 [Fig. 2(B)] are higher than expected, even though they contain fewer counts due to the breakage of larger chromosomes. The CV of these peaks is slightly less than the other peaks, therefore the log standard deviation is slightly lower, causing the peak to rise higher (see Table I). (The peaks representing chromosomes 1 and 2 have a tighter CV because they are brighter, resulting in more emission photons and, in turn, better photon statistics.) Consequently, direct visual inspection of the data on a logarithmic scale must be done with care. Another advantage of logarithmic amplification is that each peak is distributed over approximately the same number of channels, an advantage for parametric data analysis. Table I compares the statistical characteristics of both figures and illustrates why all the peaks in Fig. 2(B) are not the same height and width.

To verify the direct proportionality between the linear coefficient of variation (linear

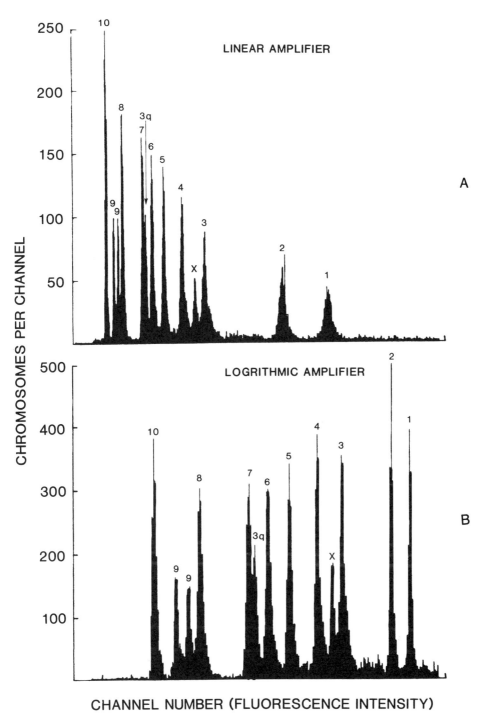

Fig. 2. Flow karyotype of CCHE 27/clB cells at passage 20 analyzed by splitting the preamplifier signal to (A) linear and (B) logarithmic amplifiers. A logarithmic amplifier (1.5 decades) was used to amplify the signal in part (B). The logarithmic transformation results in chromosome types represented in equal numbers to have the same peak height and uniform peak width. Table I illustrates why the peaks are not all of equal height.

Table I

STATISTICAL COMPARISON OF LINEAR AND
LOGARITHMIC FLOW KARYOTYPES SHOWN
IN FIG. 2

Chinese hamster chromosome	CV (%)	Peak mean	Standard deviation	Peak count (%)
Linear amplifier				
1	1.42	712	10.1	6.9
2	1.42	586	8.3	7.8
3	2.54	369	9.4	9.9
X	1.51	340	5.1	4.1
4	1.85	306	5.7	10.2
5	1.77	254	4.5	9.6
6	1.93	222	4.3	9.9
3q	1.61	204	3.3	5.7
7	1.49	195	2.9	8.4
8	2.37	140	3.3	10.2
9a	1.98	128	2.5	4.5
9b	2.08	116	2.4	4.4
10	1.85	94	1.7	8.5
Logarithmic amplifier[a]				
1	0.47	236	1.11	6.9
2	0.50	223	1.12	7.6
3	1.14	189	2.15	9.7
X	0.80	181	1.45	4.0
4	1.03	171	1.76	9.9
5	1.21	152	1.84	9.3
6	1.52	137	2.09	9.7
3q	1.36	128	1.74	5.5
7	1.32	123	1.63	8.4
8	2.47	89	2.21	10.2
9a	2.25	81	1.82	4.5
9b	2.39	73	1.74	4.4
10	2.34	57	1.34	8.8

[a] The CV in log space has very little meaning. The standard deviation in log space is important as it is proportional to the coefficient of variation in linear space.

$$s = \tfrac{1}{2}[\log(\mu + \sigma) - \log(\mu - \sigma)]$$

$$s = \tfrac{1}{2}\log\left(\frac{1 + CV}{1 - CV}\right)$$

$$s = CV + \frac{CV^3}{3} + \frac{CV^5}{5} + \cdots \text{ where } CV = \frac{\sigma}{\mu}$$

$$s \simeq CV \text{ (at low values of CV)}.$$

Therefore, peaks on a linear scale with equal count and equal CV will appear as peaks of equal height on a logarithmic scale. However, even slight variations in CV will be directly reflected as variations in heights of the peaks after logarithmic amplification [see Fig. 2(B)].

Underlying the peaks in Fig. 1 are a small number of events that are broadly distributed and referred to as background events. Background events are due to broken and aggregated chromosomes and DNA-containing nuclear fragments. Low-frequency and heterogenous aberrant chromosome types and debris that bind free fluorescent dye in the sample also contribute to the background. Background debris is typically concentrated under the midsized chromosome region of the Chinese hamster flow karyotype (51). The distribution of events in a flow karyotype also includes a large number of events that are of very low fluorescent intensity and are distributed to the far left and are of less intensity than the smallest chromosome. These events are of no consequence provided their distribution does not extend into the flow karyotype.

Peak resolution, which determines the certainty with which one can discern where one peak stops and the next one starts, is influenced by instrumental resolution and biological (chromosome) variability. Instrumental broadening is discussed in Chapter 2. Typical CVs obtained on commercial instruments when analyzing propidium-iodide-stained fibroblasts range from about 1.5 to 2.0%. Instruments optimized for chromosome analysis will yield CVs down to about 1.0%. Chromosome variability includes chromosome-to-chromosome

amplifier) and the logarithmic standard deviation, consider a normal distribution on a linear scale with a mean of μ and a standard deviation σ. The standard deviation, s, of the peak on a logarithmic scale can be *estimated* by calculating s at one standard deviation $(\mu + \sigma)$ and one standard deviation below $(\mu - \sigma)$ the mean (μ):

staining variations, perhaps some second-order effects in the degree of chromosome condensation and real differences in DNA content between what one would expect to be identical chromosomes (homologues). Two homologues with a significant difference in DNA content (chromosome polymorphisms) are discussed in detail in the applications section of this chapter.

As with any measurement, one is limited by the reproducibility of the measurement. Detecting a change in the number of chromosomes of a particular type and/or the presence of a new chromosome type is important. The reliability with which one can make such a determination for univariate flow karyotype analysis of human chromosomes was determined by Bartholdi *et al.* (9). Multiple high-resolution flow karyotypes were measured using KD cells. KD cells are normal diploid, 46 XX cells. G-banding confirms that the cells have a normal karyotype (30). The flow karyotype of these cells is illustrated in Fig. 1(B). Three separate chromosome preparations were made from these cells and two univariate flow karyotypes were taken for each preparation. Computer analysis of each flow karyotype was done by fitting a Gaussian distribution to each peak. A polynomial function was fitted to the underlying background debris distribution.

The peak areas in Fig. 1(B) and fluorescence intensity (DNA content) are identified in Table II. The relative number of chromosomes in each peak was multiplied by 46 to determine the number of chromosomes per cell. The relative DNA content of each chromosome type (percentage of total autosomal DNA content) was determined by adding each of the peak means for the autosomes and multiplying by the number of autosomes in each peak (9) and normalizing this number to 100 (see Table II). Some chromosomes were not resolved as noted in Table II and polymorphisms were observed in chromosomes 9, 13, 16 and 21. The relative DNA-content values in Table I agree very well with those determined by absorption microscopy (5,41). The standard deviations in Table II were computed from six separate flow

Table II

CHROMOSOME FREQUENCY AND RELATIVE DNA CONTENT IN KD CELLS

Chromosome	Chromosomes per cell[a]	DNA content[b]
1, 2	4.1 ± 0.1	4.22 ± 0.02
3	2.1 ± 0.1	3.52 ± 0.01
4	2.0 ± 0.1	3.34 ± 0.02
5	2.1 ± 0.2	3.18 ± 0.02
6	1.9 ± 0.1	3.00 ± 0.02
7	1.9 ± 0.2	2.82 ± 0.02
X	2.1 ± 0.1	2.70 ± 0.02
8	1.9 ± 0.2	2.56 ± 0.01
9a, 10, 11, 12	7.9 ± 0.9	2.36 ± 0.01
9b	0.7 ± 0.1	2.15 ± 0.01
13a	1.0 ± 0.1	1.96 ± 0.01
13b	0.9 ± 0.2	1.87 ± 0.01
14	2.0 ± 0.1	1.79 ± 0.01
15, 16a	3.0 ± 0.2	1.68 ± 0.01
16b	1.2 ± 0.2	1.55 ± 0.02
17, 18	3.2 ± 0.4	1.42 ± 0.01
20	2.1 ± 0.2	1.14 ± 0.01
19	2.0 ± 0.2	1.04 ± 0.01
21a	1.0 ± 0.2	0.92 ± 0.02
22	2.0 ± 0.5	0.82 ± 0.01
21b	1.0 ± 0.2	0.75 ± 0.01

[a] Relative frequency of chromosomes in each peak multiplied by 46.
[b] Percentage of total autosomal DNA content (autosomal normalization).

karyotypes. These results indicate that in order to detect a change in the number of chromosomes per cell (due to, for example, a trisomy for one chromosome type), 5–10% of the cells must contain an extra chromosome. A shift in peak position, or DNA content (e.g. due to a deletion, addition or translocation), can be detected with an accuracy of about 1%. Thus, deviations from euploidy can be readily detected. Cells that are progressively evolving can be monitored by flow karyotype analysis and monitored without detailed microscopic observation once the initial flow karyotype has been carefully determined and confirmed by banding analysis.

For biological applications, data interpretation generally consists of examining two types of aberrations: numerical and structural. Examples of both of these types of aberrations are illustrated in Section V.

IV. UNIVARIATE CHROMOSOME SORTING

Chromosome sorting has been used for confirmation and identification of normal and aberrant chromosome types and for recovery of large quantities of single types of chromosomes.

The mechanics of univariate chromosome sorting are mostly analogous to cell sorting. Analysis rates are typically 2000–4000 chromosomes/sec depending on the resolution required to distinguish the chromosome type being sorted. Well-separated chromosome types can be analyzed at the higher rates if the loss of peak resolution does not cause peak overlap with other chromosome types. Sorting rates are determined by (1) the frequency of occurrence of the chromosome type being sorted and (2) the analysis rate that is compatible with good peak resolution. For euploid female Chinese hamster chromosomes and one-direction sorting, the flow system sorts, on average, every 11th chromosome. Using an analysis rate of 3000 chromosomes/sec, about 270 chromosomes/sec will be sorted. For human cells, lower analysis rates (1000 chromosomes/sec) are needed to maintain high resolution, resulting in typical sorting rates on commercial flow cytometers of about 40–45 chromosomes/sec.

V. SPECIFIC BIOLOGICAL APPLICATIONS OF UNIVARIATE FLOW KARYOTYPE ANALYSIS

Univariate flow karyotype analysis and sorting has proven to be an effective tool for monitoring karyotype stability, detecting numerical and structural aberrations and providing purified chromosomes. The following discussion describes several studies in which

flow karyotyping and sorting have been instrumental in addressing problems involving: analysis of karyotype instability, aneuploidy, *in vivo* karyotype progression, rapid identification of "marker" chromosomes, detecting chromosome polymorphisms, sorting for gene mapping and cloning, and mutagenesis and radiation damage studies.

A. ANALYSIS OF KARYOTYPE INSTABILITY

High-resolution flow karyotype analysis has been used extensively to analyze the progressive stages of karyotype instability that occur as a population of cells spontaneously transforms from normal euploid to aneuploid and neoplastic. In contrast to the consistent and nearly homogenous chromosome translocations that often occur in hematopoetic malignancies, solid tumors exhibit a multitude of cytogenetic aberrations that can be observed to change, in part, with time. To study these progressive changes, Kraemer *et al.* (*33*) developed a Chinese hamster cell culture model of carcinogenesis based on the fact that cells from normal Chinese hamsters regularly become neoplastic when carried in culture. The spontaneous process includes the sequential acquisition of immortality, transformation and tumorigenicity, and is associated with very early, specific changes in selected chromosomes. Chromosome changes occurring at all stages of progression were characterized using a combination of univariate flow karyotypes and G-banding analysis.

Figure 3 illustrates progressive changes in the flow karyotype of a mass culture of whole Chinese hamster embryo (WCHE) cells as it progressed from a normal euploid culture of embryonic material to a cell line that demonstrated all the *in vitro* characteristics of transformed cells including tumorigenicity. At passage seven, the culture was typical of a normal euploid population with considerable cellular heterogeneity. The cells were phenotypically normal as was the flow

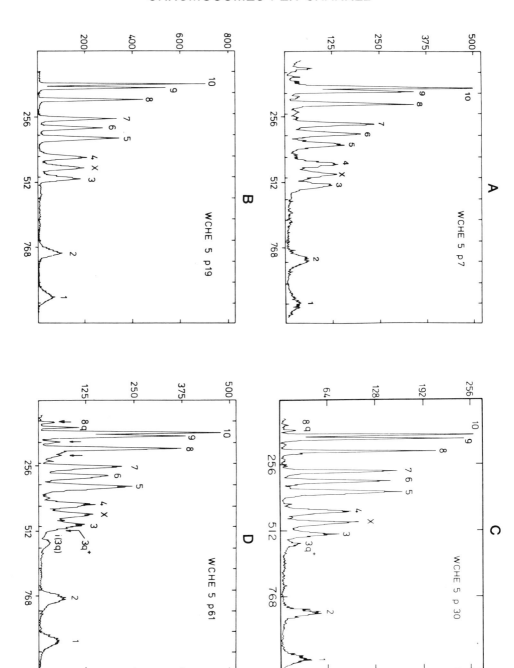

CHROMOSOMES PER CHANNEL

CHANNEL NUMBER (FLUORESCENCE INTENSITY)

Fig. 3. Flow karyotype histograms of whole Chinese hamster embryo cells (WCHE) at four successive passages in culture: (A) precrisis, passage 7; (B) postcrisis and nontumorigenic, passage 19; (C) conditionally tumorigenic, passage 30; (D) heteroploid, passage 61. The sequential series of karyotypic

karyotype. That is, all the peaks had areas and means that were within normal ranges. The background level, as noted by the number of counts occurring between the peaks, was high due to the heterogeneous cellular composition of a whole embryo culture at early passage. At passages 12–15, the culture was emerging from crisis, as characterized by slow growth and unusual cellular morphology. Following crisis, the rate of cellular proliferation increased and the flow karyotype, Fig. 3(B), indicated the first signs of karyotypic instability. This particular flow karyotype illustrates optimal performance of the instrumentation and a debris-free sample. An increase in the area of the peak identified as chromosome 5 was confirmed to be trisomy of chromosome 5 by G-banding. The frequency of trisomy of chromosome 5 was 84% of WCHE/5 passage 19 cells. Passage 19 cells were nontumorigenic and normal except for an increased frequency of clonogenic cells (14).

At passage 30, WCHE cells were partially transformed and conditionally tumorigenic. The flow karyotype [Fig. 3(C)] indicated the continued presence of trisomy 5 in 90% of the cells, the appearance of an insertion element in the long arm of chromosome 3 (3q$^+$), and an additional copy of the long arm of chromosome 8 (8q) in some of the cells. Banding analysis and flow karyotype analysis were in good agreement with the trisomy 5 frequency but not with the frequency of the insertion element. The appearance of a distinct new peak on the high-fluorescence side of chromosome 3 indicated that 55% of the cells should contain a chromosome with slightly more DNA (7%) than chromosome 3. Banding analysis directed at identifying the aberrant chromosome turned up a very small number (a few percent) of marginally identifiable chromosome 3 with an insertion element in the long arm. With continued examination, it was clear that other spreads also contained chromosomes with this same insertion element; a band that varied from moderately distinct to very slight. From the banding results it appeared that the insertion element was present in varying amounts. The value of flow karyotype analysis was therefore

two-fold; first, it provided an indication that a new marker chromosome was present at a significant level and, secondly, it confirmed that the insertion element, when present, was occurring in a uniform amount. It is interesting to note that all three examples of karyotype instability at passage 29 (3q$^+$, trisomy 5, 8q) resulted in distinctive and detectable changes in the flow karyotype.

Beyond passage 30, the degree of heteroploidy increased and the cells became highly tumorigenic with shortened latent periods (33). The combined value of flow karyotype analysis of a large number of samples and confirmatory G-banding analysis at selected points of culture progression permitted a detailed characterization that would not have otherwise been feasible. At 60–70 passages the degree of karyotype instability stabilized with the predominate aberrations being trisomy 5, trisomy 8, 8q, 3p and isochromosome i(3q). All of these aberrant chromosome types were resolved with flow karyotype analysis (Fig. 3).

Univariate flow karyotype analysis was also instrumental in exploring the universality of these findings; 21 independent lineages were derived and their karyotypes monitored. Trisomy of chromosome 5 and changes involving excess DNA of the long arm of chromosome 3 were a common finding (47). While genes located in the 3q region would appear to be implicated in the *in vitro* spontaneous neoplastic process, additional experiments indicated otherwise. Based on the idea that during midcrisis the newly immortalized cells would represent a wider range of pedigrees, numerous clones were isolated from a single lineage (WCHE/5) during midcrisis and expanded independently. These clones manifested completely new cytogenetic changes [3p$^-$, lq$^+$ or i(7q)]. One of the clones only had a small proportion of cells that were aneuploid (+10) at the time they became highly tumorigenic (32,47). Involvement of chromosome 3 at a level of instability that is undetectable at the flow karyotype or G-banding level might nevertheless be occurring, or there are more paths to the tumorigenic phenotype than involvement of chromosome 3.

B. ANEUPLOIDY

The analysis of aneuploidy by flow karyotype analysis can be influenced by the degree of cellular heterogeneity present in the culture, as illustrated in Fig. 3. This phenomenum has been demonstrated in several serially cultured Chinese hamster cells, an example of which is illustrated in Fig. 4 (7). Cloned Chinese hamster embryo cells (CCHE) were carried in culture as described above for WCHE. The CV of the flow karyotype peak is an indicator of chromosome heterogeneity; for CCHE cells the CV was observed to decrease from passage 7 to a minimum at passage 19 and to increase again as the cells became more heterogenous in their phenotype. Colony size, shape and morphology also appear to be less heterogenous at early post crisis (passage 19) stages than either before or at later stages. The debris fraction also reached a minimum at passage 19 and then increased. Repetitive analysis yielded similar results showing that instrument performance is not the limiting factor in the variable resolution of the histograms (7). This same phenomenum was also found in the WCHE/5 mass culture, as seen in Fig. 3, where the CV had a range of 1.9–2.8 at passage 7, 1.4–1.7 at passage 19 and 1.7–2.3 at passage 30.

C. IN VIVO KARYOTYPE PROGRESSION

Karyotype instability associated with *in vivo* progression of preneoplastic and neoplastic cells is often difficult to analyze because of the inherent problems of acquiring good banding of tumor cells. Primary tumor cells are often refractile to Giemsa banding. Flow karyotype analysis does not suffer from this problem and has been used to compare the degree of *in vivo* progression of Chinese hamster cells of high and low tumorigenic potential (6). Figure 5 compares the flow karyotype of two different tumors arising from cells of low tumorigenic

potential (WCHE/5 passage 33) and cells of high tumorigenic potential (WCHE/5 passage 83). In each case the flow karyotype of the cells that were injected into the athymic mouse by the pre-implanted sponge method (33) are shown in the top frame. The degree of karyotype heterogeneity in the passage 33 cell tumors is dramatic compared with passage 83 cells where no marker chromosomes (above the 5% level) were found in either tumor. This application of univariate analysis was instrumental in demonstrating that preneo-plastic cells not directly capable of tumor formation can progress *in vivo* and that karyotype instability plays an important role in providing cell variants for this tumor progression (6).

D. MARKER CHROMOSOMES

Univariate flow karyotype analysis has been used to identify and sort marker chromosomes for monitoring karyotype instability and analyzing cytogenetic rearrangements (12,13,31). The identification of aberrant chromosome types that can be considered "marker chromosomes" is difficult when extreme heterogeneity of chromosome types occurs. Cells from most solid tumors have a broad range in the number and structure of chromosome types. Figure 5 (WCHE/5 passage 33 Tl) illustrates the identification of five different marker chromosomes in a tumor of Chinese hamster origin. These peaks are considered to be produced by true marker chromosomes because they are present in statistically significant numbers and are well-defined. Low-frequency chromosome aberrations are hetero-genous in DNA content and contribute to the elevated background. Flow karyotype analysis greatly facilitates the process of discriminating these two classes of aberrant chromosomes: true marker chromosomes that are believed to replicate and expand with a clonal population, versus the large numbers of rearranged chromosomes that are heterogenous in structure and DNA content.

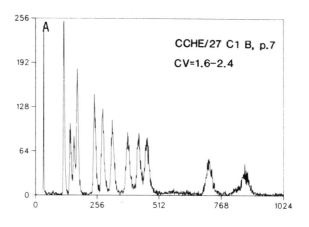

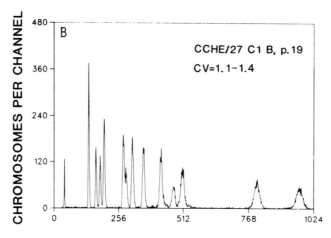

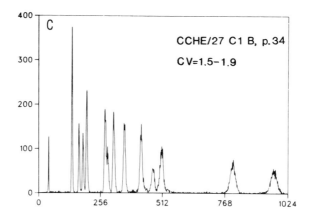

CHANNEL NUMBER (FLUORESCENCE INTENSITY)

Fig. 4. Flow karyotype analysis of cloned Chinese hamster embryo cells (CCHE/27 clone B) at three passage levels: (A), passage 7; (B), passage 19; (C) passage 34. As cellular heterogeneity decreased, there was a pronounced improvement in the coefficient of variation of the flow karyotype at passage 19. This pattern was reproduced in several cell lines.

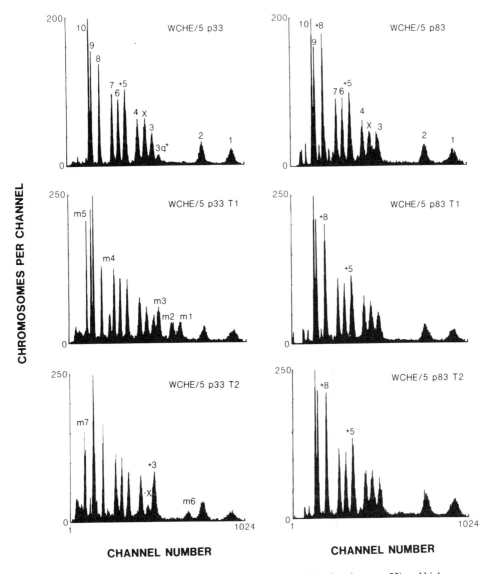

Fig. 5. Flow karyotype analysis of WCHE/5 Chinese hamster cells at low (passage 33) and high passage (passage 83) and two tumors derived from both passages. WCHE/5 passage 33 cells containing a 3q⁺ marker and trisomy 5 were injected into pre-implanted sponges in athymic mice. After considerable *in vivo* progression the resulting tumors have numerous markers as noted. WCHE/5, passage 83 cells are trisomic for chromosomes 5 and 8 and these markers are carried through to the tumors without *in vivo* selection of additional marker chromosomes.

Identification of the marker chromosomes in Fig. 5 is accomplished by flow sorting and cytogenetic analysis. Flow sorting followed by Q-banding has been reported for this purpose but is of somewhat limited utility because of the highly condensed nature of the chromosomes used for flow karyotype analysis (*11*). A flow karyotype provides a clear identification of the number of different marker chromosomes present in a population as well as quantitative information about their relative size and frequency in the population. For example, in Fig. 5, marker chromosome m1 was found to be present in about 80% of the cells and the cytogeneticist knew, based on the flow karyotype, that the chromosome to look for using the microscope would be between chromosomes 2 and 3 in size. Marker chromosome m1 was identified in this manner as a t(3;5) translocation.

The importance of combining banding analysis with flow karyotyping can be illustrated from the flow karyotype of CCHE/27 cl B cells at passage 19 [Fig. 4(B)]. From passage 7 [Fig. 4(A)] to passage 19, there was a reduction in the area of the peak for chromosome X and the simultaneous appearance of a shoulder on the peak for chromosome 7. One interpretation would be partial deletion of the lost chromosome X to generate the shoulder on chromosome 7. However, Giemsa banding confirmed a missing chromosome X and the appearance of chromosome +3q as the new peak next to chromosome 7.

E.   Chromosome Polymorphism

When the two homologues of a chromosome pair have a difference in DNA content they are said to be polymorphic (sometimes termed heteromorphic). This phenomenum is referred to as a chromosomal polymorphism to distinguish it from other types of polymorphism, such as occurs between individuals. Differences in DNA content are often due to the amount of constitutive heterochromatin on specific chromosomes. The mechanism of heterochromatin formation and its functional

significance is subject to speculation, due, in part, to the difficulty with which heterochromatin changes can be detected using banding analysis. Flow karyotype analysis has resolved several examples of chromosomal polymorphisms, some of which are observable with banding (*12*) and others that can only be detected by flow analysis (*46*).

The characteristics and inheritance pattern of a small chromosome polymorphism in Chinese hamster that could only be analyzed by flow karyotype analysis was described by Ray *et al.* (*46*). Fibroblast cultures were established using ear clippings. Samples for univariate flow karyotype analysis were prepared (using propidium iodide) within the first few passages of culture initiation. The data (see Fig. 6) indicated that three distinct peak positions for chromosome 9 were possible. Either or both of the homologues fell into the same or any two of the three positions depending on the presence of one or two discrete blocks of constitutive heterochromatin. The difference between the two peaks where only one "extra" heterochromatin block is present is equivalent to 0.15% of the DNA of the entire genome. As indicated in Fig. 6, three positions of the homologue were identified: low, (no "extra" heterochromatin); medium, (one 'extra' block); high, (two "extra" blocks). The increase in chromosome 9 fluorescence due to the presence of a single "extra" heterochromatic block is 9%. The authors also demonstrated that the polymorphism was inherited in a Mendelian fashion by mating two animals with high–low polymorphism and observing the inheritance pattern in the offspring using flow karyotype analysis (*46*). The mean and standard deviation of the autosomal normalization (see Table II for definition) of flow karyotypes from 18 individual Chinese hamster cultures with this chromosome 9 polymorphism were 1.87 ± 0.03 (low position), 2.03 ± 0.04 (medium position), and 2.23 ± 0.04 (high position). This demonstrates the remarkable reproducibility of univariate flow karyotype analysis between animals.

Green *et al.* (*27*) have characterized human

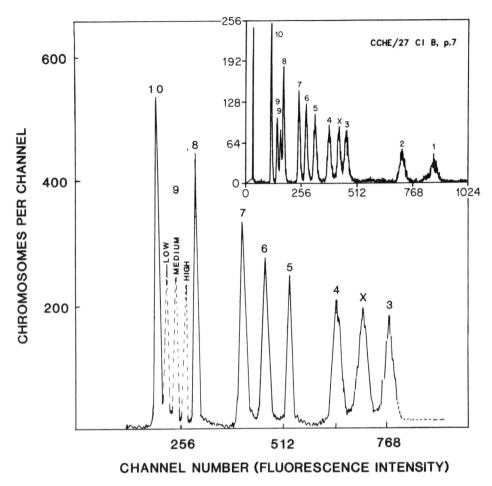

Fig. 6. Diagrammatic representation of polymorphism that can occur in the Chinese hamster chromosome 9. The two homologues can reside at the same position (low, medium or high) or in any two of the three positions shown. An example of a medium–high distribution of homologues is shown in the insert.

chromosome polymorphisms by univariate flow karyotype analysis. Nine flow karyotypes for phenotypically normal males were compared and found to differ from each other, due primarily to heteromorphisms occurring in chromosomes 1, 9, 13, 14, 15, 16, 21, 22 and Y.

F. Univariate Chromosome Sorting

Bivariate flow karyotype analysis (two-wavelength excitation) is often used to resolve

human chromosomes because of the interchromosomal variability in the AT/GC base ratio (see Chapter 8). For other species, such as mouse (4), rat, Syrian and Chinese hamsters, univariate analysis is commonly used. Univariate analysis and sorting is often selected for rodent cells because a low coefficient of variation is readily achieved and, for the most part, there is no chromosome-to-chromosome variation in AT/GC content. The Chinese hamster X chromosome is an exception as it is AT rich and can be sorted equally well by either

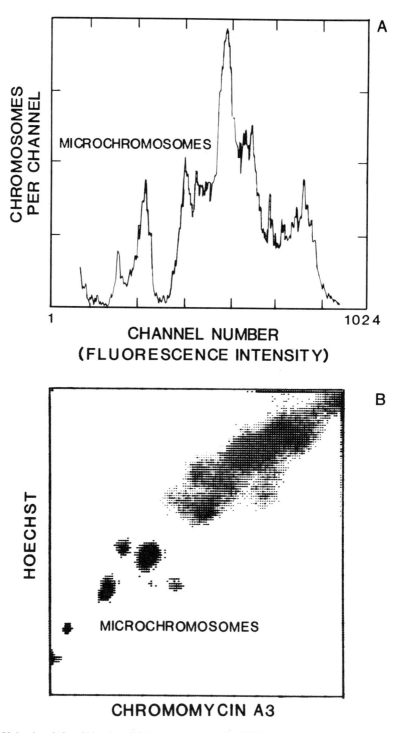

Fig. 7. Univariate (A) and bivariate (B) flow karyotypes of a 3T3M mouse cell line containing telomeric-rich microchromosomes. The univariate karyotype (A) contains two peaks that are not commonly seen in euploid mouse cells. Bivariate analysis with Hoechst 33258 and chromomycin A3 indicates that there are in fact four small chromosome types present in the population.

technique. Bivariate chromosome sorting is the method of choice for human chromosome sorting, as illustrated in Chapter 8 (36). An exception to this guideline is illustrated in Fig. 7. The availability of only a single laser can, in principle, be overcome, and Arndt-Jovin et al. (1) have demonstrated the feasibility of two-wavelength excitation with a single laser by using an all-lines excitation mirror and then spectrally and spatially separating the different wavlength beams.

Sorting applications include gene localization and mapping (10,20,21,29,35,38,54,55), specific-chromosome library construction (17,24,34,42), isolation of α-satellite repeated sequences (40) and Southern blot analysis using sorted chromosomes (23). Many of these applications have been accomplished using univariate single-laser excitation.

The univariate and bivariate distributions of mouse chromosomes obtained from a unique 3T3M cell line are illustrated in Fig. 7. The cells are from Barbara Hamkalo at the University of California, Irvine, and contain a telomeric-rich microchromosome. The univariate distribution [Fig. 7(A)] illustrates a fairly standard mouse karyotype with an extra peak about where one would expect to find a microchromosome. The bivariate flow karyotype [Fig. 7(B)] resolves the two dimmest peaks into four populations based, in part, on an altered AT/GC ratio for two of the four small chromosomes. In aneuploid karyotypes, such as often occurs in cell lines, rearrangements can influence the number and composition (AT/GC ratio) of chromosomes in unexpected ways. Another example of where two fluorochromes are useful has been found when sorting a Robertsonian 4;11 mouse translocation. The translocated chromosome is long and brighter than the normal chromosome 1 of the mouse. Univariate sorting is feasible and, depending on the application, represents a logical way to proceed. However, bivariate analysis illustrates a shift of the translocated chromosome off the 45° axis. The debris continuum remains on axis, well resolved from the 4.11 translocation.

G. Applications to Mutagenesis, Radiation Damage and Analysis of DNA Content

Univariate flow karyotype analysis has been applied to a broad variety of additional applications. Radiation damage (3,22,53) and clastogenic events (16) have been analyzed in a unique way with flow karyotype analysis. The DNA content of individual chromosome types has been precisely measured.

Fantes (22) and Welleweed (53) have used univariate flow karyotype analysis as a means to monitor and analyze chromosome damage following radiation exposure. Fantes, treating cultured lymphocytes with 50–400 rads, found a close correlation between the degree of distortion of univariate flow karyotypes and aberration frequency. The number of background counts increased, there were shifts in peak area (numerical aberrations) and shifts in peak position (deletions, translocations) corresponding to the larger chromosomes. Welleweed reported similiar results for both X-rays and α-particles. Fourier analysis was used to quantitate the level of "background" debris generated by the radiation. Flow karyotype analysis was found to be a very useful adjunct to more classical means of assessing cytogenetic analysis of radiation-induced chromosome damage. The basis for this is the ability to quantitate chromosome damage not analyzed by classical cytogenetic techniques (53). Chapters 9 and 10 describe applications of flow karyotype analysis to current problems in radiobiology and should be consulted for further details.

In a similiar fashion, Otto and colleagues have studied the effects of mutagens on flow karyotypes (43–45). High-resolution flow karyotypes were obtained using a mercury arc excitation source to excite DAPI-stained chromosomes. Increased dispersion of the largest chromosome type was found following mutagen treatment of the cell culture. The increase in CV with increasing mutagen concentration was attributed to structural aberrations associated with unequal division of the DNA at mitosis (*43*). The effect of mutagens on the uniformity of dye binding might also be a factor.

The relative DNA content of chromosomes has been precisely determined by univariate flow karyotype analysis by Bartholdi (*9*) and Harris (*28*), and by Mayall using absorption stains (*39*). These investigators report excellent agreement between the DNA content of individual human chromosomes as determined by either method. The human chromosomes with the greatest variation in DNA content are those with large centromeric heterochromatic regions, i.e. chromosomes 1, 9, 15, 16 and Y.

VI. SUMMARY AND FUTURE DIRECTIONS

Univariate flow karyotype analysis has proven to be a unique and valuable technique for a broad range of applications in cytogenetics, molecular genetics and cell biology. These techniques will continue to be of particular utility for applications involving nonhuman systems.

Rapid, quantitative analysis of single chromosomes has made it possible to sort specific chromosome types for rapid gene mapping, construction of chromosome-specific libraries and analysis of rearrangements accompanying neoplasia. In addition, the technology has made it possible to: (1) track karyotype instability of multiple cultures at sequential points throughout the multistep process, (2) discover a new polymorphism in Chinese hamster cells that controls the inheritance of a highly conserved block of heterochromatin and (3) easily identify and distinguish marker chromosomes from random aberrant chromosome types in heteroploid populations.

Future developments will likely include the successful labeling of chromosomes with DNA probes by *in situ* hybridization, the implementation of new electro-optical techniques, such as phase-sensitive detection (a technique to detect fluorochromes with overlapping emission spectra but different fluorescent life times), as well as new applications of energy transfer and emission anisotropy. New fluorochromes with different base specificity would provide alternative techniques to resolve particular chromosomes with high purity as would the availability of antibodies with chromosome specificity. New analytical capabilities are likely to be accompanied by routine implementation of four-way sorting [see Arndt-Jovin and Jovin (*2*) for a description of six-way sorting] and high-speed sorting by either the Lawrence Livermore National Laboratory technique (high pressure) or the "zapper" technique originally proposed by Martin at the Los Alamos National Laboratory and under development by Visser at TNO Rijswijk, The Netherlands (personal communication). These new concepts for increasing chromosome sorting throughput will help meet the molecular biologists needs for larger amounts (microgram quantities) of sorted chromosomes. Weeks of sorting on commercial sorters are necessary to collect microgram quantities of small chromosome types.

Flow karyotype analysis and chromosome sorting is rapidly becoming a commonplace technique for the cytogeneticist, cell biologist and molecular biologist, and the procedural options are plentiful and robust. A large frame laser is no longer required to be able to apply these techniques to a broad range of biomedical applications.

ACKNOWLEDGEMENTS

We thank Mr Kevin Albright and Mrs Carolyn Bell-Prince for their help in instrument operation and chromosome isolation and Ms Ruby Archuleta for her help in preparation of the manuscript. This work was conducted under the auspices of the Department of Energy and with support from the National Institutes of Health, Division of Research Resources, Grant RR01315.

REFERENCES

1. Arndt-Jovin, D. J., Grimwade, B. G., and Jovin, T. M. (1980). A dual laser flow sorter utilizing a CW pumped dye laser. *Cytometry* **1**, 129–131.
2. Arndt-Jovin, D. J. and Jovin, T. M. (1978). Automated cell sorting with flow systems. *Ann. Rev. Biophys. Bioeng.* **7**, 527–528.
3. Aten, J. A., Kipp, J. B. A., and Barendsen, G. W. (1980). Flow cytofluorometric determination of damage to chromosomes from X-irradiated Chinese hamster cells. In "Flow Cytometry IV, Proceedings of the IV International Symposium on Flow Cytometry", O. D. Laerum, T. Lindmo, and E. Thorud (eds). Universitetsforlaget, Oslo.
4. Baron, B., Metezeau, P., Kelly, F., Bernheim, A., Berger, R., Guenet, J. L. and Goldberg, M. E. (1984). Flow cytometry isolation and improved visualization of sorted mouse chromosomes. *Exp. Cell Res.* **152**, 220–230.
5. Bartholdi, M. F. (1985). DNA base composition of human chromosomes. *J. Colloid Interface Sci.* **105**, 426–434.
6. Bartholdi, M. F., Ray, F. A., Cram, L. S. and Kraemer, P. M. (1987). Karyotype instability of Chinese hamster cells during *in vivo* tumor progression. *Somat. Cell Molec. Genet.*, **13**, 1–10.
7. Bartholdi, M. F., Ray, F. A., Jett, J. H., Cram, L. S., and Kraemer, P. M. (1984). Flow karyology of serially cultured Chinese hamster cell lineages. *Cytometry* **5**, 534–538.
8. Bartholdi, M. F., Sinclair, D. C., and Cram, L. S. (1983). Chromosome analysis by high illumination flow cytometry. *Cytometry* **3**, 395–401.
9. Bartholdi, M. F., Travis, G. L., Cram, L. S., Porreca, P., and Leavitt, J. (1986). Flow karyology of neoplastic human fibroblast cells. *Ann. NY Acad. Sci.* **468**, 339–349.
10. Blin, N., Stohr, M., Hutter, K. J., Alonso, A., and Goerrttler, K. (1982). Assignment of snRNA gene sequences to the large chromosomes of kat kangaroo and Chinese hamster isolated by flow cytometric sorting. *Chromosoma* **85**, 723–733.
11. Carrano, A. V., Gray, J. W., and Van Dilla, M. A. (1978). Flow cytogenetics: progress towards chromosomal aberration detection. *In* "Mutagen-induced chromosome damage in man", H. J. Evans and D. C. Lloyd (eds), pp. 326–338. Edinburgh University Press, Edinburgh.
12. Carrano, A. V., Van Dilla, M. A., and Gray, J. W. (1979). Flow cytogenetics: a new approach to chromosome analysis. In "Flow Cytometry and Sorting", M. R. Melamed, P. F. Mullaney and M. L. Mendelsohn, pp. 421–451. Wiley, New York.
13. Cooke, A., Tolmie, J., Darlington, W., Boyd, E., Thomson, R., Ferguson-Smith, M. A. (1987). Confirmation of a suspected 16q deletion in a dysmorphic child by flow karyotype analysis. *J. Med. Genet.*, **24**, 88–92.
14. Cram, L. S., Bartholdi, M. F., Ray, F. A., Travis, G. L., and Kraemer, P. M. (1983). Spontaneous neoplastic evolution of Chinese hamster cells in culture: multistep progression of karyotype. *Cancer Res.*, **43**, 4828–4837.
15. Cram, L. S., Bartholdi, M. F., Wheeless, L. L., and Gray, J. W. (1985). Morphological analysis by scanning flow cytometry. *In* "Flow Cytometry Instrumentation and Data Analysis", M.A. Van Dilla, P. N. Dean, O. D. Laerum and M. R. Melamed, (eds), pp. 163–194. Academic Press, New York.
16. Cremer, C., Cremer, T. and Gray, J. W. (1982). Induction of chromosome damage by ultraviolet light and caffeine: correlation of cytogenetic evaluation and flow karyotype. *Cytometry* **2**, 287–290.
17. Davies, K. E., Young, B. D., Elles, R. G., Hill, M. E., and Williamson, R. (1981). Cloning of a representative genomic library of the human X chromosome after sorting by flow cytometry. *Nature* **293**, 374–376.
18. Deaven, L. L., Van Dilla, M. A., Bartholdi, M. F., Carrano, A. V., Cram, L. S., Fuscoe, J. C., Gray, J. W., Hildebrand, C. E., Moyzis, R. K., and Perlman, J. (1986). Construction of human chromosome-specific DNA libraries from flow sorted chromosomes. *Cold Spring Harbor Symp. in Quantit. Biol.*, **51**, 159–168.
19. Disteche, C. M., Carrano, A. V., Ashworth, L. K., Burkhart-Schultz, K., and Latt, S. A. (1981). Flow sorting of the mouse cattanach X chromosome, T(X;7) 1 Ct, in an active or inactive state. *Cytogenet. Cell Genet.* **39**, 189–197.
20. Disteche, C. M., Kunkel, L. M., Lojewski, A., Orkin, S. H., Eisenhard, M., Sahar, E., Travis, B., and Latt, S. A. (1982). Isolation of mouse X-chromosome specific DNA from an X-enriched lambda phage library derived from flow sorted chromosomes. *Cytometry* **2**, 282–286.
21. Fantes, J. A., Green, D. K., Cooke, H.J. (1983). Purifying human Y chromosomes by flow cytometry and sorting. *Cytometry* **4**, 88–91.
22. Fantes, J. A., Green, D. K., Elder, J. K., Malloy, P., and Evans, H. J. (1983). Detecting radiation damage to human chromosomes by flow cytometry. *Mutat. Res.* **119**, 161–168.
23. Fukushige, S., Murotsu, T., and Matsubara, K. (1986). Chromosomal assignment of human genes for gastrin, thyrotropin (TSH)-β subunit and C-*erb*-2 by chromosome sorting combined with velocity sedimentation and southern hybridization. *Biochem. Biophys. Res. Comm.* **134**, 477–483.
24. Griffith, J. K., Cram, L. S., Crawford, B. D., Jackson, P. J., Schilling, J., Schimke, R. T., Walters, R. A., Wilder, M. E., and Jett, J. H. (1984). Construction and analysis of DNA sequence libraries from flow-sorted chromosomes: practical and theoretical considerations. *Nucleic Acids Res.*, **12**, 4019.

25. Gray, J. W., Carrano, L. L., Steinmetz, M. A., Van Dilla, M. A., Moore II, D. H., Mayhall, B. H., Mendelsohn, M. L. (1975). Chromosome measurement and sorting by flow systesm. *Proc. Natl. Acad. Sci. USA* **72**(4), 1231–1234.

26. Gray, J. W., Carrano, A. V., Moore II, D. H., Steinmetz, L. L., Minkler, J., Mayall, B. H., Mendelsohn, M. L., Van Dilla, M. A. (1975). High speed quantitative karyotyping by flow microfluorometry. *Clin. Chem.* **21**(9), 1258–1262.

27. Green, D. K., Fantes, J. A., Buckton, K. E., Elder, J. K., Malloy, P., Carothers, A., and Evans, H. J. (1984). Karyotyping and identification of human chromosome polymorphisms by single fluorochrome flow cytometry. *Hum. Genet.* **66**, 143–146.

28. Harris, P., Boyd, E., Young, B. D. and Ferguson-Smith, M. A. (1986). Determination of the DNA content of human chromosomes by flow cytometry. *Cytogenet. Cell. Genet.* **41**, 14–21.

29. Harris, A., Young, B. D., and Griffin, B. E. (1985). Random association of Epstein–Barr virus genomes with host cell metaphase chromosomes in Burkitt's lymphoma-derived cell lines. *J. Virol.,* **56**, 328–332.

30. Kakunaga, T. (1978). Neoplastic transformation of human diploid fibroblast cells by chemical carcinogens. *Proc. Natl Acad. Sci. USA* **75**, 1334–1337.

31. Kooi, M. W., Aten, J. A., Stap, J., and Barendsen, G. W. (1984). Preparation of chromosome suspensions from cells of solid tumors for measurement by flow cytometry. *Cytometry* **5**, 547.

32. Kraemer, P. M., Ray, F. A., Bartholdi, M. F., and Cram, L. S. (1987). Spontaneous *in vitro* neoplastic evolution: selection of specific karyotype in Chinese hamster cells. *Cancer Genet. Cytogenet.* **27**, 273–287.

33. Kraemer, P. M., Travis, G. L., Ray, F. A., and Cram, L. S. (1983). Spontaneous neoplastic evolution of Chinese hamster cells in culture: multistep progression of phenotype. *Cancer Res.* **43**, 4822–4827.

34. Krumlauf, R., Jeanpierre, M., Young, B. D. (1982). Construction and characterization of genomic libraries from specific human chromosomes. *Proc. Natl Acad. Sci. USA* **79**, 2971–2975.

35. Kunkel, L. M., Tantravahi, U., Eisenhard, M., and Latt, S. A. (1982). Regional localization on the human X of DNA segments cloned from flow sorted chromosomes. *Nucleic Acids Res.* **10**(5), 1557–1578.

36. Langlois, R. G., Yu, L. C., Gray, J. W., and Carrano, A. V. (1982). Quantitative karyotyping of human chromosomes by dual beam flow cytometry. *Proc. Natl Acad. Sci. USA* **79**(24), 7876–7880.

37. Lebo, R. V., Carrano, A. V., Burkhart-Schultz, K., Dozy, A. M., Yu, L., and Kan, Y. W. (1979). Assignment of human β-, γ, and δ-globin genes to the short arm of chromosome 11 by chromosome sorting and DNA restriction enzyme analysis. *Proc. Natl Acad. Sci. USA* **76**(11), 5804–5808.

38. Lebo, R. V., Gorin, F., Fletterick, R. J., Kao, F. T., Cheung, M. C., Bruce, B. D., Kan, Y. W. (1984). High resolution chromosome sorting and DNA spot–blot analysis assign McArdle's syndrome to chromosome 11. *Science* **225**, 57–59.

39. Mayall, B. H., Carrano, A. V., Moore, D. H., Ashworth, L., Bennett, D. E., Mendelsohn, M. L. (1984). The DNA-based human karyotype. *Cytometry* **5**, 376–385.

40. McDermid, H. E., Duncan, A. M. V., Higgins, M. J., Hamerton, J. L., Rector, E., Brasch, K. R., and White, B. N. (1986). Isolation and characterization of an α-satellite repeated sequence from human chromosome 22. *Chromosoma* **94**, 228–234.

41. Mendelsohn, M. C., Mayall, B. H., Bogart, E., Moore, D. H., and Perry, B. H. (1973). DNA content and DNA-based centromeric index of the 24 human chromosomes. *Science* **179**, 1126–1129.

42. Mueller, C. R., Davies, K., Cremer, C., Rappold, G., Gray, J. W., and Ropers, H. (1983). Cloning of genomic sequences from the human Y chromosome by combined velocity sedimentation and flow sorting: applying single and dual laser cytometry. *Hum. Genet.,* **64**, 110–115.

43. Otto, F. J. and Oldiges, H. (1981). Flow cytogenetic studies in chromosomes and whole cells for the detection of clastogenic effects. *Cytometry* **1**(1), 13–17.

44. Otto, F. J., Oldiges, H., Gohde, W., Barlogie, B., and Schumann, J. (1980). Flow cytogenetics of uncloned and cloned Chinese hamster cells. *Cytogenet. Cell. Genet.* **27**(1), 52–56.

45. Otto, F., Oldiges, H., Gohde, W., and Dertinger, H. (1980). Flow cytometric analysis of mutagen induced chromosomal damage. *In* "Flow Cytometry IV, Proceedings of the IV International Symposium on Flow Cytometry" (O. D. Laerum, T. Lindmo and E. Thorud, eds). Universitetsforlaget, Oslo.

46. Ray, F. A., Bartholdi, M. F., Kraemer, P. M., and Cram, L. S. (1984). Chromosome polymorphism involving discrete heterochromatic blocks in Chinese hamster number nine chromosome. *Cytogenet. Cell Genet.* **38**, 257.

47. Ray, F. A., Bartholdi, M. F., Kraemer, P. M. and Cram, L. S. (1986). Spontaneous *in vitro* neoplastic evolution: recurrent chromosome changes of newly immortalized Chinese hamster cells. *Can. Gen. Cytogenet.* **21**, 35–51.

48. Stohr, M., Hutter, K., Frank, M., and Futterman, G. (1980). A flow cytometric study of chromosomes from rat kangaroo and Chinese hamster cells. *Histochem.* **67**, 179–190.

49. Stubblefield, E., Cram, L. S., Deaven, L. L. (1975). Flow microfluorometric analysis of isolated Chinese hamster chromosomes. *Exp. Cell Res.* **94**, 464–468.

50. Trask, B., van den Engh, G., Gray, J., Vanderlann, M. and Turner, B. (1984). Immunofluorescent

detection of histone 2B on metaphase chromosomes using flow cytometry. *Chromosoma* **90**, 295–302.

51. van den Engh, G., Trask, B., Cram, L. S. and Bartholdi, M. F. (1984). Preparation of chromosome suspensions for flow cytometry. *Cytometry* **5**, 105–117.

52. Van Dilla, M. A., Deaven, L. L., Albright, K. L., Allen, N. A., Aubuchon, M. R., Bartholdi, M. F., Browne, N. E., Campbell, E. W., Carrano, A. V., Clark, L. M., Cram, L. S., Fuscoe, J. C., Gray, J. W., Hildebrand, C. E., Jackson, P. J., Jett, J. H., Longmire, J. L., Lozes, C. R., Luedemann, M. L., Martin, J. C., McNinch, J. S., Meincke, L. J., Mendelsohn, M. L., Meyne, J., Moyzis, R. K., Munk, A. C., Perlman, J., Peters, D. C., Silva, A. J., and Trask, B. J. (1986). Human chromosome-specific DNA libraries: construction and availability. *Bio/ Technol.* **4**, 537–552.

53. Welleweerd, J., Wilder, M. E., Carpenter, S. G., and Raju, M. R. (1984). Flow cytometric determination of radiation-induced chromosome damage and its correlation with cell survival. *Radiat. Res.,* **99**, 44–51.

54. Wirschubsky, Z., Ingvarsson, S., Carstenssen, A., Wiener, F., Klein, G., and Sumegi, J. (1985). Gene localization on sorted chromosomes: definitive evidence on the relative positioning of genes participating in the mouse plasmacytoma-associated typical translocation. *PNAS* **82**, 6975–6979.

55. Wirshubsky, Z., Perlmann, C., Lindsten, J., and Klein, G. (1983). Flow karyotype analysis and fluoresence-activated sorting of Burkitt-lymphoma-associated translocation chromosomes. *Int. J. Cancer,* **32**, 147–153.

56. Young, B. D., Ferguson-Smith, M. A., Sillar, R., and Boyd, E. (1981). High resolution analysis of human peripheral lymphocyte chromosomes by flow cytometry. *PNAS* **78**(12) 7727–7731.

Bivariate Flow Karyotyping

JOE. W. GRAY, G. J. VAN DEN ENGH AND BARB J. TRASK

Biomedical Sciences Division
Lawrence Livermore National Laboratory
Livermore, CA 94550, USA

I. INTRODUCTION

Flow cytometric chromosome analysis, hereafter called flow karyotyping, is useful both for chromosome classification and for identification of chromosomes of a specific type for purification by flow sorting. Both of these applications are best served by procedures that allow the greatest possible discrimination between the chromosome types in a population. In univariate flow karyotyping (see Chapter 7), chromosome discrimination is based on the amount of fluorescent dye incorporated during equilibrium staining. Most dyes in common use today [e.g. Hoechst 33258 (HO), DAPI, propidium iodide (PI), chromomycin A3 (CA3) and mithramycin] bind only to nucleic acids so

that discrimination is largely according to DNA content. This classification procedure is limited by the fact that many chromosome types have similar DNA contents. Discrimination between such chromosome types stained with a single dye requires very high-resolution flow cytometry, if it is possible at all. Bivariate flow karyotyping (chromosome classification according to two variables) was developed to take advantage of the fact that some dyes like HO and DAPI bind preferentially to adenine-thymine (AT)-rich DNA while others like CA3 and mithramycin bind preferentially to guanine-cytosine (GC)-rich DNA (4,10,24). Measurement of the intensity with which chromosomes stain with one AT-specific and one GC-specific dye allows classification of chromosomes according to DNA content and DNA base composition (see Chapter 5 for more information on chromosome staining). This chapter focuses on techniques associated with bivariate flow karyotyping and on its application to chromo-some classification, aberrant chromosome detection and characterization, cell line characterization and chromosome sorting.

II. PRINCIPLES OF BIVARIATE FLOW KARYOTYPING

The stains HO and CA3 have been used most widely for bivariate flow karyotyping. The characteristics of these dyes are discussed in detail in Chapter 5. HO binds preferentially to AT-rich DNA, presumably in the minor groove of the double helix (*19*) and can be excited efficiently by the ultraviolet (UV) (351 + 353 nm) beam from an argon ion laser. It fluoresces around 450 nm. CA3 binds preferentially to GC-rich DNA and can be excited by the 458 nm beam from an argon ion laser. It fluoresces around 520 nm. Substantial energy transfer occurs from HO to CA3 so that most fluorescence from chromosomes stained with both dyes occurs around 520 nm (*11,25*). Neither dye binds covalently so that DNA can be isolated from chromosomes stained with these dyes; the dyes can be removed and the DNA can be processed using conventional molecular biological techniques (e.g. digested with restriction endonucleases, separated by electrophoresis and cloned into phage and cosmid vectors (see Chapters 15 and 16)).

Once stained, the chromosomes are measured using dual-beam cytometry (see Chapter 2). During flow cytometry, the chromosomes flow sequentially through laser beams adjusted to the UV to excite HO and 458 nm to excite CA3. Fluorescence resulting from the UV excitation is recorded as a measure of the HO dye content and fluorescence from the 458 nm excitation is recorded as a measure of the CA3 content. Thus, a pair of measurements are made from each chromosome.

Measurements from 10^4 to 10^6 chromosomes for each sample are accumulated to form a bivariate HO *vs* CA3 fluorescence distribution (called a bivariate flow karyotype) like that shown in Fig. 1 for normal human chromosomes.

HO *vs* CA3 flow karyotypes typically show several peaks superimposed on an underlying continuum. Each peak is produced by one or more chromosome types that have the same HO- and CA3-binding characteristics. The measurement precision (i.e. peak coefficients of variation) is 1–2% in high-resolution instruments. As a result, flow karyotypes for normal human cells typically show one peak for each chromosome type except chromosomes 9–12 and sometimes 14 and 15. In some cases, the individual homologous chromosomes can be resolved separately; probably because they differ in the amounts of centromeric heterochromatin (*7*; B. Trask *et al.* unpublished work). In other species, the chromosome discrimination is not as impressive. For example, in some mice, many of the chromosome types have the same DNA content and same DNA base composition. Thus, bivariate HO *vs* CA3 flow karyotypes for these mice show the 21 chromosome types resolved into only 12–14 peaks. The continuum that underlies the various peaks in a bivariate flow karyotype is caused by chromosome and nuclear debris fragments and chromosome clumps that are produced during the chromosome isolation process (see Chapter 4). Typically, it is highest at the origin and extends as a band of decreasing magnitude along the diagonal of the flow karyotype.

DNA-specific, base-composition-dependent dyes other than HO and CA3 also have been used successfully as stains for bivariate flow karyotyping. DAPI has proved especially useful as an alternative to HO (*12,17*). DAPI binds preferentially to AT-rich DNA as does HO. However, DAPI stains the AT-rich heterochromatic regions of some chromosomes (e.g. human chromosomes 1, 9, 16 and Y) relatively more intensely than does HO. DAPI is especially useful for distinguishing human

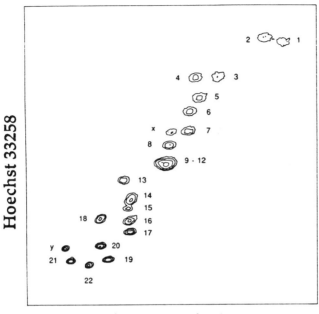

Fig. 1. This distribution shows a HO *vs* CA3 flow karyotype for a karyotypically normal male human displayed as a contour plot. The numbers indicate the identities of the chromosomes producing the peaks.

chromosome 9 from chromosomes 10–12 so that it is more clearly resolved in the flow karyotype. Nonfluorescing dyes like netropsin and distamycin that bind preferentially to AT-rich DNA also have been used with some success to enhance the discrimination of chromosomes with large heterochromatic regions (see Chapter 5 for additional details).

III. CHROMOSOME ABERRATION DETECTION

A. PEAK MEANS AND VOLUMES

The variation in the means of the peaks in bivariate flow karyotypes measured for normal individuals are usually small compared to the separation of the peaks. Thus, the peak means can be used as chromosome identifiers and indicators of chromosome structural normalcy (*5,10,13*). The volume of each peak is pro-

portional to the frequency of occurrence of the chromosome type(s) comprising that peak. Peak means and volumes can be estimated by computer analysis as described in Chapter 6.

Figure 2 illustrates the types of changes that might be produced in a flow karyotype by structural and numerical aberrations. This figure shows a bivariate flow karyotype for a hypothetical cell line with only three pairs of homologous chromosomes. Figure 2(b) shows the normal flow karyotype. Figure 2(a) shows the effect of a trisomy on peak volume. The volume of peak 3 increases by 50% relative to the volumes in peaks 1 and 2 as a result of the trisomy 3. Figure 2(c) shows the effect on peak volume and mean of a balanced, reciprocal translocation between chromosomes 1 and 3. The new peaks labeled der1 and der3 are generated by the derivative chromosomes. The difference in HO and CA3 staining between the peaks from chromosome 1 and der1 is equal and opposite to that between the peaks for

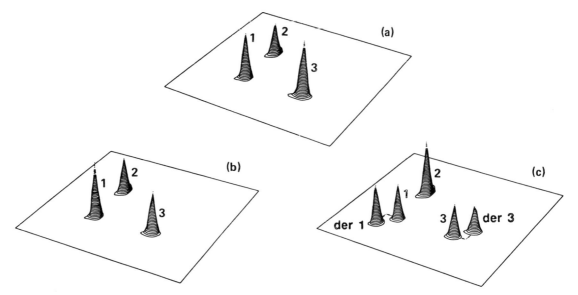

Fig. 2. Flow karyotypes for hypothetical populations carrying three pairs of homologous chromosomes. (b) The flow karyotype for the "normal" population. The hypothetical chromosomes associated with each peak are indicated by numbers on the plot. (a) The flow karyotype for a population trisomic for chromosome 3. (c) The flow karyotype for a population with a balanced reciprocal translocation between chromosomes 1 and 3. The derivative chromosomes are labeled der1 and der3.

chromosome 3 and der3 indicating that chromosomal material was not lost or gained during the exchange. The hypothetical cell now contains two copies of the normal chromosome 2 and one copy of each of chromosomes 1, 3, der1 and der3. Thus, the volumes of the peaks for the normal chromosomes 1 and 3 and those of the new peaks for der1 and der3 are half that of the peak for chromosome 2.

B. DETECTION OF NUMERICAL ABERRATIONS

Figure 3 illustrates the application of bivariate flow karyotyping to detection of a numerical aberration involving chromosome 21. Figure 3(a) shows a flow karyotype measured for a normal cell population. Figure 3(b) shows a flow karyotype for a population trisomic for chromosome 21 (5). The increased volume for the chromosome 21 peak in Fig. 3(b) is apparent. This technique allows reliable

detection of trisomies important to prenatal diagnosis (5,13). The reliability with which numerical aberrations can be detected depends on the extent to which peak volume estimates made from analysis of the flow karyotype are actually proportional to the frequency of the corresponding chromosome types in the cells from which the chromosomes were isolated. Both the chromosome isolation procedure and the mathematical procedure for estimation of peak volumes may affect this proportionality.

The chromosome isolation procedure may affect peak volume analysis in two ways: (1) Inadvertent selection of chromosomes according to size (e.g. by differential velocity sedimentation during a centrifugation or sedimentation step or by size selection during filtering) will alter the frequency distribution of the various chromosome types in the population. (2) Production of chromosome debris or clumps. The presence of a substantial debris continuum makes estimation of peak volumes difficult because the shape of the

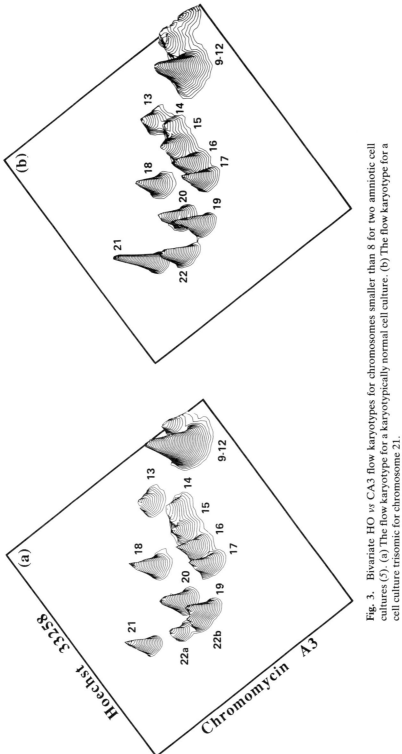

Fig. 3. Bivariate HO *vs* CA3 flow karyotypes for chromosomes smaller than 8 for two amniotic cell cultures (5). (a) The flow karyotype for a karyotypically normal cell culture. (b) The flow karyotype for a cell culture trisomic for chromosome 21.

continuum is not known with precision. As a result, it is difficult to approximate during mathematical analysis (Chapter 6).

The estimation of the volumes of well-resolved peaks in flow karyotype free of clumps and debris is straightforward and highly accurate. The accuracy is lower for peaks that overlap heavily or that are superimposed on a substantial continuum. Chromosome debris affects estimation of the volumes of the smaller chromosomes (e.g. human chromosome 21). Chromosome clumping, on the other hand, affects estimation of the volumes of the larger chromosomes (e.g. human chromosomes 1 and 2). In general, the amount of debris and clumps is sufficiently low to allow peak volume analysis with an accuracy approaching 10%. Table I, for example, shows peak volumes estimated from chromosomes 13 to 22 for flow karyotypes measured for five amniotic cell cultures that were trisomic for chromosome 21 (5). The average difference between the measured and expected peak volumes for the

nine normal and one trisomic chromosome for the five samples was 8%. However, the coefficients of variation of the volumes estimated for some chromosome types (e.g. chromosome 21) were as high as 19%. This accuracy is sufficient to allow detection of monosomies and trisomies as long as the remainder of the chromosome types in the population occur at normal or known frequencies. In this case, the relative peak volumes can be converted to frequencies per cell. This is the case for amniocytes, but may be problematic for karyotypically unstable cell populations (e.g. from solid tumors).

C. DETECTION AND CHARACTERIZATION OF STRUCTURAL ABERRATIONS

Figure 4 illustrates the use of flow karyotyping for detection of the translocation, t(9;22), in leukemic cells from an individual with chronic myelogenous leukemia (26). The two derivative chromosomes resulting from the translocation are well resolved in the flow karyotype. The change in the HO and CA3 staining from chromosome 9 to der9 is equal and opposite to that from chromosome 22 to der22, indicating that this translocation is balanced. The utility of bivariate flow karyotyping has been demonstrated for detection of structural aberrations important in prenatal diagnosis (5,14) and in detection of aberrant or marker chromosomes in tumors (26).

Flow karyotyping also allows quantitative analysis of the amount of material that may be lost from a particular chromosome as a result of a structural aberration. This application is illustrated in Fig. 5 which shows flow karyotypes measured for three human female lymphocyte cell lines. These lines were analyzed by banding analysis and found to carry deletions involving the X chromosome. The amount of chromosomal material lost to the deletion was determined by comparing the bivariate mean of the derivative chromosome to

Table I

AVERAGE PEAK VOLUMES MEASURED FROM FLOW KARYOTYPES MEASURED FOR FIVE INDIVIDUALS TRISOMIC FOR CHROMOSOME 21 (5)

Chromosome	Frequency per cell	
	Expected	Measured (SD)
13	2	1.76 (0.13)
14	2	2.42 (0.46)
15	2	1.88 (0.24)
16	2	2.12 (0.19)
17	2	1.87 (0.16)
18	2	2.04 (0.20)
19	2	1.92 (0.09)
20	2	1.95 (0.09)
21	3	3.31 (0.49)
22	2	1.77 (0.18)

The volume of each peak was converted to "frequency per cell basis" by normalizing to the volume of the peak produced by chromosomes 9–12 using the relationship:

measured frequency per cell =

(volume test peak/volume 9–12 peak) × 8.

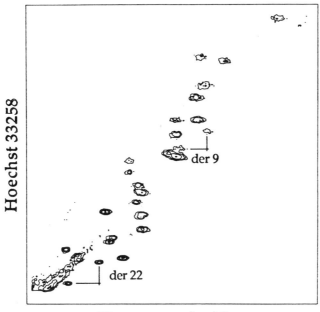

Fig. 4. Bivariate HO *vs* CA3 flow karyotype for human chronic myelogenous leukemic cells carrying t(9;22) (*26*). The derivative chromosomes resulting from the translocation are labeled der22 and der9. The horizontal and vertical lines indicate the changes in HO and CA3 content between the peaks for der22 and 22 and between the peaks for der9 and 9.

that for the remaining normal X chromosome. The flow karyotype analyses showed the deletions in X chromosomal DNA to be 27 Mbp, 15 Mbp and 11.5 Mbp, respectively. These same cell lines were estimated by G-banding to have lost 29 Mbp, 18 Mbp and 17 Mbp of their X chromosomal DNA, respectively. More extensive studies show that the amount of chromosomal material measured by flow karyotyping is linearly related to the amount of chromosomal material measured using analysis of banded metaphase spreads (*21*b).

Unfortunately, normally occurring polymorphic differences between homologs may reduce the accuracy with which structural aberrations can be detected and characterized. In particular, the accuracy with which deleted or duplicated material can be quantified depends on the accuracy of the assumption that the DNA contents of the derivative

chromosomes were the same before rearrangement as the remaining homologous chromosomes. Unfortunately, polymorphic variability may reduce the accuracy of this assumption. This variability is thought to occur primarily in the heterochromatic (non-expressed) regions of selected chromosomes. The variability is usually largest in chromosomes carrying substantial amounts of heterochromatin. In humans, chromosomes 1, 13, 14, 15, 16, 21, 22 and Y have been found to be the most variable (up to 50% in some chromosomes) while chromosomes 2, 5, 6, 7, 8 and X are the least variable (*7, 8; 21*a). Normal chromosomes may appear as structurally aberrant in flow karyotypes if they contain substantially more or less heterochromatin than the average.

Some of the complications resulting from polymorphic variability can be eliminated by measuring parental flow karyotypes.

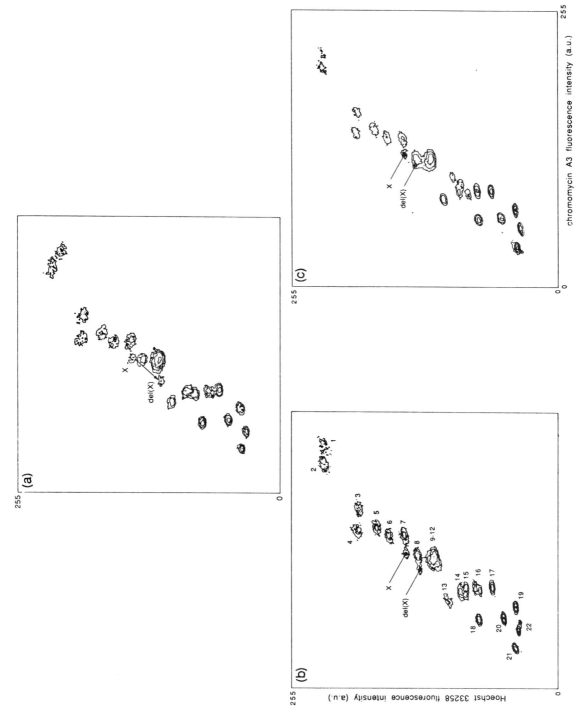

Fig. 5. Bivariate HO *vs* CA3 flow karyotypes for three human female cell lines carrying deletions involving one X homolog. The cell lines and cytogenetic classification by banding analysis are: (a) GM3923: 46, X, del(X) (pter>q13:q22>qter); (b) GM8121; 46, X, del(X) (pter:q26); (c) L2384; 46, X, del(X)

chromomycin A3 fluorescence intensity (a.u.)

Hoechst 33258 fluorescence intensity (a.u.)

Chromosomal polymorphisms seem to be inherited with high fidelity (9) so that the original DNA contents of derivative chromosomes can be inferred from the parental flow karyotypes. Figure 6, for example, shows flow karyotypes measured for the mother, father, son and daughter from one family (20a). Comparison of the flow karyotypes for the mother and father show polymorphic differences for chromosomes 1, 13, 15, 16, 21

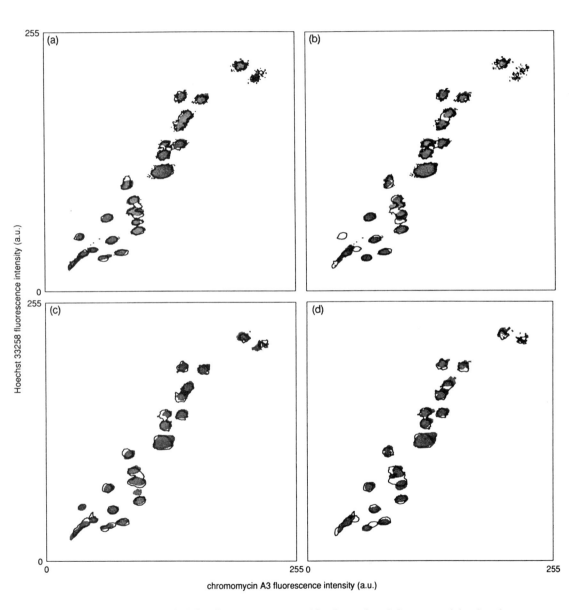

Fig. 6. Bivariate HO *vs* CA3 flow karyotypes measured for the mother, father, son and daughter from one family (20a). The contour lines and shaded areas show the portions of flow karyotypes in which the frequencies are greater than or equal to 10% of the maximum frequency in the flow karyotype. (a) Contour lines, son; shaded areas, father. (b) Contour lines, son; shaded areas, mother. (c) Contour lines, daughter; shaded areas, father. (d) Contour lines, daughter; shaded areas, mother.

and 22. The inheritance of specific polymorphic chromosomes can be traced by comparing the flow karyotypes for the offspring with the parental flow karyotypes. In this family, for example, one homolog of chromosome 21 in the paternal flow karyotype is larger than average. Examination of the flow karyotypes for the son and daughter show that this large chromosome 21 has been inherited by both. One homolog of chromosome 22 appears large in the paternal flow karyotype. However, only the daughter has inherited this homolog. In general, measurement of parental flow karyotypes allows compensation for some polymorphic variability so that changes in chromosomal DNA content resulting from structural aberrations as small as 2 Mbp may be detected. This capability may be particularly important in prenatal diagnosis where loss of chromosomal material during a *de novo* translocation has been associated with mental retardation and with other developmental abnormalities (*2*).

IV. HYBRID CELL CHARACTERIZATION

The development of hybrids of somatic human and rodent cells that carry one or a few human chromosomes has played an important role in modern molecular biology (*16*). Collections of hybrid cell lines that carry a spectrum of different human chromosomes or subregions have been used to assign hundreds of genes or DNA sequences to specific human chromosomes (*15*). Hybrids also have facilitated the isolation and cloning of specific DNA sequences. In addition, hybrids have served as useful sources of isolated chromosomes from which single human chromosomes can be sorted at high purity for production of chromosome-specific recombinant DNA libraries (see Chapters 15 and 16).

The accurate determination of the human complement of chromosomes in hybrids is essential to the proper interpretation of most hybrid-based molecular biological experiments. This is usually accomplished by screening for

human-chromosome-specific isoenzymes and/ or by cytogenetic analysis (e.g. by G-banding and/or fluorescence *in situ* hybridization). Hybrid chromosome analysis should be conducted frequently during hybrid culture because of the karyotypic instability of many hybrids. Unfortunately, isoenzyme analyses do not provide complete information about the integrity of the human chromosome(s) and cytogenetic analysis is time consuming (especially for hybrid cells). Flow karyotyping has proved to be a powerful adjunct to these hybrid classification procedures since it allows accurate and rapid assessment of the human chromosome complement of hybrids and of the structural normalcy of the human chromosomes retained by the hybrids.

Hybrid analysis by flow karyotyping is accomplished by measuring flow karyotypes for the hybrid cells, for the human cell and for a 1:1 mix of the human and hybrid chromosomes (B. Trask *et al.*, unpublished work). Ideally, the human cells analyzed are the same as those used to construct the hybrid so that normal polymorphic variability in the peak position of the human chromosome is accounted for. The human chromosome present in the hybrid can be identified and checked for normal DNA content by comparing relative peak positions in the three flow karyotypes. Figure 7, for example, shows flow karyotypes measured for the human cell line, GM131, the hybrid cell line CHH-2 carrying human chromosome 2 and a mixture of chromosomes from these two lines. The plots have been scaled so that the means of the numbers of chromosome 8 in the human and 1:1 mix flow karyotypes are the same and so that the means of a hamster chromosome reference peak in the hybrid and 1:1 mix are the same. The presence of human chromosome 2 with relatively normal DNA content in CHH-2 can be determined by comparison of the three flow karyotypes. The relative frequency of the human chromosome in the hybrid population can be estimated by analysis of the relative volumes of the human and hamster chromosomes. Flow karyotyping has been successfully applied to almost 100

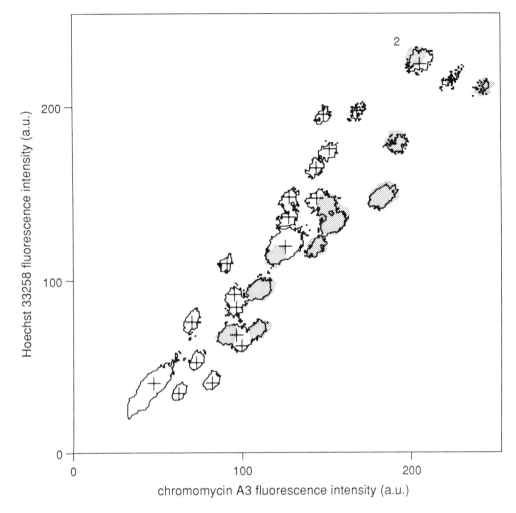

Fig. 7. HO *vs* CA3 flow karyotypes of human, hamster and human × hamster hybrid cells. The contour lines show the portion of the flow karyotype for the CHH-2 plus GM131 mixture in which the frequencies are greater than or equal to 10% of the maximum frequency in the flow karyotype. The shaded areas show the same regions of the flow karyotype for CHH-2. The crosses show the locations of the peaks for the human chromosomes determined from the flow karyotype for GM131. The only peak in which a cross, a contour and a shaded region overlap is in the region of human chromosome 2. This indicates that human chromosome 2 is the only normal human chromosome in CHH-2.

human × hamster hybrid cell lines in order to find lines carrying well-resolved normal human chromosomes that can be purified by fluorescence-activated cell sorting. The utility of bivariate flow karyotyping for hybrid analysis is illustrated by the fact that only about half of the hybrids characterized by this technique were found to contain human chromosomes

that were suitable for sorting (B. Trask *et al.* unpublished work). Approximately 50 of the cells analyzed in this study contained human chromosomes that were structurally abnormal, or were not sufficiently well resolved from the hamster chromosome. Some hybrid cell populations contained human chromosomes only in some cells so that the overall frequency of the chromosomes was too low to be useful for sorting.

V. CHROMOSOME SORTING

Chromosome purification by sorting has facilitated both gene mapping (Chapter 14) and production of chromosome-specific recombinant DNA libraries (Chapters 15 and 16). Both of these applications demand the highest possible discrimination of the desired chromosome types from each of the other chromosomes and from chromosomal debris and clumps during sorting. This is best accomplished using bivariate flow karyotyping. If small numbers of sorted chromosomes are needed, chromosomes can be processed at sufficiently low rates to maintain very high measurement resolution and high sorting purity. In this situation, populations with > 90% of one chromosome type of most human chromosomes can be sorted from chromosomes isolated from normal human cells. Unfortunately, many applications require millions of sorted chromosomes. For these applications, some resolution must be sacrificed for sorting speed (see Chapter 2). Thus, in practice, only the smaller human chromosomes (especially 13, 18, 19, 20, 21, 22 and Y) can be sorted at high purity from suspensions of chromosomes isolated from normal human cells. Human × hamster hybrid cells have proved especially useful as sources for the human chromosomes that cannot be sorted at high purity from chromosome suspensions isolated from human cells.

VI. FUTURE DIRECTIONS

A. CLASSIFICATION ACCORDING TO THE INTENSITY OF IMMUNOFLUORESCENCE STAINING

DNA-specific fluorescent dye binding is not the only basis for chromosome discrimination. Techniques also have been developed in recent years to stain chromosomes simultaneously for DNA content (e.g. using PI) and for their affinity for specific antibodies. Bivariate flow karyotypes have been measured for chromosomes that have been immunofluorescently stained with antibodies for histone H2b (21,23), for kinetochore proteins (3,6) and for bromodeoxyuridine incorporated into chromosomal DNA (20). These staining techniques have proved useful for detection of differences in chromatin structure in isolated chromosomes. They have also been used to preferentially stain the centromere multicentric chromosomes produced by exposure of cells to ionizing radiation (3). Immunofluorescent chromosome staining may become even more useful in the future if antibodies for chromosome-specific proteins or chromatin structures are discovered. Candidate proteins include those that may be associated with chromosome-specific α-satellite DNA, specific restriction endonucleases and regulatory proteins.

B. CLASSIFICATION ACCORDING TO THE INTENSITY OF STAINING USING FLUORESCENCE *IN SITU* HYBRIDIZATION

The most obvious basis for improved chromosome discrimination is according to the chromosomal DNA sequence. Fluorescence *in situ* hybridization with chromosome-specific probes permits exactly this kind of discrimination (see Chapter 17). Furthermore, large numbers of chromosome-specific probes are becoming available, making it likely that it will be possible to stain for any desired chromosome type within a few years. It remains to be seen whether hybridization can be accomplished for chromosomes in suspension so that high-quality flow karyotypes can be measured. Techniques have already been developed for fluorescence *in situ* hybridization to nuclei in suspension (22), and hybridization to chromosomes in suspension has also been reported (1). However, hybridization to isolated chromosomes generally results in excessive chromosome fragmentation and/or clumping so that flow karyotypes measured for chromosomes stained in this manner do not show distinct peaks. If more gentle techniques for chromosome denaturation, hybridization and washing can be developed, then fluorescence hybridization may become a major tool in flow karyotyping.

ACKNOWLEDGEMENTS

This work was performed under the auspices of the US Department of Energy by the Lawrence Livermore National Laboratory under Contract W-7405-ENG-48 with support from US Public Health Service grant HD17665. The authors thank Ms Lil Mitchell for expert help in the preparation of this manuscript.

REFERENCES

1. Dudin, G., Cremer, T., Schardin, M., Hausmann, M., Bies, F., and Cremer, C. (1987). A method for nucleic acid hybridization to isolate chromosomes in suspension. *Hum. Genet.* **76**, 290–292.
2. Epstein, C. (1987). *The Consequences of Chromosome Imbalance: Principals, Mechanisms and Models*, Cambridge University Press, Cambridge.
3. Fantes, J., Green, D., Malloy, P., and Sumner, A. (1989). Flow cytometry measurements of human chromosome kinetochore labeling. *Cytometry* **10**, 134–42.
4. Gray, J. W., Langlois, R. G., Carrano, A. V., Burkhart-Schultz, K., and Van Dilla, M. A. (1979). High resolution chromosome analysis: one and two parameter flow cytometry. *Chromosoma* **73**, 9–27.
5. Gray, J. W., Trask, B., van den Engh, G., Silva, A., Lozes, C., Grell, S., Schoenberg, S., Yu, L.-C., and Golbus, M. (1987). Application of flow karyotyping in prenatal detection of chromosome aberrations. *Am. J. Hum. Genet.,* **42**, 49–59.
6. Green, D. K. and Fantes, J. A. (1987). Detection of dicentric chromosomes by flow analysis. *Cytometry*, Suppl. **1**, 13.
7. Harris, P., Boyd, E., and Ferguson-Smith, M. A. (1985). Optimizing human chromosome separation for the production of chromosome-specific DNA libraries by flow sorting. *Hum. Genet.* **70**, 59–65.
8. Harris, P., Boyd, E., Young, B. D., and Ferguson-Smith, M. A. (1986). Determination of the DNA content of human chromosomes by flow cytometry. *Cytogenet. Cell Genet.* **41**, 14–21.
9. Harris, P., Cooke, A., Young, B., and Ferguson-Smith, M. (1987). The potential of family flow karyotyping for the detection of chromosome abnormalities. *Hum. Genet.* **76**, 129–133.
10. Langlois, R. G., Yu, L.-C., Gray, J. W., and Carrano, A. V. (1982). Quantitative karyotyping of human chromosomes by dual beam flow cytometry. *Proc. Natl Acad. Sci. USA* **79**, 7876–7880.
11. Latt, S. A., Sahar, E., Eisenhard, M. E., and Joergens, L. A. (1980). Interactions between pairs of DNA-binding dyes: results and implications of chromosome analysis. *Cytometry* **1**, 2–12.
12. Lebo, R., and Bastian, A. (1982). Design and operation of a dual laser chromosome sorter. *Cytometry* **3**, 213–219.
13. Lebo, R., Golbus, M., and Cheung, M.-C. (1986). Detecting abnormal human chromosome aberrations by dual label flow cytogenetics. *Am. J. Med. Genet.* **25**, 519–529.
14. Martin, A., Northrup, H., Ledbetter, D., Trask, B., van den Engh, G., LeBeau, M., Beaudet, A., Gray, J., Sikhon, G., Krossikoff, G., and Booth, C. (1988). Prenatal Detection of 46, XY, rec15, dup 8, inv (5) (p13q33) using DNA analysis, flow cytometry and *in situ* hybridization to supplement classical cytogenetic analysis. *Am. J. Med. Genet.* **31**, 643–654.
15. McKusick, V. (1986). The gene map of *Homo sapiens*: status and prospectus. In "Cold Spring Harbor Symposia on Quantitative Biology LI", pp. 15–27.
16. McKusick, V., and Roderick, T. (1987). Twenty five years in medical genetics and experimental mammalian genetics with particular reference to the gene map by mouse and man. *In "Medical and Experimental Mammalian Genetics: A Perspective"*, (V. McKusick, T. Roderick, J. Mori, and N. Paul, eds) Alan R. Liss, New York.
17. Meyne, J., Bartholdi, M. F., Travis, G., and Cram, L. S. (1984). Counterstaining human chromosomes for flow karyology. *Cytometry* **5**, 580–583.
18. Patterson, M., Schwartz, C., Bell, M., Sauer, S., Hofker, M., Trask, B., van den Engh, G., and Davies, K. E. (1987). Physical mapping studies of the human X chromosome in region Xq27-Xqter. *Genomics* **1**, 297–306.
19. Pjura, P., Grzeskowiak, K., and Dickerson, R. (1987). Binding of Hoechst 33258 to the minor groove of b-DNA. *J. Mol. Biol.* **197**, 257–271.
20. Trask, B. (1985). Studies of chromosomes and nuclei using flow cytometry. PhD Thesis, University of Leiden.
20a. Trask, B., van den Engh, G., and Gray, J. (1989). Inheritance of chromosome heteromorphisms analyzed by high resolution bivariate flow karyotyping. *Am. J. Hum. Genet.* (in press).
21. Trask, B., van den Engh, G., Gray, J., Vanderlaan,

M., and Turner, B. (1984). Immunofluorescent detection of histone 2B on metaphase chromosomes using flow cytometry. *Chromosoma* **90**, 295–302.

21a. Trask, B., van den Engh, G., Mayall, B. and Gray, J. (1989). Chromosome heteromorphism quantified by high resolution bivariate flow karyotyping. *Am. J. Hum. Genet.* (in press).

21b. Trask, B., van den Engh, G., Nussbaum, R., Schwartz, C., and Gray, J. (1989). Quantification of the DNA content of structurally aberrant X-chromosomes and X-chromosome aneuploidy using high resolution bivariate flow karyotyping. *Cytometry* (in press).

22. Trask, B., van den Engh, G., Pinke, D., Mullikin, J., van Dekken, H. G., and Gray, J. W. (1988). Fluorescence *in situ* hybridization to interphase cell nuclei in suspension allows flow cytometric analysis of chromosome content and microscopic analysis of nuclear organization. *Hum. Genet.* **78**, 251–259.

23. Turner, B., and Keohane, A. (1987). Antibody labelling and flow cytometric analysis of metaphase chromosomes reveals two discrete structural forms. *Chromosoma* **95**, 263–270.

24. van den Engh, G., Trask, B., Gray, J., Langlois, R. and Yu, L.-C. (1985). Preparation of chromosome suspensions for flow cytometry. II. Bivariate analysis of human chromosomes. *Cytometry* **6**, 92–100.

25. van den Engh, G.J., Trask, B.J. and Gray, J. W. (1986). The binding kinetics and interaction of DNA fluorochromes used in the analysis of nuclei and chromosomes by flow cytometry. *Histochem.* **84**, 501–508.

26. van den Engh, G., Trask, B., Landsdorp, P. and Gray, J. W. (1988). Improved resolution of flow cytometric measurements of Hoechst/Chromomycin-stained human chromosomes after addition of citrate and sulfite. *Cytometry* **9**, 266–70.

Relation Between Radiation-induced Flow Karyotype Changes Analysed by Fourier Analysis and Chromosome Aberrations

J. A. ATEN

Laboratory for Radiobiology
Academic Medical Center
University of Amsterdam
Amsterdam, The Netherlands

I. INTRODUCTION

Damage induced by ionizing radiation and chemical agents in the chromosomal material of mammalian cells is responsible for the induction of proliferative death, abnormal offspring and malignant transformation. At the first metaphase after the treatment these DNA lesions are expressed as chromosomal changes.

Irradiation in the G_1 phase of the cell cycle induces two main types of chromosome aberrations (9), asymmetrical translocations which lead to chromosomal fragments and symmetrical translocations which do not lead to chromosome fragments but to a redistribution of chromosomal material.

Chromosome aberrations are conventionally detected by microscopic analysis, but the limits of optical resolution are such that only a small proportion of the structural changes are recognized. Usually, only asymmetrical aberrations are scored unless chromosome banding is used.

With flow cytometry (FCM) applied to chromosomes isolated from irradiated cells, measurements can be made of changes in flow karyotypes as a function of the radiation dose.

These changes are presumably associated with a variety of types of damage, including asymmetrical and symmetrical translocations.

Several methods have been applied to derive a quantitative measure for the chromosome damage expressed as changes in the flow karyotype. Carrano et al. (5) measured the effect of irradiation on the flow karyotypes by determining the coefficients of variation of the chromosome peaks and the increase in background due to damaged chromosomes. Otto and Oldiges (8) analysed the increase in width of the peak corresponding to the chromosome with the largest DNA content. Fantes et al. (6) applied this technique to flow karyotypes of human lymphocytes. A measure of chromosome damage was the ratio of the volume of the peak of the largest chromosome and the level of the debris background. Aten et al. (1,2) and Welleweerd et al. (10) employed Fourier analysis to quantify the changes observed in the flow karyotypes.

We shall discuss the application of Fourier analysis to derive a quantitative measure of the chromosome damage expressed in the flow karyotypes from irradiated cultures, in combination with model calculations that simulate the effect of chromosomal changes on flow karyotypes. The results are compared with data on microscopically detected chromosome aberrations.

II. RADIATION-INDUCED CHANGES IN FLOW KARYOTYPES

A. FLOW KARYOTYPES FROM IRRADIATED CELL CULTURES

V79 Chinese hamster fibroblasts were irradiated with various doses of 200 kV X-rays. Cell cultures were irradiated in plateau phase containing more than 90% $G_0 + G_1$ cells, as determined by flow cytometric analysis, with the aim to treat a homogeneous population of cells and to produce only chromosome-type aberrations in metaphase.

Measurements of chromosome damage by FCM and visual analysis were performed by methods described by Aten et al. (1). The cells were accumulated in mitosis by the addition of vinblastine and collected by shake-off. The time of application of vinblastine depended on the duration of the mitotic delay caused by the radiation treatment. It was adjusted for each radiation dose to give an optimal mitotic yield.

Part of each metaphase cell suspension was analyzed by microscope for asymmetrical chromosome aberrations. After hypotonic treatment, these metaphase cells were fixed in methanol:acetic acid (1:3), pipetted onto slides and stained with Giemsa. Well-spread metaphases were scored for dicentrics, tricentrics and centric rings.

To liberate the metaphase chromosomes for FCM analysis the cell suspension were swollen and syringed in the presence of $50 \mu g/ml$ propidium iodide, (added for chromosome stabilization and staining), and 1% v/v Triton X-100, (added to partially dissolve the cell membranes (4)). The chromosome suspensions were analyzed using an ORTHO system 30 flow cytometer equipped with a Spectra Physics 164-05 argon ion laser operated at 250 mW and at 488 nm wavelength.

Unspecific background in the flow karyotypes, not due to chromosome damage, should be avoided in these experiments where random chromosome aberrations are analysed through changes in peak distributions of flow karyotypes. We therefore tested cells from irradiated cultures for membrane damage which, through the chromosome isolation procedure, could cause an increase in the amount of debris in the chromosome suspensions. This check was performed by adding the fluorescent DNA stain propidium iodide to all the cultures just before the disruption of the metaphase cells. Cells with damaged membranes are permeable to propidium iodide and could thus easily be detected. Inspection by fluorescence microscopy of irradiated cultures showed no increase in the frequency of fluorescent cells when compared to control cultures, indicating the absence of radiation-induced membrane

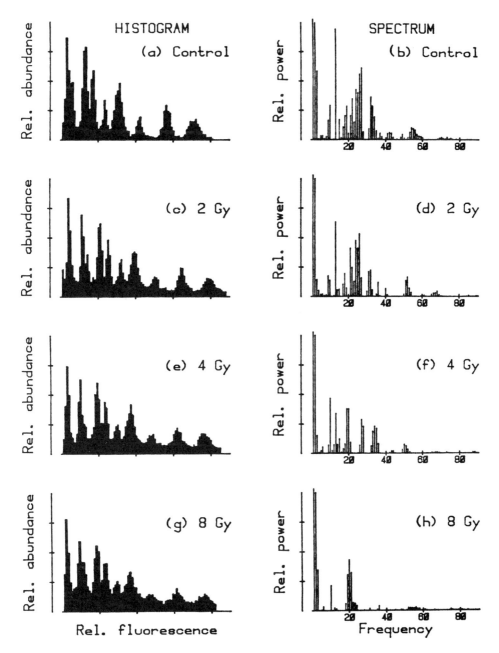

Fig. 1. DNA histogram of V79 Chinese hamster chromosomes, irradiated with 0, 2, 4 and 8 Gy of X-rays. In each of the histograms 2×10^4 chromosomes were analysed [(a),(c),(e),(g)]. The effect of radiation on the Fourier cosine power spectra of the chromosome histogram [(b),(d),(f),(h)].

damage at the time of the chromosome isolation.

Figure 1 shows examples are shown of flow karyotypes from a control culture and from cultures treated with 0, 2, 4 and 8 Gy of X-rays. In the flow karyotypes of chromosome suspensions prepared from irradiated cells, the shapes of the peaks are altered. The heights of the peaks representing intact chromosomes are reduced and their widths are increased. In addition, the areas between the peaks are partly filled with signals from damaged chromosomes.

B.   MODEL CALCULATIONS SIMULATING THE EFFECT OF CHROMOSOMAL CHANGES ON FLOW KARYOTYPES

To interpret the changes observed in flow karyotypes from irradiated cultures, we have developed a computer model to simulate the effects of various chromosomal changes on peak shapes in the histograms. The simulation procedure is illustrated in a qualitative way in Fig. 2. In this schematic example, the effects of the induction of chromosome breaks, dicentrics and deletions on a flow karyotype are represented. It was assumed that all chromosomes in the schematic flow karyotype contained equal amounts of DNA.

In the actual model calculations, we applied this method to flow karyotypes from unirradiated cultures, see Fig. 3(a), with the aim to simulate the effects of radiation-induced chromosomal changes on flow karyotypes. In the calculations it was assumed that the relative probability for damage induced in individual chromosomes is proportional to their DNA content and that it is produced at random along the chromosome.

In Fig. 3(b), a flow karyotype is shown from a culture irradiated with 6 Gy of X-rays. The peaks on the right side of the histogram have been affected more strongly by the irradiation than the peaks on the left side, which appears to be in agreement with the random distribution of radiation damage.

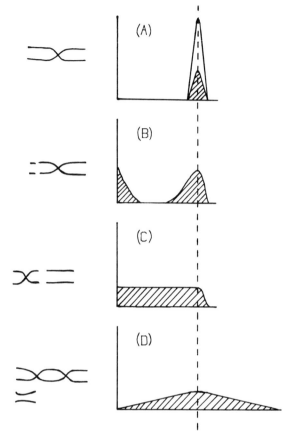

Fig. 2. Schematic representation of the effect of chromosome aberrations on a flow karyotype. (A) Original chromosome peak and chromosomes undamaged after radiation treatment (shaded area). (B) Hypothetical distribution of chromosomes with deletions. (C) Chromosome breaks. (D) Dicentrics and acentric fragments.

A result of the model calculations is shown in Fig. 3(c). In this example the effect is simulated that would have been observed in Fig. 3(a) if 20% of the chromosomes analysed had been damaged by radiation. In the simulated flow karyotype of Fig. 3(c), the same type of changes can be observed as in Fig. 3(b), i.e. broadening and lowering of the peaks and an increase in the background level.

(a) Control

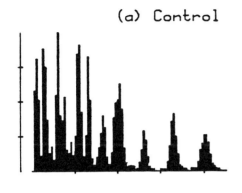

(b) 6 Gy X-Rays

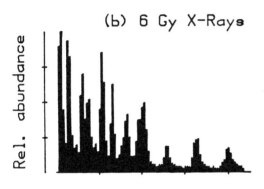

Rel. abundance

(c) Model calc.

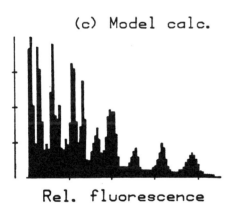

Rel. fluorescence

Fig. 3. Model calculation simulating radiation damage in DNA histograms. (a) Chromosome histogram from control cells. (b) Histogram from cells irradiated with 6 Gy of X-rays. (c) The histogram from part (a), changed by simulating damage in 20% of the chromosomes.

III. APPLICATION OF DISCRETE FOURIER ANALYSIS TO FLOW KARYOTYPES

A. DISCRETE FOURIER REPRESENTATION OF HISTOGRAMS

In the flow cytometric analysis of chromosome suspensions, the DNA content of a large number of chromosomes is determined individually as they pass through the laser beam. The distribution of DNA values is recorded in a histogram where chromosomes with identical DNA content are grouped as peaks. In the flow karyograms of chromosome suspensions prepared from irradiated cells, the shapes of the peaks are altered, c.f. Fig. 1. To quantify the changes observed in the histograms, the distribution of peaks can be analysed by assuming that it represents an oscillating function.

An efficient way to analyse oscillating functions is by Fourier representation. With this technique, graphs or histograms are analysed, not as a composition of independent peaks, but as the superposition of sinusoids with different frequencies.

To explain the procedure a schematic flow karyotype is presented in Fig. 4(a), as a histogram h_n ($n = 0, N - 1$) composed of rapid oscillations representing the chromosome peaks superimposed on a continuum. By using a Fourier representation, all oscillations are extracted from the histogram and sorted in order of increasing frequency. The result is given in the form of a second histogram H_k ($k = 0, N - 1$), Fig. 4(b), representing the power, or the square, of the amplitude of the extracted oscillations, in order of increasing frequency. In our procedure the data were symmetrically extended, i.e. $h_{-n} = h_n$ ($n = 0, N - 1$). In this way, the convergence of the series is increased and the representation reduces to the cosine term only. The relation between the flow karyotype and its Fourier

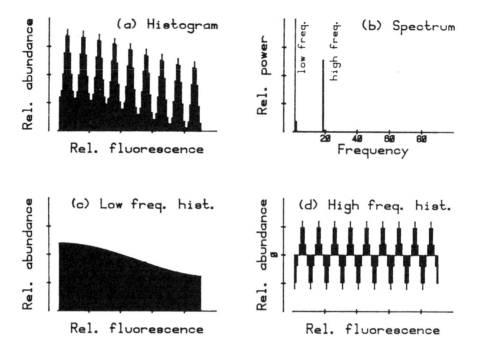

Fig. 4. The application of Fourier analysis to DNA histograms of chromosomes. (a) Schematic chromosome histogram. (b) Power spectrum of the cosine Fourier series. (c) Result of the back transform of the low-frequency part of the Fourier series. (d) Back transform of the high-frequency part of the series.

representation is given by

$$H_k = (2N-1)^{-1}$$

$$\times \left\{ h_0 + 2 \sum_{n=1}^{N-1} h_n \cos(2\pi nk/(2N-1)) \right\}$$

$$\text{for } k = 0, N-1 \quad (1)$$

and

$$h_n = H_0 + 2 \sum_{k=1}^{N-1} H_k \cos(2\pi nk/(2N-1))$$

$$\text{for } n = 0, N-1. \quad (2)$$

The low and high frequencies are located on the left and right sides of the spectrum, respectively. The chromosome peaks are represented by the high-frequency part and the slow variation in the histogram by the low-frequency part, as can be seen from Fig. 4(c) and (d) where the low- and high-frequency parts of the Fourier spectrum have been transformed back into a "slow variation" and into a

"chromosome peak" histogram. Together they reconstitute the original histogram.

As a measure for the size and shape of the chromosome peaks in the flow karyotypes we have taken the ratio of the high-frequency contribution in the power spectrum to the zero-frequency component in the power spectrum, the latter being equal to the square of the surface of the flow karyotype, including the debris background.

The argument for the application of this procedure to measure changes in chromosome peaks is provided by Parseval's theorem for Fourier transforms (9),

$$h_0^2 + 2 \sum_{n=1}^{N-1} h_n^2 = (2N-1)$$

$$\times \left\{ H_0^2 + 2 \sum_{k=1}^{N-1} H_k^2 \right\}, \quad (3)$$

where h_n are the function values in the histogram and H_k the values of the frequency

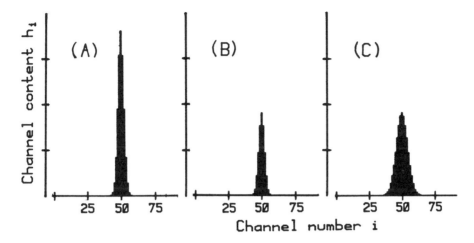

Fig. 5.   Effects of peak reduction and peak broadening in the histogram h_n ($n = 0, N - 1$) on the sum Σh_n^2. (A) Gaussian distribution: $\Sigma h_n^2 = S$. (B) Reduction of histogram volume by 50%: $\Sigma h_n^2 = S/4$. (C) Broadened distribution with volume equal to the distribution in part (A): $\Sigma h_n^2 = S/2$.

components in the spectrum. Through the application of this relation to the high-frequency part of the spectrum, c.f. Fig. 4(b) and (d), it can be shown that increases in the background, reduction in peak size and peak broadening in the flow karyotype all contribute to a relative decrease in volume of the high-frequency part of the power spectrum.

This is illustrated in Fig. 5 where schematic examples are given of a reduced and a broadened peak. When the sum of the squared histogram values is calculated for these examples, a decrease in peak volume by 50% reduces both terms in Eq. (3) by a factor of 4. Doubling of the peak width while keeping the peak volume constant results in a reduction of the volume of the power series by a factor of 2. Moreover, by using only the high-frequency part of the spectrum the result of the calculation is not affected by the background, except through a relative increase in the zero component of the power spectrum.

B.   APPLICATION TO FLOW KARYOTYPES

We have applied the procedure described above to quantify the changes observed in flow karyotypes caused by various amounts of chromosome damage. Figure 6 shows the application of Fourier analysis to a flow karyotype from a control culture. In the Fourier spectrum the low-frequency "slow variation" and the high-frequency "chromosome peaks" distributions are wider than in the example of Fig. 4, but they can still be separated, as is demonstrated in the two bottom histograms. In order to obtain a measure for the size of the chromosome peaks, the volume of the high-frequency part of the power spectrum was divided by the zero component, as discussed before.

In the Fig. 1(a), (c), (e) and (g), chromosome histograms are shown for cells irradiated with 0, 2, 4 and 8 Gy of X-rays. The corresponding Fourier spectra are shown in Fig. 1(b), (d), (f) and (h). The effects of the irradiation on both the flow karyotypes and on the Fourier power spectra can be clearly distinguished. With increasing dose the high-frequency components in the spectra are progressively suppressed. This corresponds to the reduction in the chromosome peaks in the flow karyotypes, which Fourier analysis reveals as a damping of

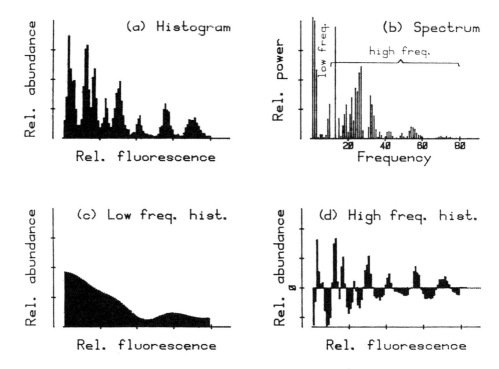

Fig. 6. Application of Fourier analysis to flow karyotypes. (a) Flow karyotype of V79 cells. (b) Power spectrum of the cosine Fourier series. (c) Back transform of the low-frequency part of the Fourier series. (d) Back transform of the high-frequency part of the series.

the high-frequency oscillations in the DNA histograms.

It should be noted, however, that the decrease in the area of the high-frequency part of the power spectrum is not necessarily directly proportional to the number of damaged chromosomes observed in the flow karyotypes. The actual relationship between changes in the flow karyotypes and the number of damaged chromosomes can only be established through comparison with results from the model calculations described above, simulating the effects of chromosome damage on flow karyotypes.

We therefore applied the Fourier analysis procedure to flow karyotypes which were affected by simulated chromosome damage. By comparing the decrease in high-frequency contributions of the power spectra from the experimental flow karyotypes and from the

flow karyotypes containing various amounts of simulated chromosome damage, we obtained quantitative estimates for the frequencies of chromosome damage associated with the flow karyotype changes observed in the radiation experiments.

IV. CHROMOSOME DAMAGE OBSERVED IN FLOW KARYOTYPES IN RELATION TO ASYMMETRICAL CHROMOSOME ABERRATIONS

Quantitative comparison of the radiation effects detected microscopically or by flow cytometry requires that the data be presented in a common format. Figure 7 shows the experimental results expressed as frequencies of lesions characteristic of each of the two end points, as a function of dose. For each data

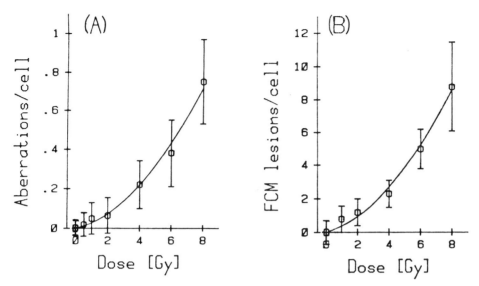

Fig. 7. Dose–response relations for two end points of radiation damage in V79 fibroblasts. The coefficients of the fitted dose–response curves, c.f. Eq. (4), are listed in Table I. (A) Frequency of chromosome aberrations, corrected for the background in control cultures. Dicentrics, tricentrics and centric rings. (B) Frequency of chromosomal lesions determined through the analysis of the flow karyotypes.

point the standard deviation is given, based on values obtained in three or more independent experiments. Linear-quadratic curves have been fitted to the data using a least squares fitting procedure.

The data on chromosome aberrations are shown in Fig. 7(A) fitted by a linear-quadratic dose–response curve,

$$A(D) = a_1 D + a_2 D^2, \qquad (4)$$

where a_1 and a_2 (Table I) are parameters depending on the cell type, on the type of radiation and on the conditions of exposure (3); $A(D)$ is the frequency of aberrant events and D is the radiation dose. The experimental values represent the average induced frequencies of dicentrics, tricentrics and centric rings, obtained from three to five irradiated cultures. At least 50 metaphases were analysed. For doses $\leqslant 2$ Gy, 100 metaphases were analysed for each culture. The aberration frequencies were corrected for the background value obtained from metaphase spreads of the control cultures.

Table I

COEFFICIENTS OF THE DOSE–RESPONSE CURVES, EQ. (4), FOR THE FREQUENCY OF LESIONS PER CELL

	$a_1/(Gy)$	$a_2/(Gy)$
Aberrations	0.019	0.0088
FCM lesions	0.27	0.100

Figure 1 shows flow karyotypes from a control culture and from cultures treated with 2, 4 and 8 Gy of X-rays. The frequencies of chromosomal lesions registered by changes in the flow karyotypes for various dose values were derived using the methods described in the sections on computer analysis of the karyotypes. The results of the analysis are shown in Fig. 7(B) together with a linear-quadratic dose–response curve, c.f. Eq. (4) and Table I.

Figure 7 and Table I show that the frequency of events calculated from the changes in flow karyotype is an order of magnitude larger than the frequency of asymmetrical chromosome translocations determined by microscopic analysis. The large discrepancy in detection efficiency seems to imply that the two end points analysed do not measure the same type of chromosome damage.

It is generally accepted that only a fraction of the primary changes caused by a dose of ionizing radiation results in effects observable as asymmetric chromosomal exchanges. In flow karyotyping analysis, on the other hand, it is to be expected that both asymmetrical and symmetrical translocations are detected and it has been suggested by Welleweerd *et al.* (*10*) that damage in chromosomes, normally not expressed as breaks, might be exposed by the isolation of the metaphase chromosomes from the protective environment of the cell. Changes in the degree of condensation of the chromatin induced by radiation treatment, moreover, might affect the interaction of the intercalating fluorochrome propidium iodide with the DNA. This could result in a change in fluorescence intensity. These different factors could offer an explanation for the relatively large number of radiation-induced events deduced from the flow karyotypes.

ACKNOWLEDGEMENTS

The author wishes to thank Professor Dr G. W. Barendsen for stimulating this research and for his help in preparing the manuscript. Dr J. Grasman and Professor Dr J. Strakee are acknowledged for their comments, Mr. J. Stap for preparing the drawings and Mrs G. van Vuuren-Bakvis for typing the manuscript.

REFERENCES

1. Aten, J. A., Kipp, J. B. A., and Barendsen, G. W. (1980). Flow cytofluorometric determination of damage to chromosomes from X-irradiated Chinese hamster cells. *Acta Microbiol. Scand., Suppl.* 1980, 287–292.
2. Aten, J. A., Kooi, M. W., Bijman, J. Th., Kipp, J. B. A., and Barendsen, G. W. (1984). In "Biological Dosimetry: Cytometric Approaches to Mammalian Systems" (W. G. Eisert, and M. L. Mendelsohn, eds), pp. 51–59. Springer-Verlag, Berlin.
3. Barendsen, G. W. (1979). Influence of radiation quality on the effectiveness of small doses for induction of reproductive death and chromosome aberrations in mammalian cells. *Int. J. Radiat. Biol.* **36**, 49–63.
4. Buys, C. H. C. M., Koerts, T., and Aten, J. A. (1982). Well-identifiable human chromosomes isolated from mitotic fibroblasts by a new method. *Hum. Genet.* **61**, 157–159.
5. Carrano, A. V., Gray, J. W., and Van Dilla, M. A. (1978). In: "Mutagen-induced Chromosome Damage in Man" (H. J. Evans, and D. C. Lloyd, eds), pp. 326–338. Edinburgh University Press, Edinburgh.
6. Fantes, J. A., Green, D. K., Elder, J. K., Malloy, P., and Evans, H. J. (1983). Detecting radiation damage to human chromosomes by flow cytometry. *Mutat. Res.* **119**, 161–168.
7. Lanzos, C. (1966). "Discourse on Fourier Series," pp. 51–54. Oliver & Boyd, Edinburgh.
8. Otto, F. J., and Oldiges, H. (1980). Flow cytogenetic studies in chromosomes and whole cells for the detection of clastogenic effects. *Cytometry* **1**, 13–20.
9. Savage, J. R. K. (1975). Classification and relationships of induced chromosomal structural changes. *J. Med. Genet.* **12**, 103–122.
10. Welleweerd, J., Wilder, M. E., Carpenter, S. G., and Raju, M. R. (1984). Flow cytometric determination of radiation-induced chromosome damage and its correlation with cell survival. *Radiat. Res.* **99**, 44–51.

Detection of Randomly Occurring Aberrant Chromosomes as a Measure of Genetic Change

DARYLL K. GREEN, JUDITH A. FANTES and JOHN H. EVANS

MRC Human Genetics Unit
Western General Hospital
Crewe Road
Edinburgh EH4 2XU, UK

I. UNSTABLE CHROMOSOME ABERRATION SCORING

The accepted method of measuring clastogenic effects on human peripheral blood cells is to count the number of unstable damaged chromosomes per cell (7). The unstable chromosome aberrations which can be observed include acentric fragments, dicentric chromosomes and centric ring chromosomes and these are easily recognized and counted on a microscope slide preparation. A similar, dose-related quantity of stable chromosome aberrations also occur and these include reciprocal and nonreciprocal translocations or inversions, for example. Such stable aberrations are not usually counted in routine investigations of suspected clastogen exposure because their detection is difficult and time consuming. The close relationship between the clastogen dose and the number of unstable chromosome aberrations induced has been very comprehensively described in the literature (3,8). In order to reach the stage of counting damaged chromosomes, peripheral blood lymphocytes are cultured for 48 h, which is the average time for cells to reach their first division, and the metaphase cells are harvested, spread on a microscope slide and stained with an absorbtion stain such as orcein or Giemsa. Skilled technicians can maintain a scoring rate for metaphase spreads with low levels of damage of around 50 cells/h, approximately 1 cell/min. A single high-confidence measurement of a low dose of mutagen, 0.15 Gy of X-radiation for example, requires at least 500 cells to be scored and so would amount to a day's work involving high levels of skill and a fair degree of tedium.

Had the metaphase chromosomes been released into suspension, stained with a fluorescent dye, such as ethidium bromide or Hoechst 33258, and passed through a flow cytometer to produce a "flow karyotype", (*4,11*) chromosomes from 500 cells could be scored in 0.5 min. For this reason a great deal of research effort has been spent attempting to extract reliable dose response information from flow karyotypes. The analysis and sorting of chromosomes by flow methods was heralded as a new approach which could remove much of the tedium associated with scoring damaged chromosomes (*18*).

At a radiation dose of 0.5 Gy, which is considerably greater than safety standards allow for occupational exposure of individuals, the average frequencies of unstable aberrations is around 0.2/cell or 4/1000 chromosomes. These aberrations mainly appear as acentric fragments or dicentric chromosomes. Aberrant chromosomes are randomly sized, covering the whole range of normal chromosome sizes, fragments tending to be smaller than dicentrics for example. Assuming that a flow karyotype covers 512 channels of fluorescence intensity and that aberrant chromosomes from 0.5 Gy radiation are spread evenly throughout the intensity range, a sample of 10^5 chromosomes would include approximately one abnormal chromosome per channel.

II. SINGLE-FLUOROCHROME FLOW KARYOTYPES

A. Observed Clastogenic Effects

A flow karyotype produced from mitotic chromosomes from cultured peripheral blood lymphocytes of a normal individual (female) is shown in Fig. 1. There are two main components to the distribution of fluorescence. The pattern of peaks and troughs arise out of the homogeneous groups of unbroken metaphase chromosomes. Chromosome groups associated with each peak have been labeled following identification by flow sorting. Individual human donors may be identified by this unique invariant pattern of their chromosome profile (*12*). Underlying the peaks there is a smooth distribution of background fluorescent objects making up the second component of the picture. Broken chromosomes, nuclear and cytoplasmic debris, and unstable heterogenous aberrations give rise to the background, the area of which is sensitive to the source of cells, the culture conditions and the chromosome preparation technique. The background distribution in Fig. 1, shown as the lower of the two smooth curves, was somewhat higher than average but within the normal range. A magnified (1000 ×) distribution of dicentric chromosomes induced with 0.5 Gy X-irradiation is also shown in Fig. 1. It is clear that the dicentrics would have little or no effect on the background area at this level of chromosome damage. Only at much higher doses would the number of dicentrics approach the magnitude of background debris and in those cases damage could be quickly recognized under the microscope. In spite of these practical arguments, which seemingly predict a negligible dose response at relatively low, nonlethal doses, many researchers observed that the level of background was increased when cells had been exposed to clastogenic agents prior to culture (*14–6,9*). Much of the work was reported from experiments using Chinese hamster cell lines. Clastogens have other degrading effects on flow karyotypes, namely the broadening of chromosome peaks. Quantitative studies of these effects have been reported by Otto and Oldiges (*15*), who related the coefficient of variation (CV) of a large chromosome peak to dose and Aten *et al.* (*2*); Chapter 9 who related dose to the power of a flow karyotype Fourier spectrum at frequencies close to that produced by chromosome peaks and troughs. Results of studies involving Chinese hamster or human chromosomes have not as yet produced high-confidence dose responses for less than the equivalent of 1.0 Gy X-irradiation.

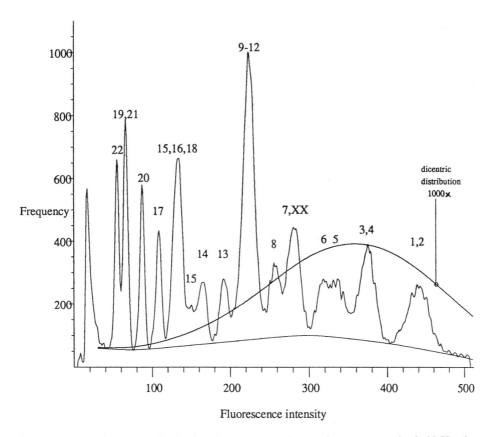

Fig. 1. Fluorescence intensity distribution of normal female human chromosomes stained with Hoechst 33258. The distribution of background events (lower of the two smooth curves under peaks) was constructed from a tensioned spline fit to a number of fixed points chosen by hand. The theoretical distribution of dicentric chromosomes produced by 50 rads of X-irradiation is shown, magnified 1000 times.

B. SOURCE OF BACKGROUND

In PHA-stimulated peripheral blood lymphocyte cultures harvested approximately 52 h after exposure to X-rays, increasing the dose of radiation usually leads to a decrease in mitotic index because of mitotic delay and an increase in the number of dead cells. We have found that chromosome samples prepared from cultures with a low mitotic index will give flow karyotypes with poor resolution of peaks and high background; this is probably due to nuclear degeneration and an increase in chromosome stickiness, associated with an increase in debris. This may contribute to the increased

background found with irradiated peripheral blood lymphocytes in culture.

However, similar results have been reported by Aten et al. (2) and Welleweerd et al. (19) after exposing Chinese hamster cells to X-rays or α-particles and using mitotic shake-off cells as a source of chromosomes. With this system the effects described above should be less important, as most of the cells used to prepare the chromosome sample will be in mitosis, and it is therefore necessary to look for further explanations. Conventional slide-based aberration scoring is usually confined to scoring dicentrics, centric rings and fragments. An equivalent number of symmetrical aberrations, such as translocations, inversions and deletions, are also produced but rarely scored. The limits of optical resolution are such that only a small proportion of structural changes are actually detected (*16*); the undetected changes may contribute to the background of a flow histogram. An additional factor, suggested by Welleweerd *et al.* (*19*), is that areas of damage in the chromosome which might not be expressed as breaks in a metaphase spread might be broken in the flow cytometry preparation procedure.

C. DOSE RESPONSE

Several approaches to quantify a dose response using the degradation of a univariate flow karyotype have been mentioned. The level of fluorescent background for human chromosomes, harvested from peripheral blood cells, is not only dose dependent but also dependent on the donor, the blood sample storage time, the culture medium and other factors. Attempts have been made at this laboratory to generate for each experimentally irradiated sample, a control chromosome suspension. This was similar in every way to the primary suspension except for the culture time, which was made sufficiently long to ensure an order of magnitude reduction in unstable chromosome aberrations and hopefully a significant reduction in other background objects related to clastogen dose. The control suspension was used to generate a flow karyotype for direct comparison with the flow karyotype produced from the normal 48 h culture. T-cell growth factor was used to extend the control culture to 120 h leading to six cell divisions in a proportion of the cells thereby culturing-out of unstable chromosome damage. Figure 2 shows the comparison between short-term (48 h) and long-term (120 h) flow karyotypes for a given experiment.

Several experiments similar to the one illustrated in Fig. 2 produced sufficient data to estimate a dose response. Each flow karyotype was first normalized along the intensity axis

such that the chromosomes 9–12 peak was positioned at channel 225 and the chromosomes 1 and 2 peak was positioned at channel 440 and normalized in total area between channels 40 and 511 to 100 000 units. An operator, using an interactive graphics screen, defined points corresponding to the deepest part of the initial valley, the valley between chromosomes 19 and 21 and chromosome 20, chromosomes 20 and 17, chromosomes 13 and 14, chromosomes 3 and 4 and chromosomes 1 and 2, and the trailing background at channel 500. A tensioned spline fit was drawn automatically through the defined points and the area beneath, between channels 40 and 511, was deemed the background area. At each radiation dose the short-term background:long-term background ratio was calculated. Figure 3(a) shows the mean values and standard deviations of this ratio for doses up to 1.0 Gy. The average CVs for the chromosomes 1 and 2 and chromosomes 3 and 4 peaks for each flow karyotype shown in Fig. 2 are plotted in Fig. 3(b). It would appear from both these results that the expected trend of increasing background and increasing CVs with increasing X-ray dose, though present, has little significance below 1.0 Gy.

Positive dose relationships have been reported by Otto and Oldiges (*15*) for the effect of clastogens on the CV of large Chinese hamster chromosomes and by Aten *et al.* (*2*) for the degrading effect of clastogens on the flow karyotype Fourier power spectrum for Chinese hamster chromosomes. These results are reproduced here in Fig. 4(a) and (b). Chinese hamster cell lines were used to generate chromosomes for these experiments and, clearly, the customary high yield of metaphase chromosomes harvested from cell cultures of this type gives rise to a reliable test system for dosimetry studies. Human peripheral blood T-cells, which must be artificially induced to divide, often deliver a poor and variable harvest of chromosomes (mitotic index < 30%) by comparison with immortalized cell lines, of human or Chinese hamster origin. Although the ultimate aim of dosimetry research is to quickly and reliably measure doses to man, the clear Chinese hamster results shown in Fig. 4 cannot

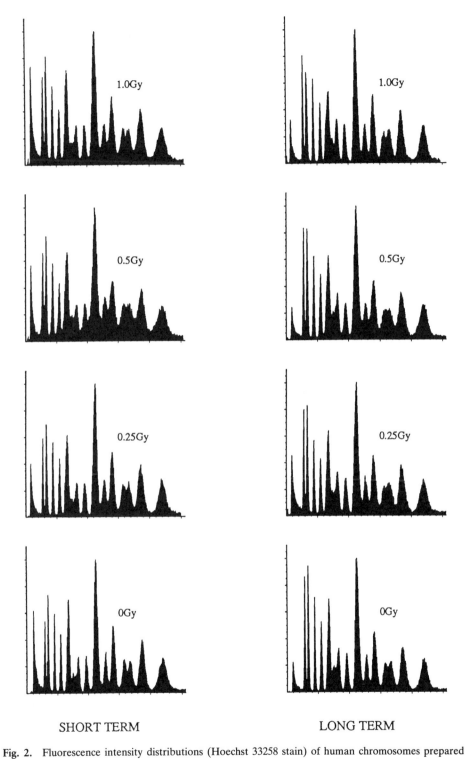

SHORT TERM   LONG TERM

Fig. 2. Fluorescence intensity distributions (Hoechst 33258 stain) of human chromosomes prepared from X-irradiated peripheral blood lymphocytes. The distributions in the short-term column were obtained from 48 h cultures and those in the long-term column were obtained from 120 h cultures. T-cell growth factor was added to the long-term cultures to prolong the number of cell divisions.

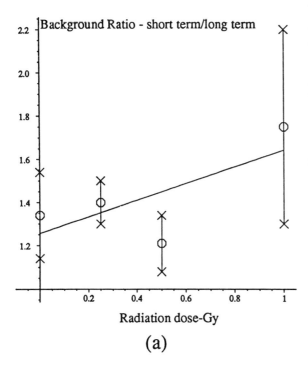

(a)

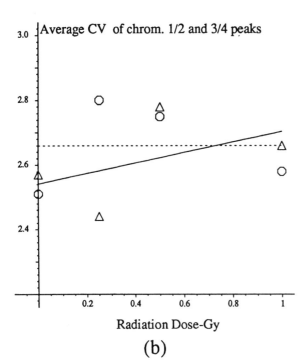

(b)

be unambiguously extrapolated to human chromosomes prepared from peripheral blood cells.

III. SCORING DAMAGED CHROMOSOMES FROM SORTED FRACTIONS

Another approach to extracting dose response information from univariate flow karyotypes was to use the ability of flow cytometers to sort chromosomes in preset regions of the fluorescence intensity distribution into separate fractions (4). It can be seen in Fig. 1 that in the region of the chromosomes 3 and 4 peak there is also a peak in the dicentric chromosome distribution, and hence it might be expected that a sorted fraction of chromosomes from this part of the flow karyotype would be enriched in dicentric chromosomes and would therefore produce a microscope slide preparation which would be less tedious to score than normal metaphase spreads. Experimental data from this approach to aberration scoring is sparse. The results of an experiment performed at this laboratory are shown in Fig. 5. Here, human peripheral blood lymphocytes were irradiated with 2.5 Gy X-radiation and prepared for flow cytometry.

Fig. 3. The effects of radiation dose on the area of background and the coefficient of variation (CV) of each peak in a univariate flow karyotype of human chromosomes. In part (a) the results of several experiments are combined, where the average ratio of background areas from short-term and long-term cultures at each radiation level are shown as open circles. Standard deviations above and below each average value are shown as crosses and a solid line shows the least squares linear fit to the circled points. In part (b) the average CVs of the chromosomes 1 and 2 and chromosomes 3 and 4 peaks are shown for the experiment illustrated in Fig. 2. CVs were in all cases computed using a least squares Gaussian fitting algorithm. A dotted line is drawn through the long-term culture points (open circles) and a solid line is drawn through the short-term culture points (open triangles). A least squares linear fit was used in both cases.

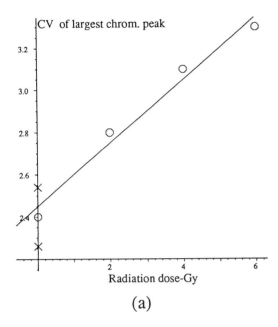

(a)

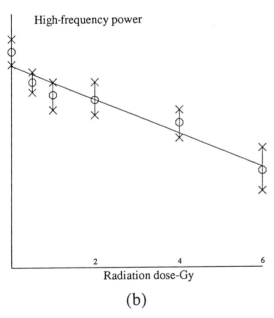

(b)

Fig. 4. The effects of large radiation doses on mammalian chromosome flow karyotypes. (a) Some results reported by Otto and Oldiges (*15*), demonstrating the effects of X-irradiation on the CV of the large chromosomes of Chinese hamster cells. (b) The broadening of Chinese hamster chromosome peaks through the decreasing power of the high-frequency Fourier spectrum; results from Aten *et al.* (*2*).

Objects in four regions of the intensity distribution were sorted and spun down onto microscope slides. These slides, together with a slide made from the whole chromosome suspension and a traditional metaphase spread preparation, were scored for dicentric chromosomes. Each sample from the flow cytometer involved depositing 60 000 sorted fluorescent objects into 0.25 ml of polyamine buffer, fixing with formaldehyde and spinning onto a microscope slide. This protocol has been shown to produce a microscope slide sample on which the sorted chromosomes are identifiable from their fluorescent band structure and are sufficiently elongated to reveal the centromeres. However, 60 000 sorted objects do not lead to the same number of objects deposited on a slide, since chromosomes adhere to every surface throughout the slide preparation. The cytogeneticist sees a sparsely populated slide, where it is difficult to discriminate between dicentric chromosomes and chromosome aggregates, since there is no dense background of normal randomly grouped chromosomes. Sorted fractions from the valleys of the fluorescence distribution were the most difficult to score because of the near absence of any normal chromosomes. Only one of the sorted fractions, the one from the chromosomes 1 and 2 peak, showed a significant enrichment of dicentric chromosomes.

Although worth mentioning, sorting and scoring damaged chromosomes can be more tedious and time consuming than scoring by traditional methods, and hence this approach has been quickly abandoned at many laboratories.

IV. DIRECT DETECTION OF DICENTRIC CHROMOSOMES

Another promising approach toward the detection of aberrant chromosomes in flow relies on the recognition and enumeration of chromosome centromeres. There are two

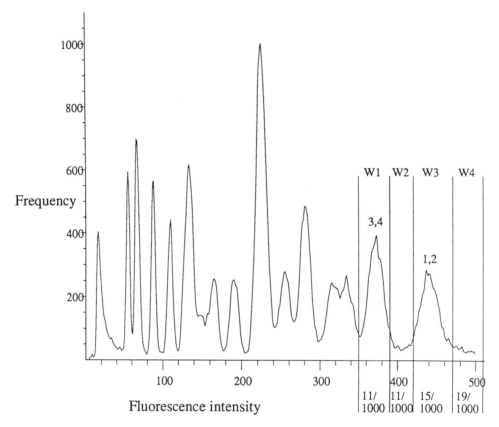

Fig. 5. Vertical lines are drawn through the flow karyotype of a male donor, whose blood lymphocytes were irradiated with 2.5 Gy X-rays (*in vitro*) prior to culture, to mark the placing of four sorting windows W1–4. The ratio of dicentric chromosomes per 1000 normal chromosomes in each sorted fraction is shown at the bottom of each window. The corresponding count of dicentric chromosomes for a parallel metaphase spread preparation from the same source was 12/1000 chromosomes.

approaches: slit scanning DNA-stained chromosomes and counting the number of centromeric dips in the fluorescence profile, or staining centromeres and measuring the total amount of centromere-linked fluorescence. Slit scanning is discussed in Chapter 11. Here, the specific labeling of chromosome centromeres (sometimes called kinetochores) is discussed. Methods have been developed for immuno-fluorescently staining isolated chromosomes (*17*). In this study, chromosomes were incubated first with a mouse monoclonal antibody to the histone chromosomal protein H2B. The bound anti-H2B was visualized with a second-stage fluorochrome-labeled antimouse antibody. Here the uniform distribution of histone along

the chromosome produced uniformly fluorescing chromosomes. We have adapted these methods and those of Gooderham and Jeppeson (*10*) to stain the centromeres of isolated chromosomes. CREST serum from scleroderma patients binds to the kineto-chores of human and several other mammalian chromosomes (*13,14*). The centromeres are then visualized following incubation with fluorochrome-labeled antihuman antibody. Human metaphase spreads stained with Hoechst 33258 are shown in Fig. 6 (upper plate of each pair of photographs). The same metaphase spreads were treated with CREST serum and then FITC-labeled antihuman antibody; the FITC labeling is shown in the

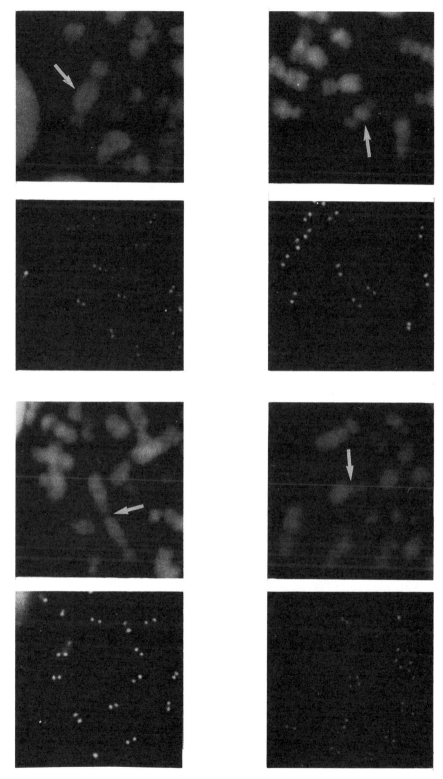

Fig. 6. Counterstained human metaphase chromosomes. Four pairs of photomicrographs show, in the upper plate of each pair, chromosomes stained with Hoechst 33258, a DNA-specific stain and, in the lower plate, the corresponding chromosomes stained with FITC at the centromeres by indirect immunofluorescence using the scleroderma antibody. Arrows indicate a dicentric chromosome in each pair.

lower plate of each pair of photographs. Chromosomes have clearly marked kinetochore regions, and particular attention is drawn to the dicentric chromosomes (arrowed) which show two sets of active kinetochores.

When the above staining technique can be accomplished satisfactorily for chromosomes in suspension the recognition of dicentric chromosomes or accentric fragments in flow cytometry should become a matter of: (i) detecting the small amount of kinetochore-bound fluorochrome at each centromere, (ii) establishing confidence in the measurement of zero, one or two increments of centromere fluorescence and (iii) measuring the percentage of zero-centromere fluorescence objects which are genuine chromosome fragments and the percentage of 2 units of centromere fluorescence objects which are genuine dicentrics. In an actual flow experiment it will be useful to have the whole chromosomes counterstained either with a fluorochrome which can be excited with the same focused laser beam as the indirectly bound fluorochrome, but having a different emission spectrum (FITC plus ethidium bromide, for example), or with a fluorochrome which is excited with a second focused laser beam tuned to a different wavelength. The dual-beam approach has the advantage that because the beams are physically separated, energy transfer between the stains is kept to a minimum and the stain emission spectra can overlap since each colour is detected quite independently. Hoechst 33258 for DNA staining and FITC for antibody staining combined with an ultraviolet (350 nm) and a visible (488 nm) laser beam provide a good working combination since Hoechst 33258 is recognized for low levels of background staining and FITC is readily available conjugated with many useful antibodies.

V. FUTURE PROSPECTS

It is quite clear that flow cytogenetics has not produced a viable, fast method for scoring unstable aberrant chromosomes resulting from clastogenic effects; however, hidden some-where within the technology is the prospect of a dosimeter capable of scanning many thousands of chromosomes in a few minutes. For this reason research will probably continue until a solution is found.

Three main approaches have been discussed here, the analysis of a flow karyotype, the counting of aberrant chromosomes in sorted fractions and the prospect of scoring staining chromosome centromeres in flow. The quality of the results seen in the first approach depends on both the source of lymphocytes and the exact sample preparation conditions. Until such time as human peripheral blood lymphocytes from different donors become uniformly similar in growth performance, and preparation standards are exactly reproducible, the effects of low doses of clastogen will be impossible to identify. The second approach shows no promise of being either reliable or faster than normal scoring methods. Concentrating attention on the scoring of actual damaged chromosomes provides the greatest promise of a flow dosimeter and we are left with the technical problems of guaranteeing bright centromere fluorochromes and of being assured that chromosomes are nearly perfectly dispersed in suspension such that each event observed in flow is a single piece of intact or damaged DNA.

REFERENCES

1. Aten, J. A., Kipp, J. B. A., and Barensden, G. W. (1980). Flow cytometric determination of damage to chromosomes from X-irradiated chinese hamster cells. *In* "Flow Cytometry IV," (O. Laerum, T. Lindmo and E. Thorud, eds), pp. 287–292. Univeritetsjerlagat, Oslo.
2. Aten, J. A., Kooi, M. W., Bijman, J. Th., Kipp, J. B. A., Barensden, G. W. (1984). Flow cytometric analysis of chromosome damage after irradiation: Relation to chromosome aberrations and cell survival. *In* "Biological Dosimetry," (W. G. Eisert and M. L. Mendelsohn, eds), pp. 51–60. Springer-Verlag, Berlin.
3. Bauchinger, M. (1984). Cytogenetic effects in human lymphocytes as a dosimetry system. *In* "Biological Dosimetry," (W. G. Eisert and M. L. Mendelsohn, eds), pp. 15–24. Springer-Verlag, Berlin.

4. Carrano, A. V., Gray, J. W. and Van Dilla, M. A. (1978). Flow cytogenetics: progress towards chromosomal aberration detection. *In* "Mutagen Induced Chromosome Damage in Man," (H. J. Evans and D. C. Lloyd, eds), pp. 326–338. University Press, Edinburgh.

5. Cram, L. S., Bartholdi, M. F., Ray, F. A., Travis, G. L. and Kraemer, P. M. (1983). Spontaneous neoplastic evolution of chinese hamster cells in culture: multistep progression of karyotype. *Cancer Res.* **43**, 4828–4837.

6. Cremer, C., Cremer, T., and Gray, J. W. (1982). Induction of chromosome damage by ultraviolet light and caffeine: correlation of cytogenetic evaluation and flow karyotype. *Cytometry* **2**, 287–290.

7. Evans, H. J. (1984). Human peripheral blood lymphocytes for the analysis of chromosome aberrations in mutagen tests. *In* "Handbook of Mutagenicity Test Procedures," (Kilbey *et al.*, eds). Elsevier, Amsterdam.

8. Evans, H. J., and Lloyd, D. C. (1978), "Mutagen Induced Chromosome Damage in Man." Edinburgh University Press, Edinburgh.

9. Fantes, Judith A., Green, D. K., Elder, J. K., Malloy, Patricia and Evans, H. J. (1983). Detecting radiation damage to human chromosomes by flow cytometry. *Mutat. Res.* **119**, 161–168.

10. Gooderham, K., and Jeppeson, P. (1983). Chinese hamster metaphase chromosomes isolated under physiological conditions. *Exp. Cell Res.* **144**, 1–4.

11. Green, D. K., Fantes, J. A., and Spowart, G. (1984). Radiation Dosimetry using the methods of flow cytogenetics. *In* "Biological Dosimetry," (W. G. Eisert and M. L. Mendelsohn, eds), pp. 67–76. Springer-Verlag, Berlin.

12. Green, D. K., Fantes, Judith A., Buckton, Karin E., Elder, J. K., Malloy, Patricia, and Evans, H. J. (1984). Karyotyping and identification of human chromosome polymorphisms by single fluorochrome flow cytometry. *Hum. Genet.* **66**, 143–146.

13. Moroi, Y., Peebles, Carol, Fritzler, M. J., Steigerwald, J., and Tan, Eng M. (1980). Autoantibody to centromere (kinetochore) in scleroderma sera. *Proc. Natl Acad. Sci. USA* **77**, 1627–1631.

14. Moroi, Y., Hartman, A. L., Nakane, P. K. and Tan, Eng M. (1981). Distribution of kinetochore (centromere) antigen in mammalian cell nuclei. *J. Cell Biol.* **90**, 254–259.

15. Otto, F. J., and Oldiges, H. (1980). Flow cytogenetic studies in chromosomes and whole cells for the detection of clastogenic effects. *Cytometry* **1**, 13–17.

16. Savage, J. R. K. (1976). Annotation: classification and relationships of induced chromosomal structural changes. *J. Med. Genet.* **13**, 103–122.

17. Trask, B., Van den Engh, G., Gray, J. W., Vanderlaan, M., and Turner, B. (1984). Immunofluorescent detection of histone 2B on metaphase chromosomes using flow cytometry. *Chromosoma* **90**, 295–302.

18. Van Dilla, M. A., Carrano, A. V., and Gray, J. W. (1976). Flow karyotyping: current status and potential development. *In* "Automation of Cytogenetics, Asilomar Workshop," (M. L. Mendelsohn, ed.), pp. 145–164. NTIS, Springfield.

19. Welleweerd, J., Wilder, M. E., Carpenter, S. G., and Raju, M. R. (1984). Flow cytometric determination of radiation induced chromosome damage and its correlation with cell survival. *Radiat. Res.* **99**, 44–51.

Chromosome Classification by Scanning Flow Cytometry

JOE LUCAS

Biomedical Sciences Division
Lawrence Livermore National Laboratory
Livermore, CA 94550, USA

MARTY F. BARTHOLDI

Los Alamos National Laboratory, US Department of Energy
Los Alamos, NM 87545, USA

I. INTRODUCTION

Flow cytometry and sorting have been used for classification and purification of chromosomes of a single type since the report in 1975 that chromosomes could be isolated from cells, stained with DNA-specific dyes, and processed flow cytometrically (*4,7–9,14,15, 26,27*). For flow cytometric analysis of chromosomes, the chromosomes are isolated from mitotic cells, stained with one or more fluorescent dyes, and classified one by one according to their dye contents(s). Before 1979, chromosomes were stained with one dye and the flow karyotypes were univariate. In those studies, the fluorescence from each chromosome was approximately proportional to its dye content so that the chromosomes were classified according to their DNA contents. However, some chromosome types have nearly identical DNA contents and cannot be distinguished in univariate flow karyotypes. Beginning about 1979, Gray *et al.* (9) started using pairs of DNA-specific dyes to generate bivariate flow karyotypes. In this approach, chromosomes usually are classified according to their DNA content and DNA base composition. Most human chromosomes are better resolved in bivariate flow karyotypes than in univariate flow karyotypes. Still, not all human chromosomes can be distinguished. For example, chromosomes 9–12 have roughly the same DNA content and DNA base composition, and thus generate a single peak in the bivariate flow karyotype. In addition, bivariate flow karyotypes provide very limited information about the frequency of randomly occurring aberrant chromosomes, such as might result from radiation exposure.

In 1975, Van Dilla *et al.* (28) proposed that chromosome shape parameters may be measured using scanning flow cytometry to improve chromosome discrimination and allow detection of distinctively shaped aberrant chromosomes such as dicentrics. During scanning flow cytometry, stained chromosomes are forced to flow with their long axes parallel to the direction of flow through a laser beam. Chromosome scanning may be accomplished using image plane scanning, object plane scanning, or a combination of the two. In object plane scanning [Fig. 1(a)], the exciting laser beam is focused so that it is thin compared to the length of the object being scanned. The time-dependent variation in fluorescence intensity emitted as each chromosome passes through the beam is recorded as a measure of its shape. Chromosomes also can be scanned in the image plane as illustrated in Fig. 1(b). In this approach, fluorescence collected from the chromosome laser intersection point is imaged onto a mask with a thin slit perpendicular to the axis of flow. Only the light from a small section of the chromosome passes through the slit. Again, the time-dependent variation in the intensity of fluorescence reaching detection is recorded as a measure of the chromosome profile. The centromere location can be easily derived from fluorescence profiles measured for chromosomes stained with DNA-specific dyes since the chromosome centromere, with its reduced DNA content, produces a characteristic dip in the fluorescence profile. Thus, chromosome centromeric indices (CI = fluorescence from the long arm of the chromosome divided by the total chromosome fluorescence) can be calculated from the relative location of the dip and used for chromosome discrimination (*11*). Dicentric chromosomes can be recognized as producing profiles with two centromeric dips (*21*).

In the past few years the resolution of slit-scan instruments has been improved to about $1.0–1.5\,\mu m$ so that scanning flow cytometry can now be applied to some of the smaller human chromosomes. Scanning flow cytometry has been used successfully for chromosome classification (*10,11,16,17*), for detection of homogenously occurring structural chromosome aberrations (J. Lucas *et al.*, unpublished work) and for measurement of the frequency of radiation-induced dicentric chromosomes as an indication of genetic damage. In addition, scanning flow cytometry is beginning to be used as the basis for improved chromosome sorting (M. Bartholdi, unpublished work). The effectiveness of flow cytometric techniques for chromosome analysis and sorting depends on the discrimination of the desired chromosome type during sorting. The techniques commonly

Fig. 1. Slit-scan flow cytometry. (a) During object plane scanning, chromosomes stained with a DNA-specific dye are forced to flow lengthwise through a thin ($1.0–1.5\,\mu m$ thick) laser beam. The resulting time-varying fluorescence intensity is recorded as a measure of the distribution of fluorescent dye along the length of the chromosome. Recorded profiles typically dip when the centromere with its reduced DNA content crosses the laser beam. (b) Image plane scanning is accomplished by measuring only light intensity that passes through a narrow slit placed in the image plane of the fluorescence collection lens.

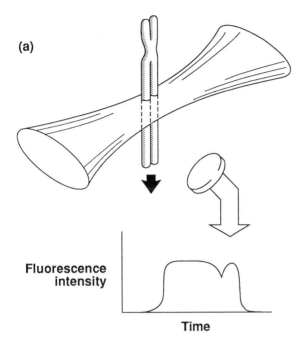

(a)

Fluorescence
intensity

Time

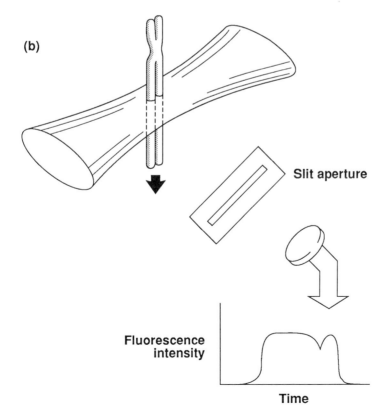

(b)

Slit aperture

Fluorescence
intensity

Time

uscd for resolving chromosomes by flow cytometry, however, typically rely only on relative DNA content and base composition. Scanning flow sorting promises to increase the purity of sorting by increasing the discrimination between chromosomes of similar DNA content and DNA base composition and by allowing discrimination against DNA fragments. This chapter is designed to introduce the reader to techniques for measuring chromosome shape and to prepare chromosomes for shape analysis. The utility of these techniques in flow karyotyping is also discussed.

II. CHROMOSOME ISOLATION AND STAINING FOR SLIT SCANNING

Estimation of chromosome shape information during scanning cytometry requires that chromosomal shape be preserved during chromosome isolation and preparation for scanning. The accuracy of shape analysis can be increased, however, if the chromosomes can be enlarged or decondensed so that the chromosome features to be measured are large compared to the spatial resolution of the scanning system. Several procedures to accomplish this have now been developed (8,16,17). Chromosomes prepared for stretching were isolated by the magnesium sulfate procedure detailed in Chapter 4. The stretching procedure is simple. To enlarge the chromosomes, the isolated chromosomes are first treated with RNase, then stained with propidium iodide (0.05 mg/ml) and placed on ice. After 30 min, the ionic strength of the chromosome suspension is increased from 0.075 to $0.5 M$, which causes the chromosomes to swell to about twice their previous size. Figure 2 shows that chromosome length as measured by scanning flow cytometry may be increased two-fold by increasing the ionic strength of salt in the isolation buffer from 0.075 to $0.5 M$. This has led to a more accurate estimation of the centromere location, particularly for the smaller human chromosomes.

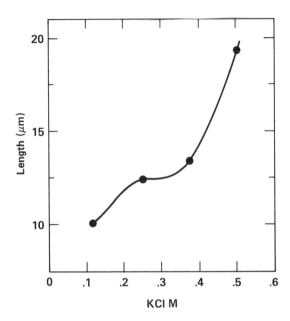

Fig. 2. The average length of human chromosome 1 *vs* KCl concentration. The chromosomes expand with increasing KCl concentration.

Other chromosome features, such as bands, also may be made sufficiently large and distinct that they can be resolved. Preliminary studies suggest that these features can be enlarged by altering the ionic strength and composition of the isolation buffer. Ohnuki's buffer, for example, seems to alter the tight packing of condensed metaphase chromosomes, leaving them in an extended, partially uncoiled state. This buffer is an equimolar solution of 55 nM KCl, $NaNO_3$ and $NaC_2H_3O_2$ (24). Figure 3 shows that chromosomes isolated using polyamine isolation buffer and swollen in Ohnuki's buffer produce well-resolved peaks in the flow karyotypes after chromomycin A3 staining. Scans of these chromosomes show that a large fraction exhibit structural features. Figure 4, for example, shows profiles of chromosomes isolated in Ohnuki's buffer that exhibit a number of peaks and valleys with a less well-differentiated centromere. So far, conditions have not been developed so that the profile shapes are highly reproducible for

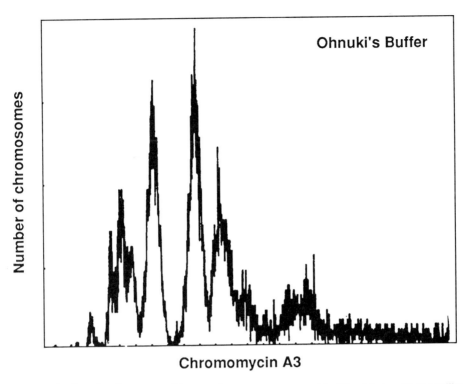

Fig. 3. Univariate flow karyotype of human chromosomes prepared in Ohnuki's hypotonic swelling buffer. The chromosomes were stained with chromomycin A3.

chromosomes of one type. However, the spacing between the peaks is not random, suggesting that the peaks may be representative of an underlying structural unit. In another approach, agents can be introduced during cell culture to partially inhibit chromosome contraction and to induce fluorescence-staining differences with the dye chromomycin A3.

III. INSTRUMENTATION

Two high-resolution scanning cytometers have been reported to date that are suitable for chromosome shape analysis. One, at the Lawrence Livermore National Laboratory (LLNL) is an object plane scanner. The other at the Los Alamos National Laboratory (LANL) is a combined image–object plane scanner. The details of these systems are presented below.

A. OBJECT PLANE SCANNING

During object plane scanning, chromosomes are scanned as they flow lengthwise through a thin excitation laser beam. In this scheme, the scanning resolution is determined solely by the scanning beam thickness. The slit-scan flow cytometer in operation at the LLNL is an object plane scanner (10). It consists of three major components: the fluidics, optics for light focusing and light collection, and electronics.

a. Fluidics

Precise and stable control of the sheath flow rate and sample injection rate is essential for slit scanning since the objects to be scanned must be constrained to flow within a region no larger than 3 μm in diameter to insure that they pass through the thin part of the scanning beam. In the LLNL system, both the sheath and sample

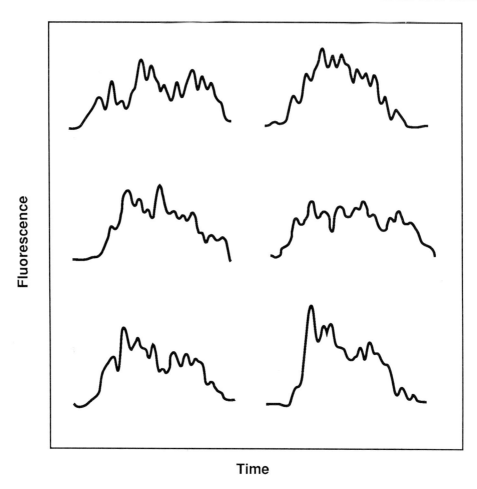

Fig. 4. Chromosome scans from a preparation in Ohnuki's swelling buffer and polyamine swelling buffer. The chromosomes were stained with chromomycin A3.

are driven through the flow system by regulated, compressed air. The sheath fluid passes through a flow restrictor made of 29 gauge cannula to decouple slight pressure changes in the pressure vessel and changes in the fluid level. The flow rate of the sheath fluid is maintained at 10 ml/sec which produces a fluid velocity of 3.9 m/sec at the center of a 0.25×0.25 mm chamber. The sheath fluid enters the flow chamber along the axis of the measurement channel to reduce the turbulence any sharp turns might introduce. The sample is injected into the flow chamber from a coil of

0.015 in. (ID) polyethylene tubing which serves as the storage vessel. The sample is pushed out of the polyethylene tubing by a liquid driver fluid separated from the sample by a bubble to prevent mixing between the driver and the sample. The driver fluid traverses a 1 mm diameter flow restrictor prior to entering the polyethylene tube to damp out small fluctuations induced by the driving air pressure. The presence of the flow restrictor allows precise control of the sample flow at rates as low as 0.001 ml/min. The sample is injected at 0.0017 ml/min to produce a 3μm diameter sample stream. The tubing reservoir contains enough sample fluid for a 15 min run at a flow rate of 0.001 ml/min.

Orientation of chromosomes in flow is an important feature of the flow chamber. Chromosomes which are not properly oriented cannot be properly scanned. Rod-like objects such as chromosomes in a small accelerating stream orient with their long axis parallel to the direction of flow (18). The orienting force is greatest near the sample injection point and decreases along the sample stream. Interpretation of the data generated from slit-scan profiles of improperly oriented chromosomes is difficult, and dips corresponding to the centromere may be completely obliterated. However, the system described here allows excellent orientation for all of the human chromosomes (19); even the smallest can be forced to properly orient in most cases.

b. Optics

A Spectra Physics ceramic tube, water-cooled laser (Model 2025-05) generates the 1–1.5 W of 488 nm light used for slit scanning. The beam is spatially filtered and expanded to a diameter of 2.5 mm $1/e^2$. It then passes through a long focal length cylindrical lens and is focused to form a 1.5-μm-thick ellipse at the flow stream using a Leitz 32 ×, 0.6 NA long working distance microscope objective corrected for the 1.8 mm thickness of the flow chamber walls. The fluorescence from the slit-scan region is collected by a 40 ×, 0.75 NA microscope objective, focused through a 1 mm pinhole, spectrally filtered to pass only the fluorescence, and projected into the slit-scan photomultiplier tube (PMT).

c. Electronics

The signals from the slit-scan PMT are amplified and directed along two pathways. Pulses proceeding along one signal path are integrated with a shaping time constant of 3 μsec. Those proceeding along the other are digitized at a sample rate of 50 MHz and eight-bit resolution, using a Biomation 8100 waveform recorder. When the peak value of the integrated pulse falls within a predetermined range, a gate pulse is generated which stops the digitization, thereby freezing the profile already stored in the memory of the waveform recorder. The recorded profile is then read by a PDP 11-23 computer and stored on disk. These data are transferred to a VAX 750 computer for profile analysis.

B. COMBINED IMAGE–OBJECT PLANE SCANNING

a. Fluidics

The scanning flow cytometer at LANL is a combined image–object plane scanner (6) that allows very high-resolution total fluorescence analysis (1). This instrument was designed to overcome the limitation inherent in the short depth of field of high-numerical optics by constricting the flow stream to a very fine diameter. The performance of high-precision total-fluorescence measurements (CV ≃ 1%) are achieved while maintaining a slit-scan resolution of 1 μm. A triple-nozzle flow chamber maintains the stable, narrow sample stream diameter flow (5). This chamber consists of a square cross-section outer nozzle (2 mm inner dimension) and two cylindrical inner nozzles, each 50 μm in diameter. The chromosome suspensions are introduced into the innermost nozzle from a motor-driven syringe capable of delivering sample volumes at rates as low as 0.0021 μl/min. Typically, sample streams around 0.7 μm in diameter are generated at flow rates of 0.006 μl/min. The dual sheath flows are driven by regulated pressurized air. The sample stream velocity at the laser beam intersection is determined for each run by analysis of the width of profiles from 10.26 μm diameter microspheres. Flow velocities around 5.4 m/sec are typical. This information is used to calibrate chromosome length measurements.

b. Optics

Chromosome excitation is accomplished by focusing a 488 nm, 1–1.5 W laser beam onto the sample stream using a $10\times$, 0.4 NA objective partially filled to 0.9 NA by a 90 nm diverging toric lens and a 630 mm focal length cylindrical lens. This system generates an elliptical beam profile at the sample stream whose major and minor dimensions are $27\,\mu$m and $2\,\mu$m, respectively. The peak irradiance of the system is ~ 2 MW/cm^2. Fluorescence from the system is collected by a $40\times$, 0.75 NA microscope objective coupled to the chamber using glycerol. The scanning resolution of the system is determined by a slit placed in the image plane whose narrow dimension is perpendicular to the direction of flow. The slit width is set at $30\,\mu$m to give a theoretical scanning resolution of $0.75\,\mu$m. This system is able to resolve doublets from $0.91\,\mu$m diameter microspheres, indicating that, while the resolution is not as high as that expected theoretically, it is adequate for resolution of chromosome features such as centromeres and large bands.

To maintain a constant, large depth of field for integrated fluorescence measurements while maintaining high scanning resolution, a beam splitter was placed before the slit, reflecting about 10% of the fluorescence into a second PMT. In this way the fluorescent light from the entire illuminated section of each chromosome is measured along with the corresponding chromosome scan measured through the slit. The precision of microsphere fluorescence measurement is about 1% CV, compared to 0.8% CV with a full reflector in place of the beam splitter (2). Fluorescence distributions from microspheres determined from integrated fluorescence were equivalent to the distributions reconstructed from the waveform areas. Distributions from many chromosome preparations, however, were often of higher precision measured with the beam splitter, showing the effect of the short depth of field used in slit scanning.

c. Electronics

The analog signal from the PMT detecting light deflected from the beam splitter placed in front of the slit is amplified and integrated for pulse height and area analysis. The signal pulse is delayed $0.25\,\mu$sec after crossing the trigger threshold for accurate integration of the entire pulse. The integral signal is digitized with ten-bit resolution and the pulse height with eight-bit resolution by the LANL cell analysis data acquisition system (12). The signals are acquired in correlated list mode files with the digitized waveforms.

The pulse shape signal from the second PMT placed to the rear of the slit is digitized by a Biomation 8100 waveform recorder with sample intervals of 10 or 20 nsec. Usually, 128 bytes in a single waveform are acquired through a modification of the LACEL (Los Alamos Cell Analysis) system. The data acquisition system is interfaced to a PDP 11/34 computer. The LACEL software (25) was slightly modified to accept the 128-byte waveform and to place the Biomation 8100 recorder under digital control.

IV. SCANNING THEORY

Only morphological information along the chromosome axis perpendicular to the excitation (scanning) beam is provided. Consider, for example, a hypothetical chromosome of length L flowing with velocity v through a scanning beam of negligible width [see Fig. 5(a)]. The lengths of the short and long arms of this hypothetical chromosome are P and Q, respectively, and the centromere has a length C. Thus, a chromosome flowing with its long axis perpendicular to the scanning beam would produce a profile whose total, short-arm and long-arm durations are L/v, P/v and Q/v, respectively, and whose centromeric dip duration is C/v. However, a chromosome traversing the scanning beam at an angle, σ, to the direction of flow, would have total profile, short-arm and long-arm durations of $L\cos\sigma/v$, $P\cos(\sigma)/v$ and $Q\cos(\sigma)/v$, respectively. The duration of the centromeric dip would be $C\cos(\sigma)/v$ [see Fig. 5(b)]. These equations only hold for reasonably small values of σ, and for an excitation beam of negligible width.

Laser beams, however, have finite widths.

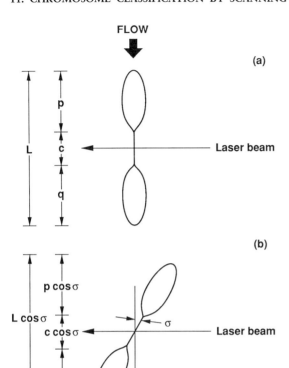

Fig. 5. (a) Depiction of a profile of a chromosome scanned with a very thin laser beam. The profile is essentially the true chromosome profile. (b) The effect of scanning with the chromosome oriented at angle σ with respect to the flow axis.

This causes the observed chromosome profile to be somewhat blurred. The chromosome measured during scanning may be characterized as a convolution of the instrument excitation beam profile or image plane scanning aperture with the true chromosome shape along the axis of flow. Figure 6(a) depicts a hypothetical profile of a chromosome scanned with a very thin laser beam. The profile is essentially the true distribution of dye along the chromosome $c(x)$. The blurring due to the finite laser beam width can be described mathematically as a convolution. That is, the measured profile $i(t)$ can be expressed mathematically as

$$i(t) = \int_{-\infty}^{\infty} a(x)\, c(x - vt)\, dx, \qquad (1)$$

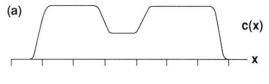

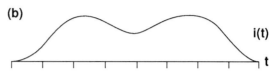

Fig. 6. An illustration of the "blurring" that occurs as a result of the finite width of the scanning beam. (a) Hypothetical "true" chromosome profile. (b) The blurred profile that would be measured during passage through a beam (whose thickness is comparable to the centromere width) with a normally distributed intensity profile.

where $a(x)$ is the scanning aperture (i.e. the distribution of the laser beam along the chromosome flow axis in object plane scanning) and v is the velocity of the chromosome through the scanning region. The blurring effect can be seen schematically in Fig. 6(b).

V.   SHAPE ANALYSIS

In this section we focus on three aspects of chromosome classification: (1) classification by the location of the centromere, (2) classification according to the number of centromeres and (3) classification according to "banding pattern."

A.   CLASSIFICATION ACCORDING TO CENTROMERE LOCATION

a.   CI Calculation

The centromeric dip or depression in the profile for a DNA-stained chromosome separates the profile into two segments. The CI is determined from the fluorescence profile as the ratio of the area of the larger segment of the profile to the total profile area. The algorithm used for CI calculation from slit-scan profiles is conceptually similar to one developed by

Mendelsohn *et al.* (*21*) for CI calculation from chromosome images measured by scanning cytophotometry. In this approach, the profile to be analyzed is subtracted from a similarly sized standard profile lacking a centromeric depression. The centromere location is defined to be the point of maximum difference between the standard and the experimental profile. To be useful, the algorithm for calculating the CI should locate the position of centromeres and chromosome ends accurately, and perform this task in a relatively short amount of time. Lucas *et al.* (*16*) evaluated two algorithms for this purpose. In one, error functions of the form

$$\text{Erf}(x) = A \int_0^x e^{-t^2/\sigma^2}\, dt \qquad (2)$$

(where x is position along the chromosome profile) of equal amplitude A are fitted to the left and right edges of the experimental profile

by adjusting A and σ. The standard profile is completed by joining the error functions at their maximum amplitude. Error functions were chosen because they approximate the shape of the pulse to be expected from the passage of the sharp edge of a chromosome end through a Gaussian-shaped laser beam. Evaluation showed this method to be accurate, but slow. The time-consuming step involves calculation of the error functions.

The second algorithm (see Fig. 7) is termed the reflection algorithm because it is based on a symmetrical reflection of half of the experimental profile about the profile midpoint. The half of the profile containing the maximum is reflected about the profile midpoint, producing a symmetrical profile. The standard profile is completed by joining the two points of maximum amplitude. Evaluation showed the reflection algorithm to be accurate

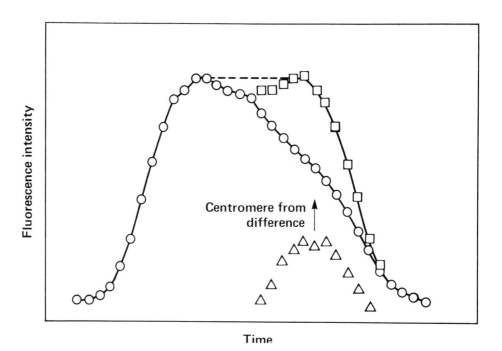

Fig. 7. Reflection algorithm for centromere location. The midpoint of the experimental profile is located, and the half of the profile containing the profile maximum is reflected about the midpoint to create a symmetrical profile. The two high points in this profile are connected with a horizontal dashed line to generate a standard "centromere-free" profile. The centromere is located at the point of maximum difference (arrowed) between the measured pulse profile and the standard.

when applied to human chromosome profiles. The reflection algorithm was selected as the procedure of choice because of its conceptual simplicity and speed.

Chromosome classification according to CI and DNA content offers promise for discrimination between chromosomes 10–12, which cannot be distinguished by other flow procedures (16). This approach has been successfully applied to normal (19) and homogeneous abnormal chromosomes (J. Lucas et al., unpublished work) isolated from established human cell lines and from human amniocytes. Figure 8 shows a bivariate DNA vs CI distribution measured for human chromosomes (19) isolated from a human foreskin cell line (LLL811). The peak identities in this figure have been determined by comparing the peak location in this distribution with peak locations in DNA vs CI distributions

measured for chromosomes from human/rodent hybrid cell lines (23) containing a few known human chromosomes. Many of the larger chromosomes, e.g. 1 and 2, 3 and 4, and 9–12, are much better resolved with this technique than in bivariate Hoechst 33258 vs chromomycin A3 distributions. Figure 9 shows a bivariate DNA vs CI distribution measured for chromosomes isolated from a human amniocyte containing a homogeneous occurring translocation between chromosomes 2 and 11. Comparing Figs 8 and 9, one sees a decrease in the number of contours for chromosomes 2 and 11. In fact, at the contour level displayed in Fig. 9, the contour of chromosome 2 does not appear. This is not surprising, because chromosomes 11 and 2 each have only one copy present in their normal respective peak positions. The other copies of chromosomes 11 and 2 are represented in the translocated peaks

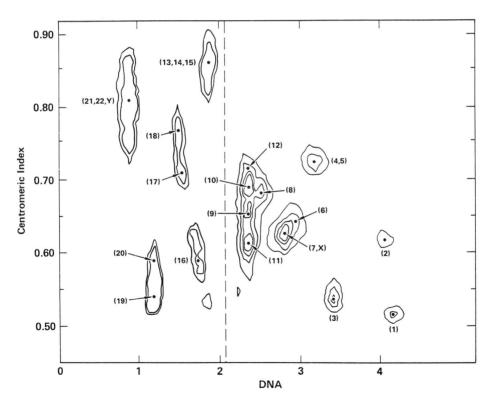

Fig. 8. CI vs DNA flow karyotype for human chromosomes. In this normal flow karyotype, the solid dots in the contours denote human chromosomes whose CIs have been measured in hybrids.

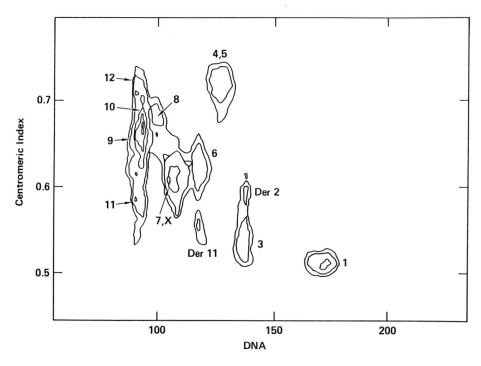

Fig. 9. CI *vs* DNA flow karyotype for an abnormal amniocyte containing a translocation between chromosomes 2 and 11.

labeled Der 11 and Der 2. Other flow methods are able to detect translocations involving chromosomes 9–12, but they are unable to determine which of the chromosomes are involved.

B. DICENTRIC CHROMOSOME DETECTION ALGORITHM

The frequency of dicentric chromosomes has been shown to be a useful indicator of radiation-induced genetic damage, and has been used to estimate radiation exposure levels in humans (20). The development of fast, reliable methods to detect dicentrics is at the center of much ongoing research. Detection of dicentric chromosomes in flow is complicated by their infrequent occurrence and their random size and shape. The utility of slit scanning in estimating the frequency of dicentric chromosomes was demonstrated using

irradiated populations of cultured Chinese hamster M3-1 cells (11,17). In that study, profiles were recorded from the larger chromosomes. Dicentric chromosomes were recognized as producing pulse profiles having two distinct and approximately equal centromeric dips, while normal metaphase chromosomes produced pulse profiles having only one centromeric dip. Chinese hamster M3-1 cells exposed to 0, 0.5, 1.0, 1.5, 2.0 and 4.0 Gy of X-irradiation were harvested 20 h after irradiation. The frequency of dicentric chromosomes with total fluorescence higher than that for chromosome 3 was estimated both by slit-scan analysis of isolated chromosomes and by visual analysis of metaphase spreads. Figure 10(a) shows the frequencies of dicentric chromosomes estimated by slit scanning at each dose level during three different experiments. These data show a linear-quadratic increase in dicentric frequency with increasing dose. The

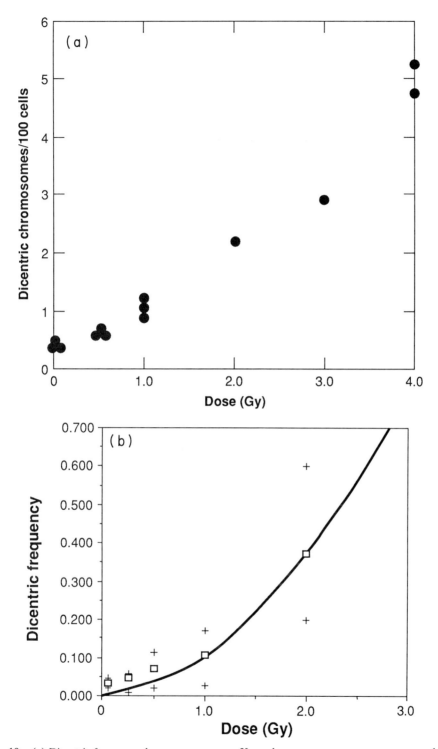

Fig. 10. (a) Dicentric frequency dose-response curve. X-ray dose-response curves were measured by scanning flow cytometry for Chinese hamster M3-1 cells on three separate occasions. Doses ranged from 0 to 4 Gy. The individual frequency estimates for each experiment are shown separately as solid dots. (b) Dicentric dose-response curve. Scanning flow cytometry was used to measure pulse profiles for chromosomes isolated from human lymphocytes following exposure to 0, 0.25, 0.5, 1.0 and 2.0 Gy of [137]Cs. The solid line is a least squares fit to the data. One standard deviation of the data is indicated by +.

close agreement between the three different experiments shows high reproducibility for the slit scanning approach. Figure 10(b) shows the application of slit scanning to the quantification of the dicentric frequency in human lymphocytes exposed to 0, 0.2, 0.5, 1.0 and 2.0 Gy of ^{137}Cs γ-rays. The cells were harvested 14 h after irradiation, and the frequency of dicentric chromosomes with total fluorescence higher than that for chromosome 2 was estimated by slit-scan analysis of isolated chromosomes. The data show a linear quadratic increase in dicentric frequency with increasing dose. Over 5×10^6 chromosomes were analyzed in this study. Dicentric frequencies reported in the literature for irradiated human lymphocytes (20) correlated well with the slit-scan-generated frequencies. However, slit-scan detection efficiency was only about 1/300. Figure 10(b) shows a significant increase in the estimated dicentric frequencies even at low doses (0.25–0.5 Gy).

C. CHROMOSOME STRUCTURE ANALYSIS

The capability of measuring simultaneous fluorescence and light scattering waveforms was developed to obtain information on chromosome structure (13). The two simultaneous waveforms were detected separately by two PMTs and digitized in a single-memory recorder after delaying one of the waveforms. Significant differences were observed between the fluorescence and light scattering profiles (Fig. 11). The centromeric dip appeared more often in the light scattering profiles than fluorescence profiles. The 90° light scattering depends on the mass or volume of chromosome, but because the dependence is unlikely to be linear, light scattering may elicit enhanced resolution of physical constrictions. Chromosome structures also have been detected in scans of partially decondensed chromosomes (see Fig. 4). The challenge now is to develop algorithms that allow classification of chromosomes using these profiles.

Fluorescence   **Light Scatter**

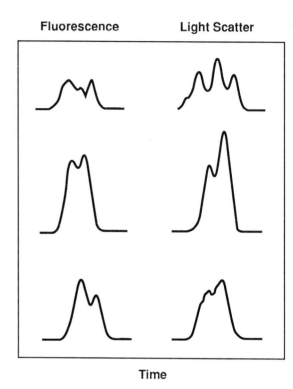

Time

Fig. 11. Correlated fluorescence and 90° light scattering scans for three Chinese hamster chromosomes (6, 2 and 4) stained with propidium iodide. Even though the fluorescence and light scanner scans were obtained simultaneously, substantial differences are seen between fluorescence and light scattering profiles.

VI. FUTURE DIRECTIONS

Fringe-scan flow cytometry and trivariate slit-scan flow cytometry are now being developed to further improve chromosome classification based on shape.

A. FRINGE-SCAN FLOW CYTOMETRY

Fringe-scan flow cytometry (FSFCM) has been developed to further increase the spatial resolution available during scanning (22). Fringe-scanning takes advantage of the constructive and destructive interference that occurs when two laser beams intersect. This interference produces a set of parallel planes

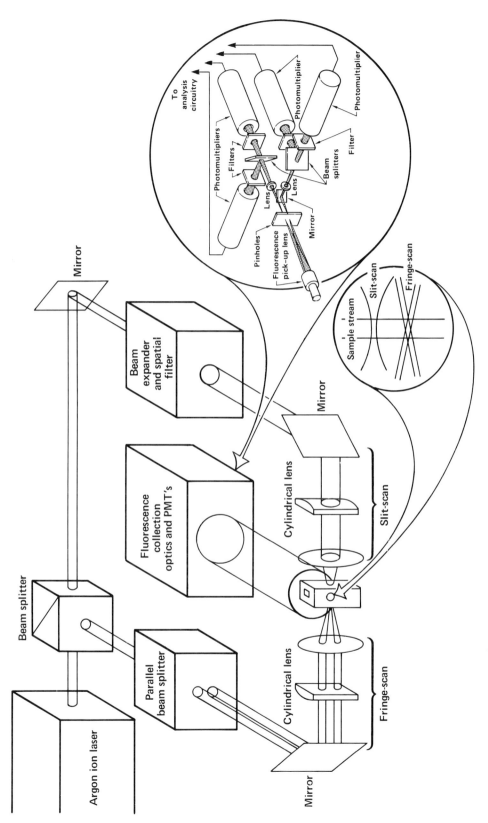

Fig. 12. Schematic illustration of the optical components of the slit-/fringe-scan flow cytometer.

of intense illumination, the spacing of which can approach $0.3\,\mu$m. High-resolution scans of chromosomes and other objects can be calculated from the fluorescence profiles measured as chromosomes flow lengthwise through the planes of illumination.

As discussed earlier, the signal $i(t)$ that is measured is the convolution of the fringe field $a(x)$ with the fluorescent intensity profile of the chromosome $c(x)$ [see Eq. (1)]. Software has been developed to calculate the original chromosome profile $c(x)$ from the measured fluorescence $i(t)$ and the fringe field $a(x)$. This is accomplished by taking the Fourier transform of Eq. (1). For detailed description of the deconvolution process, see Bracewell (3) and Mullikin et al. (22). This transformation is illustrated in Eq. (3) where $I(w)$, $A(w)$ and $C(w)$ are the transforms of $i(t)$, $a(x)$ and $c(x)$, respectively.

$$I(\omega) = A(\omega)\,C(\omega). \qquad (3)$$

In the spatial domain, the convolution is represented by an integral, see Eq. (1). In the spatial frequency domain this convolution integral becomes a multiplication. By rearranging Eq. (3), multiplying numerator and denominator by $A^*(\omega)$, and adding K^2 to the denominator, an approximation to the transform of the original chromosome profile is generated in the spatial frequency domain:

$$C'(\omega) = \frac{I(\omega)\,A^*(\omega)}{|A(\omega)|^2 + K^2}. \qquad (4)$$

The value K keeps Eq. (4) in bounds when $A(\omega)$ becomes small. To transform $C'(\omega)$ to $c'(x)$, the inverse Fourier transform is applied.

The latest generation FSFCM, developed at LLNL, allows sequential fringe-scanning and slit-scanning in the same flow cell (see Fig. 12). Using that system, the two scanning methods have been directly compared on the same objects. Figure 13 shows three human chromosome 4 flourescence profiles sequentially measured by the slit and fringe systems. The profiles in Fig. 13(a) are the slit-scan profiles and those in Fig. 13(b) are the deconvolved fringe-scan profiles. Both sets show clearly resolved centromeric dips as

expected for this chromosome. Moreover, CIs calculated from deconvolved profiles of the larger human chromosomes (10 000 profiles) were in close agreement with corresponding CI values calculated from slit-scan profiles of the same chromosomes.

B. TRIVARIATE SLIT-SCAN FLOW CYTOMETRY AND SORTING

Work is now underway to classify chromosomes according to three variables, Hoechst 33258 fluorescence, chromomycin A3 fluorescence and shape. This approach should allow chromosomes that bind equal amounts of Hoechst 33258 and chromomycin A3 but differ in the position of the centromere to be distinguished. Thus, chromosomes that are difficult to separate by dual-beam flow cytometry (e.g. the larger human chromosomes, especially chromosomes 9–12) can be distinguished by SSFCM. Further, SSFCM allows detection of morphologically aberrant chromosomes such as dicentrics.

ACKNOWLEDGEMENTS

Support for Joe Lucas was performed under the auspices of the US Department of Energy by the Lawrence Livermore National Laboratory under contract number W-7405-ENG-48 with support from the USPHS Grant GM 25076. Marty Bartholdi was supported by the Los Alamos National Flow Cytometry Resource sponsored by the Division of Research Resources of the National Institutes of Health (Grant RR01315) Department of Energy.

REFERENCES

1. Bartholdi, M. F., Sinclair, D. C., and Cram, L. S. (1983). Chromosome analysis by high illumination flow cytometry. *Cytometry* 395–401.
2. Bartholdi, M. F., Ray, F. A., Cram, L. S., and Kraemer, P. M. (1984). Flow karyology of serially cultured Chinese hamster cell lineages. *Cytometry* 5, 534–538.

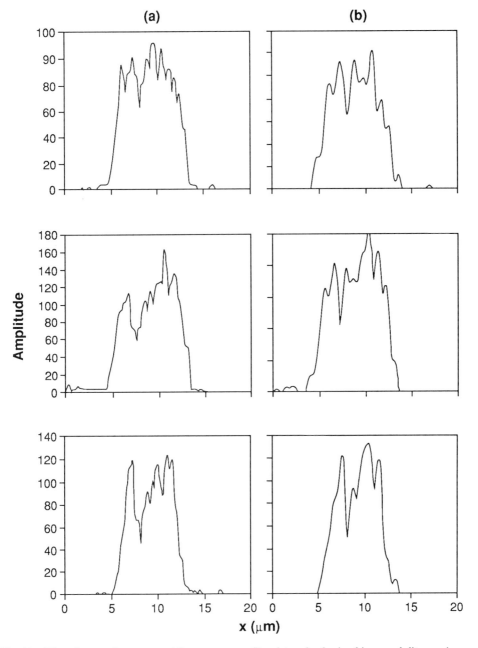

Fig. 13. Three human chromosome 4 fluorescence profiles determined using fringe- and slit-scanning. (a) Slit-scan profile estimates and (b) fringe-scan profiles.

3. Bracewell, R. (1965). "The Fourier Transform and its Applications." McGraw-Hill, New York.

4. Carrano, A. V., Gray, J. W., Langlois, R. G., Burkhart-Schultz, K., and Van Dilla, M. A. (1979). Measurement and purification of human chromosomes by flow cytometry and sorting. *Proc. Natl Acad. Sci. USA.* **76**, 1382–1384.

5. Cram, L. S., Arndt-Jovin, D. J., Grimwade, B. G., and Jovin, T. M. (1980). Low resolution image analysis of spheres and Chinese hamster cells in a flow cytometer/cell sorter. *In* "Flow Cytometry" (O. D. Laerum, T. Lindmo, and E. Thorud, eds), Vol. IV, pp. 256–259. Universitetsforlaget, Oslo.

6. Cram, L. S., Bartholdi, M. F., Wheeless, L. L., and Gray, J. W. (1985). Morphological analysis by scanning flow cytometry. *In* "Flow Cytometry, Instrumentation and Data Analysis" (M. Van Dilla, P. N. Dean, O. D. Laerum, and M. N. Melamed, eds), pp. 163–194. Academic Press, New York.

7. Davies, K., Young, B., Elles, R., Hoill, M., and Williamson, R. (1981). Cloning of a representative genomic library of human X chromosome after sorting by flow cytometry. *Nature* **293**, 364–366.

8. Gray, J. W., Carrano, A. V., Steinmetz, L., Van Dilla, M., Moore, D., Mayall, B., and Mendelsohn, M. (1975). Chromosome measurement and sorting by flow systems. *Proc. Natl Acad. Sci. USA* **72**, 1231–1234.

9. Gray, J. W., Langlois, R. G., Carrano, A. V., Burkhart-Schultz, K., and Van Dilla, M. A. (1979). High resolution chromosome analysis: one and two parameter flow cytometry. *Chromosoma* **73**, 9–27.

10. Gray, J. W., Lucas, J. N., Pinkel, D., Peters, D., Ashworth, L., and Van Dilla, M. A. (1980). Slit-scan flow cytometry: analysis of Chinese hamster M3-1 chromosomes. *In* "Flow Cytometry IV," p. 49. Universitetsforlaget, Bergen.

11. Gray, J. W., Lucas, J. N., Yu, L.-C., and Langlois, R. G. (1984). Flow cytometric detection of human aberrant chromosomes. *In* "Biological Dosimetry" (W. G. Eisert and M. L. Mendelsohn, eds), pp. 25–35. Springer-Verlag, Berlin.

12. Hiebert, R. D., Jett, J. H., and Salzman, G. C. (1981). Modular electronics for flow cytometry and sorting: the LACEL systems. *Cytometry* **1**, 337–341.

13. Johnston, R. G., Bartholdi, M. F., Hiebert, R. D., Parson, J. D., and Cram, L. S. (1985). Slit-scan flow cytometer for recording simultaneous waveforms. *Rev. Sci. Instrum.* **56**, 691–695.

14. Kunkel, L., Tantravahi, U., Eisenhard, M., and Satt, S. (1982). Regional localization on the human X of DNA segments cloned from flow sorted chromosomes. *Nucleic Acids Res.* **10**, 1557–1579.

15. Lebo, R. (1982). Chromosome sorting and DNA sequence localization. *Cytometry* **3**, 145–152.

16. Lucas, J. N., Gray, J. W., Peters, D. C., and Van Dilla, M. A. (1983). Centromeric index measurement by slit-scan flow cytometry. *Cytometry* **4**, 109–116.

17. Lucas, J. N., and Gray, J. W. (1983). Quantification of the frequency of occurrence of dicentric chromosomes by slit-scan flow cytometry. *In* "Radiation Research, Somatic and Genetic Effects" (J. J. Broerse, G. W. Barendsen, H. B. Kal, and A. J. Van der Kogel, eds), pp. B4–21. Martinus Nijhoff, Amsterdam.

18. Lucas, J. N., and Pinkel, D. (1986). Orientation measurements of microsphere doublets and metaphase chromosomes in flow. *Cytometry* **7**, 575–581.

19. Lucas, J. N., and Gray, J. W. (1987). Centromeric index versus DNA content flow karyotypes of human chromosomes measured by means of slit-scan flow cytometry. *Cytometry* **8**, 273–279.

20. Lloyd, D. C., Purrott, R. J., and Reeder, E. J. (1980). The incidence of unstable chromosome aberrations in peripheral blood lymphocytes from unirradiated and occupationally exposed people. *Mutation Res.* **72**, 523–532.

21. Mendelsohn, M. L., Mayall, B. H., Bogart, E., Moore, D. H., and Perry, B. H. (1973). DNA content and DNA based centromeric index of the 24 human chromosomes. *Science* **179**, 1126–1129.

22. Mullikin, J., Norgren, R., Lucas, J., and Gray, J. W. (1988). Fringe-scan flow cytometry. *Cytometry* **9**, 111–120.

23. Ohnuki, Y. (1965). Demonstration of the spiral structure of human chromosomes. *Nature* **208**, 916–917.

24. Ohnuki, Y. (1968). Morphological studies of the spiral structure of human somatic chromosomes. *Chromosoma* **25**, 402–428.

25. Salzman, G. C., Wilkins, S. F., and Whitfill, J. A. (1981). Modular computer programs for flow cytometry and sorting: the LACEL system. *Cytometry* **1**, 325–326.

26. Sillar, R., and Young, B. (1981). A new method for the preparation of metaphase chromosomes for flow analysis. *J. Histochem. Cytochem.* **29**, 74–78.

27. Stubblefield, E., Deavan, L., and Cram, L. S. (1975). Flow microfluorometric analysis of isolated Chinese hamster chromosomes. *Exp. Cell Res.* **94**, 464–468.

28. Van Dilla, M. A., Carrano, A. V., and Gray, J. W. (1975). Flow karyotyping: current status and potential development. *In* "Automation of Cytogenetics, US Energy Research and Development Administration Conf-751158" (M. L. Mendelsohn, ed.), pp. 145–164. National Technical Information Service, Springfield, Va.

Prepurificiation of Chromosomes by Velocity Sedimentation and Applications

J. G. COLLARD, A. TULP AND J. W. G. JANSSEN*

Antoni van Leeuwenhoekhuis, The Netherlands Cancer Institute
Division of Cell Biology, 121 Plesmanlaan,
1066 CX Amsterdam, The Netherlands

I. INTRODUCTION

Flow cytometry is uniquely suited to the task of distinguishing between individual isolated mammalian metaphase chromosomes differing in DNA content (*5,11,33*). For analytical purposes this technique can be applied to flow karyotyping of short-term peripheral blood cultures (*29,59*) and thus may improve diagnosis by testing chromosomal rearrangements of genes. On a preparative scale, flow cytometry allows the isolation of individual metaphase chromosomes by fluorescence-activated flow sorting. DNA isolated from fractionated or sorted chromosomes can be used for gene mapping studies (*7,9,35,36*) as well as for the construction of chromosome-specific DNA libraries (*14,16,22,27,28*). Chromosome banks have the advantage that the number of manipulations needed for

* Present address: Deutsches Rotes Kreuz, Blutspendezentrale, Ulm, Oberer Eselsberg 10, Postfach 1564, 7900 Ulm/Donau, FRG.

FLOW CYTOGENETICS
ISBN 0-12-296110-2

191

screening specific DNA sequences is reduced considerably. These libraries are, therefore, a valuable aid to human gene mapping and molecular analysis of human genetic diseases (3,6,18,43).

For the construction of chromosome-specific libraries, rather large amounts of DNA derived from sorted chromosomes are required, forming one of the main impediments to the construction of chromosome banks. This is particularly serious if complete libraries with overlapping DNA inserts are desired. One of the possibilities to facilitate or to speed up the sorting of individual chromosomes is to pre-enrich the desired chromosomes prior to flow sorting. This may be performed by making use of the differences in size between metaphase chromosomes present in cells under investigation. In the past, chromosome separations according to size have been performed either by swinging-bucket centrifugation (40), zonal rotor centrifugation (51,52) or velocity sedimentation in specially designed sedimentation chambers (8–12). Velocity sedimentation using a chamber as designed in our laboratory has proved to be a simple and fast technique to separate cells, nuclei and chromosomes in different size classes (8,53,54). This chapter emphasizes use of the Tulp velocity sedimentation chamber for application to chromosome research.

II. RATIONALE FOR PRE-ENRICHMENT PROCEDURES PRIOR TO FLOW SORTING

The yield of any sorting method, that processes particle by particle, is limited by time and by dead time of the sorting decision. Quantitative yield benefits greatly from pre-enrichment measures. These pre-enrichment steps are generally batchwise and lend themselves to scaling up of the dimensions whenever massive amounts are needed.

If in a mixture there are N particles, of which n are desirable for flow sorting, then contamination of the sorted pool by unwanted

particle is given by

$$f = \frac{N - n}{N} \frac{3vt}{1 + v(3t - \tau)}, \qquad (1)$$

in which f denotes the fraction of contaminants, t the drop generation time, τ the dead time of analog electronics and v the flow rate (particles/sec).

Equation (1) takes into account: (1) Poisson statistics, (2) dead time of the analog electronics, (3) that three droplets are deflected per desired particle and (4) absence of overlap rejection (39). It follows from Eq. (1) that a number of particles is never detected nor analyzed and that there is always a controversy of purity versus recovery. Under typical working conditions of a commercial flow sorter ($\tau = 20\ \mu sec$;* $t = 25\ \mu sec$ and $v = 1000$ particles/sec) Eq. (1) yields

$$f = 0.0714 \frac{N - n}{N}.$$

A numerical example, for which $n/N = 0.1$, gives the contamination as 6.4%; when a pre-enrichment step augments n/N to 0.8, calculation using Eq. (1) shows that a similar purity is obtained if the particles flow serially at 5456 particles/sec so that $0.8 \times 5456\ (= 4365)$ desired particles/sec are collected instead of $0.1 \times 1000\ (= 100)$ particles/sec from a nonenriched suspension. In this theoretical case, the sorting process is speeded up 44-fold. The lower the initial percentage of desired particles in the starting suspension, the more impressive the speeding factor. In general, when desired particles (at initial fraction a) are sorted at V_1 particles/sec, then pre-enrichment of these particles to fraction b permits sorting at V_2 particles/sec to obtain a similar purity, and thus the speeding-up factor K can be written as

$$K = \frac{4(b - a) + \tau v_1(1 - b)}{V_1} \frac{a}{b}.$$

Theoretically, it follows that the sorting of a particular chromosome (at $a = 0.0435$ and

---

* Editors note: The electronic deadtime in modern instruments is considerably lower than 20 μsec (e.g. ~1 μsec). In these instruments high purity is maintained at the cost of lower sorting efficiency.

pre-enriched to $b = 0.5$) can be speeded up 482-fold. Apart from considerable speeding up, pre-enrichment steps generally remove debris, large aggregates and dead cells. In the case of chromosome pre-enrichment, cell debris, chromosome clumps and nuclei are removed, thus facilitating chromosome analysis and sorting. However, it is desirable that the particles which generally become diluted by the pre-enrichment procedure(s) can be easily washed and concentrated by sedimentation (centrifugation, or at $1g$) without major loss, damage due to pelleting or clump formation. In the case of sticky particles like chromosomes this is obviously difficult and depends on the chromosome isolation and fractionation buffers used.

III. THE LOW-GRAVITY-FORCE SEDIMENTATION CHAMBER (TULP CHAMBER)

A. CONSTRUCTION AND HANDLING OF THE CHAMBER

The sedimentation chamber, originally made of Perspex and commercially available through Brand & Co. (Postfach 310, 6980 Wertheim, FRG) is described in detail in the literature (8,56). Figure 1(A) shows a separation chamber made of the nylon "Macrolon" that is sterilizable. A diagram of the chamber is depicted in Fig. 1(B). The effective internal cross sectional area is 50 cm^2 and the effective internal height is 2 cm. The flow deflectors in the top and bottom cones guarantee undisturbed thin sample layering and undisturbed rapid fractionation. The stainless-steel antivortex cross aborts eddy formation upon acceleration (linear in 4 min to 500 rpm) and deceleration (to 0 rpm in 4 min) (56).

The chamber is first completely filled with a 15% sucrose solution cushion via the inlet (i) (see Fig. 1) using a peristaltic pump or a simple gravity feed from a reservoir. A linear sucrose gradient ranging from 5 to 15% (100 ml) is introduced into the chamber within 10 min via the outlet (o) by reversing the flow from the peristaltic pump. The sucrose gradient is made

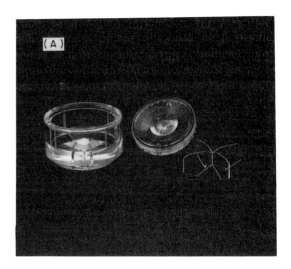

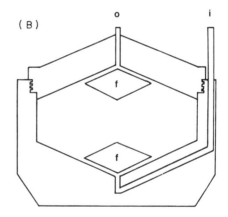

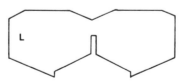

Fig. 1. Photograph and diagram of the Perspex sedimentation chamber (effective height 2 cm; internal diameter 8 cm; surface 50 cm^2). (A) The top cone (middle) can be placed on the base cone (at left). The antivortex cross (at right) is placed in the chamber. The chamber can be lifted out of the centrifuge bucket by a lid. (B) i, inlet port; o, outlet port; f, flow deflectors; L, one of the lamellae of the antivortex cross.

using 0.1 *M* Tris, 0.1 *M* NaCl, 5 m*M* MgCl$_2$ pH 7.4. The chamber is then placed in position in a bucket of a speed-controlled Kontron TGA-6 centrifuge to avoid blurring of the thin sample that is introduced next. The crude chromosomal preparation (2–3 ml) followed by an overlay (28 ml) of Tris buffer without sucrose is introduced into the chamber via the outlet (o) as above. At this stage the whole chamber is filled with fluid and the band of chromosomes will have reached the cylindrical part of the device, resulting in a thin film (thickness 0.4 mm) of undisturbed chromosome material sandwiched between the overlay and the 2-cm-high density gradient. Chromosomes derived from 10^7 mitotic cells can be layered on the gradient for optimal separation. The centrifuge is run at room temperature with linear acceleration to 500 rpm in 4 min, a constant run for 50 min at 500 rpm (52 *g*) and an almost linear deceleration in 4 min. Under these conditions the whole gradient is used for the separation of chromosomes. The contents of the chamber can be fractionated within 5 min in 3–5 ml fractions via the outlet (o) of the chamber by introducing the cushioning solution via the inlet (i). Fractions can be stored at −20°C until use for analysis or sorting.

B. Chromosome Isolation Prior to Chromosome Velocity Sedimentation

Various procedures have been developed to isolate metaphase chromosomes for flow analysis. Each method appears to have advantages and disadvantages, as discussed elsewhere in this volume. Generally, isolation procedures which cause contraction of the chromosomes are very suitable for subsequent flow analysis. A drawback of these methods is that chromosome identification by banding is made difficult (*50,60*).

For fractionation of chromosomes by velocity sedimentation we used citric acid as the chromosome isolation solution (*8*). Briefly, mitotic cells were accumulated with vinblastine

(0.02–0.1 µg/ml). After hypotonic treatment (0.075 *M* KCl) the mitotic cells were suspended in chromosome isolation solution (2% citric acid, containing 1 m*M* CaCl$_2$, 1 m*M* MgCl$_2$ and 0.1 *M* sucrose) at a concentration of 5–10 × 10^6 cells/ml. Subsequently, the metaphase chromosomes were isolated by shearing the cells several times through a 22 gauge needle. The crude chromosomal preparation could be layered without further purification on the neutral pH sucrose gradient in the sedimentation chamber. The short exposure to citric acid (usually 15 min) fixed the chromosomes and kept them in good condition for sedimentation. To prevent effects of prolonged acid treatment on the DNA of the chromosomes, a neutral sucrose gradient was used, into which all the chromosomes, including the small ones, sedimented within 30 min after isolation. High molecular weight DNA (40–60 kbp) could be isolated from the chromosomal fractions. This DNA appeared to be suitable for restriction enzyme analysis (hybridization studies) (*9*) and for molecular cloning in plasmids, phages or cosmids (*22*).

C. Advantages of the Sedimentation Chamber

There are several factors that render high resolving power to the device:

(i) The flow deflectors mounted in the top and bottom cones permit samples of 0.3 mm thickness to be distributed uniformly over a surface of 50 cm^2. This is of importance since resolution in zonal velocity sedimentation is in part determined by the initial width of the sample layer.

(ii) At speed and upon complete swing-out of the sedimentation chamber, the sample layer is at 21 cm from the axis of rotation. Since the chamber is only 2 cm high, it follows from geometry that there is almost no wall sedimentation (loss of particles) (*57*). In contrast, metaphase chromosomes separated in ordinary cylindric centrifugal tubes suffer from considerable losses due to severe wall sedimentation (40). Comparing zonal rotors (1,4,48) and the re-orienting density gradient zonal rotor (21,51,52) with the present small sedimentation chamber points to an equal or even higher degree of resolving power for the chamber (8).

(iii) The liquid content of the chamber is prevented from swirling during acceleration and deceleration. The four perpendicular lamellae that form the antivortex cross (Fig. 1) are sufficient to suppress eddy generation completely, being aided by a moderate (but absolutely necessary) acceleration – deceleration schedule ($\pm \Delta\omega = 200$ rpm).

(iv) The chamber being loaded at normal gravity, liquid isodense layers, initially flat, turn into paraboloids of revolution upon centrifugation. This reorientation of isodense layers occurring twice (during acceleration and deceleration) is only marginal and is easily tolerated by the density gradient used. It was found that chromosome separations at higher gravity forces are consistently better than those performed at unit gravity (8). The short duration (1 h versus 24 h at unit gravity) is beneficial for structural intactness of the chromosomes. Secondly, so-called "anomalous" band broading (41,42,55) is strongly reduced at higher acceleration. At unit gravity, initially thin zones of particles (even at very low concentrations of 10^4/ml) are rather unstable in density gradients and are subject to considerable band broadening by a still unexplained mechanism. Miller (41,42) has studied these 1g-driven band instabilities and concluded that these are excessively large in weak density gradients, but relatively small in strong macromolecular gradients (55). Our experience is that band broadening also occurs in strong sucrose gradients at 1g but is reduced at higher g forces. Therefore, reduction of the extent of "anomalous" band broadening at higher g forces is highly beneficial to the resolution of some of the more complex metaphase karyotypes separated in sucrose density gradients.

IV. APPLICATIONS OF THE LOW-GRAVITY-FORCE CHAMBER FOR ENRICHMENT OF CHROMOSOMAL POPULATIONS

A. FRACTIONATION OF HUMAN CHROMOSOMES

Operating the chamber at 1g, chromosomes derived from Muntjac (10) and Chinese hamster cells (11) have been separated in 24 h. However, operating the chamber at 30–50g, chromosomes derived from rat (9), human (8,12) and mouse (22) cells were separated with good resolution within 1 h. Figure 2 illustrates the separation of human chromosomes by velocity sedimentation at 52g. The upper-left histogram (HFL) represents the flow karyotype of the unfractionated human chromosomes. The histograms A–O show the chromosomal composition of the fractions after sedimentation in a 5–15% sucrose gradient. Chromosomes were stained with a mixture of mithramycin and ethidium bromide. Mithramycin preferentially stains guanine-cytosine GC-rich chromosomes (30). The combined use of mithramycin and ethidium bromide results in a transfer of excitation energy from mithramycin to ethidium bromide. This combination of stains exhibits a low-fluoresence background, especially when compared to ethidium bromide alone. As can be seen after chromosome fractionation, the first two fractions, A and B, contain primarily the smallest chromosomes Y, 21 and 22. Deeper in the gradient (fractions C, D and E) chromosomes 18–20 appear, and chromosomes Y, 21 and 22 disappear. In the middle of the gradient, fractions highly enriched with chromosomes 13–17 were obtained (especially fraction G). The chromosomes 9–12 are dominant in fractions I and J, while fraction L, M and N contain mainly chromosomes 3–8. The last fraction, O, contains virtually only the largest chromosomes 1 and 2. Thus, all fractions are highly enriched with certain chromosomes.

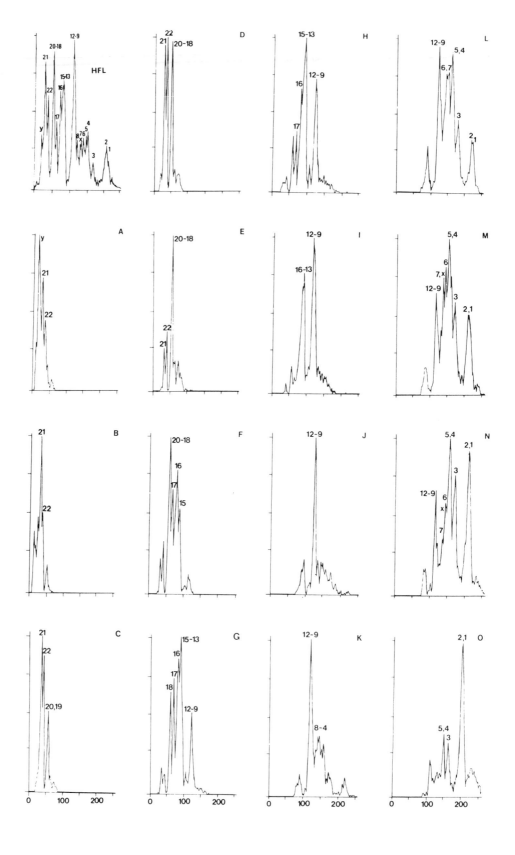

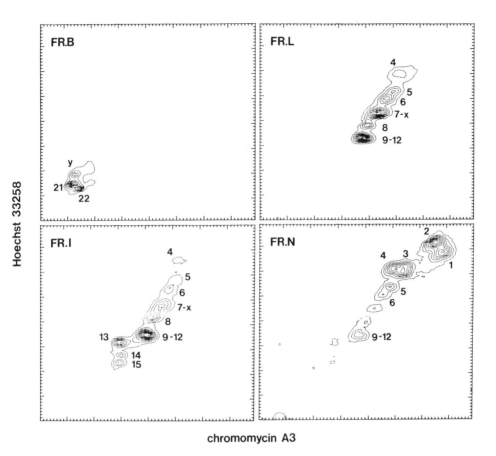

Fig. 3. Bivariate flow karyotypes of size-separated chromosomes similar to the ones depicted in Fig. 2 and stained with Hoechst 33258 ($2\,\mu$g/ml) and chromomycin A3 ($20\,\mu$g/ml). Chromosomes were analyzed with a FACSIV dual-laser flow sorter. The positions of the human chromosomes present are indicated and the height of each peak in the distributions is indicated by contours. Fractions B, I, L, N are not identical with, but are comparable to, the fractions B, I, L and N presented in Fig. 2. For details see Collard *et al.* (*8*).

Fig. 2. Separation of human chromosomes by velocity sedimentation at $52g$ using the Tulp chamber. Chromosomes were analyzed with a FACS IV flow sorter equipped with one 4 W argon ion laser and stained with mithramycin ($25\,\mu$g/ml) and ethidium bromide ($10\,\mu$g/ml). The histograms presented are directly derived from the FACS flow sorter without further mathematical manipulation or application of a curve-fit program. The maximum peak height is at 512 chromosomes per channel. Histogram HFL, unfractionated chromosomes; histograms A–O, the chromosomal composition of the different 4 ml fractions after $52g$ sedimentation. Numbers 1–22, X and Y indicate the position of the different human chromosomes in the histograms. For details see Collard *et al.* (*8*).

As discussed previously (*8*), by using different DNA-specific stains (mithramycin, ethidium bromide or Hoechst 33258) certain human chromosomes can be resolved in clear, single peaks using single-laser flow cytometry. By applying dual-laser flow cytometry with two different stains, almost all human chromosomes can be made visible as individual populations (*8,34,38,59*). Some representative examples of dual-laser analysis of chromosomal fractions are presented in Fig. 3. Fraction B contains only chromosomes Y, 21 and 22. Fraction I is highly enriched with chromosomes 9–12 and 13 while chromosomes 7, 8, 14 and 15 are present in lower amounts. Fraction L contains mainly chromosomes 9–12 with decreasing amounts of the larger chromosomes 3–8. Note that chromosomes 13–22 are completely absent in this fraction. Fraction N contains mainly the largest chromosomes 1–4.

As illustrated in Figs 2 and 3, velocity sedimentation makes it possible to enrich individual human chromosomes. The fractions thus obtained may be used either directly or as starting material for additional chromosome purification by sorting.

B. CHROMOSOME VELOCITY SEDIMENTATION AS A PRE-ENRICHMENT STEP PRIOR TO FLOW SORTING

It is evident that the chromosomal fractions, obtained by velocity sedimentation and stained with different fluorochromes, are ideal starting material for the sorting of specific chromosomes by fluorescence-activated flow sorting. The fractions are virtually free of cell debris, nuclei and chromosome clumps, which facilitates flow sorting. Furthermore, the desired chromosomes are present in relatively large amounts, which speeds up the sorting times considerably.

A representative example of flow sorting of various chromosomes using unfractionated or fractionated chromosomes is depicted in Fig. 4. All flow sortings are performed at a sample rate of 1000 chromosomes/sec. Using unfractionated chromosomes stained with a mithramycin/ethidium bromide mixture, chromosomes 21 and 22 (and in a following sort chromosomes 18–20 and 17) can be sorted, each at a rate of 15–40 chromosomes/sec (see Fig. 4). However, the sorting rate for chromosomes 21 and 22, predominantly present in the pre-enriched chromosome fraction D, can be increased up to 300 and 200 chromosomes/sec, respectively. The pre-enrichment step speeds up the sorting of these chromosomes by a factor of 10. The same result is obtained with chromosomes 18–20, which can be sorted at a rate of 40 chromosomes/sec from the unfractionated chromosome sample. Using fraction F, stained with the mithramycin/ethidium bromide mixture, chromosomes 18–20 can be obtained at a sort rate of 350 chromosomes/sec. This fraction contains hardly any chromosome 18 as a result of the prior separation on the basis of size. After staining fraction F with Hoechst 33258 (*32*), chromosomes 19 and 20 appear as single peaks in the histogram and can each be sorted at a rate of 150 chromosomes/sec. To check the purity after sorting, the deflected chromosomes are analyzed again as shown in the lower part of Fig. 4. Thus, velocity sedimentation of chromosomes appears to be an effective means to reduce the time needed to isolate a certain number of individual chromosomes. Unavoidably, velocity sedimentation will result in some dilution of the separated chromosomes and a too high dilution will limit the sorting speed. Chromosome fractions can be stored at −20°C so that chromosomes present in the additional fractions of the velocity sedimentation gradient may also be used for the purificatior of other desired chromosomes.

C. GENE MAPPING BY MOLECULAR HYBRIDIZATION WITH GENE-SPECIFIC DNA PROBES

Besides the classical gene mapping procedure with somatic cell hybrids, four other different

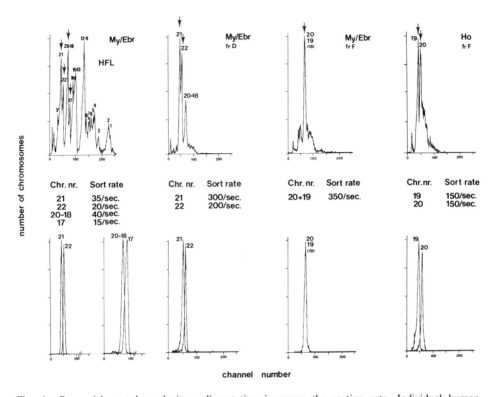

Fig. 4. Pre-enrichment by velocity sedimentation increases the sorting rate. Individual human chromosomes were sorted from unfractionated (HFL) and fractionated (fractions D and F) chromosomes. Sorting was performed at a sample rate of 1000 chromosomes/sec. Abscissa, channel number x, relative number of chromosomes per channel. The particular chromosomes which were sorted are arrowed. The sorting rate for each chromosome is presented under the upper panels. The corresponding lower panels represent the flow analysis of the deflected chromosomes. My/Ebr and Ho, chromosomes stained with mithramycin/ethidium bromide and Hoechst 33258, respectively. Numbers in the histograms denote the chromosome numbers. For details see Collard *et al.* *(8)*.

methods, all utilizing molecular hybridization techniques, localize structural genes or introduced genes at the chromosomal level.

The first and perhaps most direct approach involves *in situ* hybridization with radioactive or fluorescent-labeled probes to detect certain DNA sequences on individual metaphase chromosomes *(20,23)*. However, due to the limitations of sensitivity frequently experienced with single-copy genes, this method is rather time consuming and tedious by the necessity to score each chromosome in many karyotypes.

In a second approach, localization of structural genes can be performed by using DNA from a wide range of somatic cell hybrids in filter hybridization experiments with suitable c-DNA probes *(47)*. If the appropriate cell hybrids are available, structural genes can be localized on specific chromosomes in certain cell types, but the localization of rearranged or transfected genes requires the formation of a new panel of somatic cell hybrids.

The third approach involves the combination of molecular hybridization to DNA separated on agar gels and blotted on nitrocellulose filters with large-scale chromosomal fractionation by velocity sedimentation or by fluorescence-activated flow sorting. This approach permits the localization of structural genes or introduced specific DNA sequences on groups of chromosomes, if not on individual chromosomes (*9,37*).

These three straightforward physical methods have the advantage of circumventing the requirement of genetic manipulation (i.e. somatic cell hybridization). In addition, the DNA of fractionated chromosomes can be used for gene-enrichment purposes in molecular cloning experiments.

The fourth method, avoiding DNA isolation and agar gel electrophoresis, is the recently developed chromosome spot hybridization technique (*2,7,36,38*, see Chapter 14). Small numbers of chromosomes are sorted directly onto nitrocellulose filters and the chromosomal DNA is subsequently hybridized with gene-

specific radioactively labeled DNA probes. In the following sections examples of the latter two techniques are given. Of these, the chromosome spot hybridization method is the most powerful technique for gene mapping studies on individual chromosomes. In addition, this method lends itself to the identification of individual chromosomes or chromosome translocation corresponding with the peaks of a flow karyotype histogram.

a. Gene Mapping on DNA Isolated from Fractionated Rat Chromosomes

DNA of size-separated chromosomes may be used to localize structural genes on the

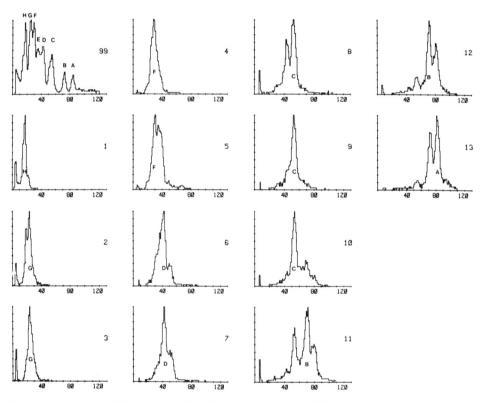

Fig. 5. Flow cytometric histograms of rat chromosomes separated by velocity sedimentation at 30*g*. Histogram 99, unfractionated sample; histograms 1–13 relate to fraction numbers, the chromosomal composition of the fractions after 30*g* sedimentation. The different peaks denoted A–H in the unfractionated histogram represent different groups of chromosomes: A, chromosome 1; B, chromosome 2; C, chromosomes 3–5; D, chromosomes 6, 7 and X; E, chromosomes 8–10; F, chromosomes 11, 13, 14; G, chromosomes 15–18; H, chromosomes 12, 19, 20, Y and an extra chromosome. Chromosome analysis was performed with an ICP11, equipped with a mercury arc and a sheath flow cell (*9*).

chromosomal level. In Fig. 5, a representative example of the separation of rat chromosomes by velocity sedimentation is presented. Flow cytometric analysis of the unfractionated rat chromosomes reveals a histogram with eight distinct peaks, designated A–H (Fig. 5, histogram 99). The different peaks represent the various groups of rat chromosomes with overlapping DNA content and size. Histograms 1–13 show the different fractions obtained after velocity sedimentation. The first fractions contain only the small chromosomes belonging to peak H. Going to higher fraction numbers, the larger chromosomes appear while at the same time the smaller chromosomes disappear. The last fractions contain virtually only the largest rat chromosomes. For gene mapping purposes, DNA from the chromosomal fractions was purified, digested with restriction endonuclease, and after agarose gel electrophoresis transferred to nitrocellulose filters (9). These filters were then used in hybridization studies with radioactively labeled gene-specific DNA probes. As shown in Fig. 6

the rat immunoglobulin genes were detected on the DNA of the fractionated rat chromosomes by cross hybridization with a mouse immunoglobulin light-chain kappa c-DNA probe. The hybridization pattern on the DNA of the chromosomal fractions is similar to that obtained with DNA of unfractionated chromosomes (lane T). After chromosome fractionation, the hybridization signals are found in the chromosomal fractions 8–10, being predominant in fraction 9, which contain mainly the chromosomes of peak C. Fractions enriched in the adjacent peaks (B, D) show a lower degree of hybridization. Consequently, the rat genes coding for the immunoglobulin light-chain kappa are located on chromosomes 3, 4 or 5; the chromosomes belonging to peak C. Thus, chromosome velocity sedimentation and subsequent analysis of chromosomal DNA by filter hybridization allows one to localize structural genes on the chromosomal level of cells. The same procedure may be followed when studying chromosomal or extrachromosomal localization of transfected DNA sequences in cells. The advantage of this procedure is that besides the gene mapping possibilities, DNA of the chromosomal fractions is also available for cloning purposes (see Section IV.D). Obviously, in general, the method is limited to localization of genes within groups of chromosomes and seldom is direct assignment to a single chromosome possible.

b. Gene Mapping by Chromosome Spot Hybridization

Recently, a more direct and highly preferential method has been developed to map genes on isolated metaphase chromosomes (2,7,36,38) (see also Chapter 14). Since virtually all human chromosomes can be resolved in a flow karyotype these chromosomes may be isolated by fluorescence-activated flow sorting. The chromosome spot hybridization technique is based on the sorting of individual chromosomes directly onto nitrocellulose filters. By mild aspiration of the filters during sorting (2,7) the chromosomes are collected and confined to a small spot and sorting of a small number of

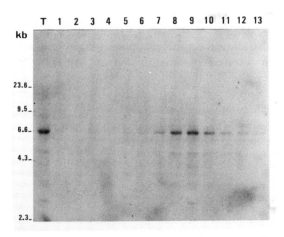

Fig. 6. Filter hybridization of a mouse immunoglobulin light-chain kappa c-DNA probe with DNA derived from size-separated rat chromosomes; lane T, 4 μg DNA of unfractionated chromosomes; lanes 1–13, 1 μg DNA of each chromosomal fraction. The markers indicate the position of the fragments of Hind III digested lambda phage DNA. The DNA in lanes 1–13 corresponds to the flow cytometric histograms in Fig. 5, fractions 1–13. For details see Collard *et al.* (9).

chromosomes is sufficient to detect gene-specific hybridization with single-copy DNA probes. Besides use for gene mapping, chromosome spot hybridization may also be used to identify individual chromosomes or chromosomal aberrations in a flow karyotype (*2,7*). Chromosomes present in a certain peak of a flow karyotype can be sorted onto filters and identified by hybridizing the sorted chromosomal DNA with chromosome-specific DNA probes. In this way, chromosome spot hybridization circumvents the problem of chromosome contraction, interfering with the identification of sorted chromosomes by banding techniques.

A representative example of the chromosome spot hybridization technique is depicted in Fig. 7, underlining the high detection level of the method. Chromosomes pre-enriched by velocity sedimentation were used as starting material for chromosome sorting (see Sections IV.A and B). From velocity sedimentation fraction I, highly enriched with the human chromosomes 9–12 and 13–16, increasing amounts of chromosomes 9–12 and 13–16 were sorted onto nitrocellulose filters. Subsequently, the chromosomal DNA on the filters was hybridized with the c-KI-ras-2 DNA probe. This single-copy gene has been mapped earlier to chromosome 12 (*44*). As expected, gene-specific hybridization was found only on filters containing chromosomes 9–12. Sorting of 40 000 chromosomes 9–12 was sufficient, indicating that 10 000 c-K-ras gene copies (chromosome 12) were required to detect the gene. In order to demonstrate that all filters contain chromosomal DNA, the filters were rehybridized with the human ALU probe (*46*). This probe recognizes human repetitive sequences present on all human chromosomes. Hybridization was found in all spots on the filters and was proportional to the number of chromosomes sorted as determined by spot size measurements. From the velocity sedimented pre-enriched chromosome fraction N, chromosomes 1–2 and 3–5 were sorted onto filters and hybridized with the N-ras probe (Fig. 7). The N-ras gene, localized to chromo-some 1 (*19*), was detected after sorting of 40 000 chromosomes 1 and 2, indicating that 20 000 chromosomes 1 were sufficient to detect the gene. As before, the filters were re-hybridized with the ALU-repeat probe to visualize the chromosomal DNA on all filters.

Chromosome spot hybridization is a rapid and simple method to detect gene-specific hybridization to flow sorted chromosomes. Sorting of 10 000–30 000 chromosomes (1–15 ng DNA) is sufficient to detect single-copy probe hybridization, depending on the quality of the probes used. Spot hybridization to chromosomes derived from the pre-enriched velocity sedimentation fractions can quickly narrow the choice of chromosomes which need to be sorted (*7*). Gene mapping can be performed directly on chromosomal DNA without time-consuming DNA extraction from flow-sorted (*37*) or fractionated chromosomes (*9*). Since resolution of virtually all human chromosomes in a flow karyotype can be achieved by applying dual-laser flow cytometry, filter panels of individual human chromosomes can be made for gene mapping purposes, as shown recently (*36,38*). If suitable cell lines are available with specific translocated chromosomes, the spot hybridization method also allows regional mapping of DNA probes to flow-sorted or fractionated chromosomes as reported for the C-myb oncogene (*7*). Although chromosome spot analysis requires considerable investment to assemble a sorting system, an operational system can localize genes rapidly and at low cost.

D. CHROMOSOME FRACTIONATION AND THE CONSTRUCTION OF CHROMOSOME-SPECIFIC RECOMBINANT DNA LIBRARIES

Chromosome-specific recombinant DNA libraries are a valuable aid to human gene mapping and to the molecular analysis of human genetic diseases. Such libraries may facilitate the acquisition of probes used in

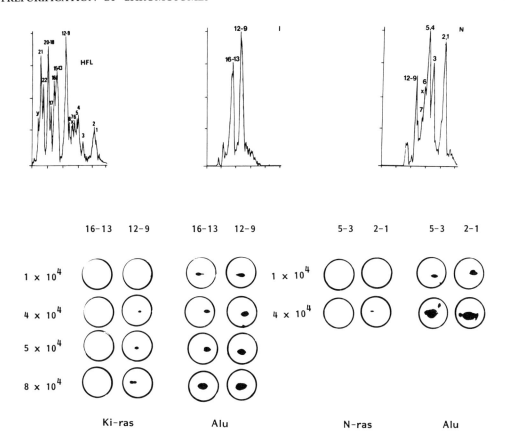

Fig. 7. (Top) Flow histograms (FACS IV) of human chromosomes stained with mithramycin and ethidium bromide. Abscissa, channel number, the relative amount of DNA per chromosome; ordinate, the relative number of chromosomes per channel. Histogram HFL, unfractionated chromosomes; histograms I and N, chromosomal composition of two fractions obtained by velocity sedimentation. Numbers in the histogram indicate the position of the different human chromosomes as identified previously (8). (Bottom) Chromosome spot hybridization. The eight filters on the left contain various numbers of sorted chromosomes 9–12 and 13–16 (as indicated) derived from fraction I. The filters were hybridized with c-Ki-ras probe (Ki-ras, exposure time 5 days) and subsequently with the human ALU-repeat probe (ALU, exposure time 12 h). The four filters on the right contain various numbers of sorted chromosomes 3–5, 1 and 2 (as indicated) derived from fraction N. The filters were hybridized with the N-ras probe (N-ras, exposure time 5 days) and subsequently with the ALU probe (ALU, exposure time 1 day). For details see Collard et al. (7).

the investigation of restriction-enzyme-length polymorphisms linked to human genetic diseases (3,6,18,43); see also Chapters 15 and 16.

Obviously, the most direct approach to the purification of chromosome-specific DNA sequences involves the isolation of the desired chromosomes by fluorescence-activated flow sorting. In this way, several chromosome-specific or chromosome-enriched recombinant DNA libraries have been constructed using human or rodent cell lines with normal and abnormal chromosome numbers (13,16,26,28) or cell lines with marker chromosomes (14,28,31). However, flow sorting alone is still very limited in providing the large quantities of chromosomes required for cloning purposes. Therefore, virtually all chromosome-specific or chromosome-enriched recombinant DNA libraries have been made from DNA digested to completion because of the difficulty in controlling partial restriction enzyme distribution. This approach does not provide complete representation of chromosome-specific DNA sequences, because of the limited acceptance range of the cloning vector and the spacing of the restriction enzyme recognition sites. This problem may perhaps be obviated by constructing separate chromosome-specific libraries from total restriction digests of DNA by using various vectors with different acceptance ranges and for vectors that accept different restriction endonuclease cleavage fragments. Alternatively, if sufficient DNA is available, chromosome-specific recombinant DNA libraries may be constructed from partially digested DNA, thus providing a complete representation of the DNA sequences on a chromosome.

As discussed in Sections IV.A and B, chromosome separation by velocity sedimentation may be applied to pre-enrich chromosomes prior to flow sorting in order to obtain relatively large amounts of chromosomal DNA within a reasonable time for molecular cloning. In the near future, high-speed flow sorters may also facilitate the acquisition of large amounts of chromosomal DNA from unique human chromosomes. Alternatively, it is possible to use mouse–human hybrid cell lines, containing one human chromosome of interest only, as starting material for the construction of chromosome-specific libraries. Instead of cloning total hybrid cell DNA, the DNA of the desired human chromosome can be considerably enriched by chromosome velocity sedimentation. The large amount of DNA

available then allows a partial restriction enzyme digestion and DNA-size fractionation prior to cloning. In this way, a library can be constructed highly enriched with DNA sequences derived from one unique human chromosome. An example of this approach is given below.

For the construction of a human chromosome-21-specific recombinant phage library we used a mouse–human hybrid cell line (SCC 16.5) in which chromosome 21 is the only human chromosome present. Chromosomes of SCC 16-5 cells were isolated and separated according to their differences in size by velocity sedimentation (Fig. 8). The first fractions were highly enriched with small chromosomes; the medium-sized and larger chromosomes being absent. Since human chromosome 21 was one of the smallest chromosomes present in the cells, it was expected that this chromosome would appear in one of the early fractions of the gradient. As shown in Fig. 9, fraction 1 and, especially, fraction 2 are highly enriched with human DNA as assessed by dot hybridization with the human "ALU" repeat probe. By comparing the dot hybridization results (Fig. 9) with the flow histograms (Fig. 8) it is concluded that peak B contains human chromosome 21. Optimal chromosome fractionation in four separation chambers and careful pooling of fractions highly enriched with chromosome 21 (peak B) resulted in chromosomal material consisting of approximately $2\mu g$ DNA (50% human chromosome 21 DNA). Due to the fortunate size difference of chromosome 21 relative to that of the mouse chromosomes we were able to enrich chromosome 21 approximately 70-fold (from 0.7% per SCC 16-5 genome to 50% in the chromosomal

Fig. 8.   Flow cytometric histograms (FACS IV) of the chromosomes from the hybrid cell line SCC 16-5 separated by velocity sedimentation at $52g$ for 50 min at 22°C. Abscissa, channel number x, the relative amount of DNA per chromosome; ordinate, distribution $f(x)$, the relative number of chromosomes per channel. Histogram SCC, unfractionated chromosomes analyzed before layering onto the gradient; histograms 1–15, chromosomal composition of the different fractions after velocity sedimentation. A–L represent different chromosome peaks in the histograms. A mixture of mithramycin and ethidium bromide was used for chromosome staining.

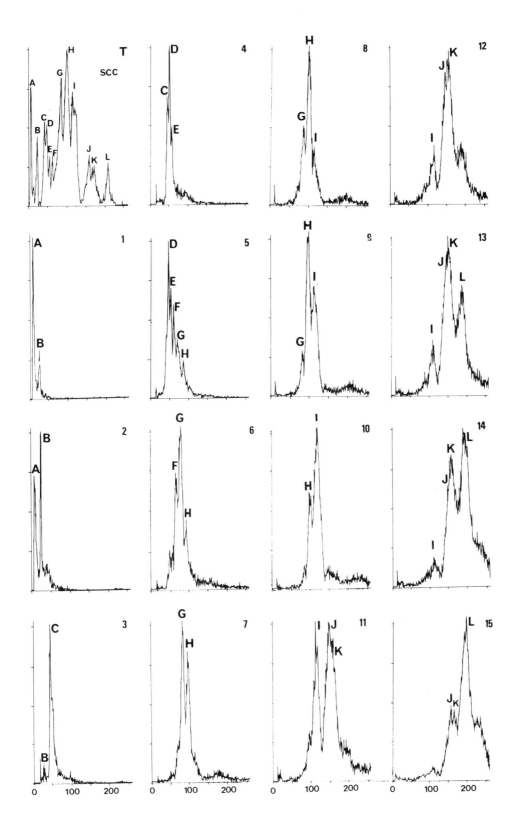

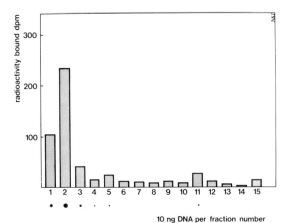

Fig. 9. Dot hybridization of DNA from the different chromosomal fractions with the human ALU-repeat probe, BLUR-8. Numbers 1–15 represent 10 ng DNA of the different chromosomal fractions as presented in Fig. 8. The radioactivity bound to the DNA filters was counted by liquid scintillation counting.

fractions). In general, the size position of the desired chromosome in hybrid cells will be less ideal so that additional purification by chromosome sorting will still be required. The DNA derived from the chromosomal fractions highly enriched for chromosome 21 was subsequently used for the construction of the library. DNA (2 μg) was partially digested with MbO I, size fractionated on a NaCl gradient, and fragments between 15 and 25 kbp in size were cloned in the EMBL-3 phage vector (*15*). Of the approximately 21 000 phage clones obtained at least 99% appeared to be recombinants. Following phage plaque filter hybridization and Southern blotting it was found that half of the recombinants were positive for human repetitive DNA (*22*). All 12 human recombinants studied thus far contain DNA inserts ranging in size between 15 and 20 kbp and originate from chromosome 21 only (*22*).

In principle, the presented strategy meets the criteria of purity, quality and completeness of chromosome-specific recombinant DNA libraries and circumvents the drawbacks of libraries made from DNA digested to completion. Since the isolated DNA derived from the velocity sedimentation fraction appeared to be at least 40–60 kbp in size, it should be feasible to clone this DNA directly into a cosmid vector (*17*). Furthermore, improvement of the packaging procedure (*45*) will considerably reduce the amount of DNA required for cloning. With regard to the cells used, it is of interest that different methods have been developed to obtain stable hybrids retaining unique human chromosomes. Hybrids carrying human chromosomes 11 and 12 only have been made by fusion of human cells with auxotrophic Chinese hamster cells (*24*). Human cell harboring transfected selectable marker genes on specific chromosomes may be used in microcell transfer, or cell fusion experiments, in order to obtain hybrids containing unique human chromosomes (*25,58*). Newly arisen cell surface antigens in combination with fluorescence-activated cell sorting may be used to select hybrids carrying specific human chromosomes (*49*). Chromosomes isolated from such stable hybrids either by chromosome velocity sedimentation, by fluorescence-activated chromosome sorting or by a combination of both techniques may be used to construct recombinant DNA sequences specific for any individual human chromosome.

ACKNOWLEDGEMENTS

J.W.G. Janssen has been supported by the Netherlands Cancer Foundation (Koningin Wilhelmina Fonds).

REFERENCES

1. Benz, R. D., and Burki, J. (1978). Preparation and identification of partially fractionated Chinese hamster chromosomes. *Exp. Cell Res.* **112**, 143–153.

2. Bernheim, A., Metezeau, P., Guellaen, G., Fellous, M., Goldberg, M. E. and Berger, R. (1983). Direct hybridization of sorted human chromosomes: localization of the Y chromosomes on the flow karyotype. *Proc. Natl Acad. Sci. USA* **80**, 7571–7575.

3. Bhattacharya, S. S., Wright, A. F., Clayton, J. F., Price, W. H., Phillips, C. I., McKeown, C. M. E., Jay, M., Bird, A. C., Pearson, P. L., Southern, E. M., and Evans, H. J. (1984). Close genetic linkage between X-linked retinis pigmentosa and a restriction fragment length polymorphism identified by recombinant DNA probe L 1.28. *Nature* **309**, 253–255.

4. Burki, J. H., Regimbal, T. J., and Mel, H. C. (1973). Zonal fractionation of mammalian metaphase chromosomes and determination of their DNA content. *Prep. Biochem.* **3**, 157–182.

5. Carrano, A. V., Gray, J. W., Langlois, R. G., Burkhart-Schultz, K. J., and Van Dilla, M. A. (1979). Measurement and purification of human chromosomes by flow cytometry and sorting. *Proc. Natl Acad. Sci. USA.* **76**, 1382–1384.

6. Cavenee, W. K., Drya, T. P., Phillips, R. A., Benedict, W. F., Godbout, R., Gaillie, B. L., Murphree, A. L., Strong, L. C. and White, R. L. (1983). Expression of recessive alleles by chromosomal mechanism in retinoblastoma. *Nature* **305**, 779–784.

7. Collard, J. G., de Boer, P. A. J., Janssen, J. W. G., Schijven, J. F., and De Jong, B. (1985). Gene mapping by chromosome spot hybridization. *Cytometry* **6**, 179–185.

8. Collard, J. G., Philippus, E., Tulp, A., Lebo, R. V., and Gray, J. W. (1984). Separation and analysis of human chromosomes by combined velocity sedimentation and flow sorting applying single- and dual-laser flow cytometry. *Cytometry* **5**, 9–19.

9. Collard, J. G., Schijven, J. F., Tulp, A., and Meulenbroek, M. (1982). Localization of genes on fractionated rat chromosomes by molecular hybridization. *Exp. Cell Res.* **137**, 463–469.

10. Collard, J. G., Tulp, A., Hollander, J. H., Bauer, F.W., and Boezeman, J. (1980). Separation of large quantities of chromosomes by velocity sedimentation at unit gravity. *Exp. Cell Res.* **126**, 191–197.

11. Collard, J. G., Tulp, A., Stegeman, J., Boezeman, J., Bauer, F. W., Jongkind, J. F., and Verkerk, A. (1980). Separation of large quantities of chinese hamster chromosomes by velocity sedimentation at unit gravity followed by flow sorting (FACS IV). *Exp. Cell Res.* **130**, 217–227.

12. Collard, J. G., Tulp, A., Stegeman, J., and Boezeman, J. (1981). Separation of human metaphase chromosomes at neutral pH by velocity sedimentation at twenty times gravity. *Exp. Cell Res.* **133**, 341–346.

13. Davies, K. E., Young, B. D., Elles, R. G., Hill, M. E., and Williamson, R. (1981). Cloning of a representative genomic library of human X chromosome after sorting by flow cytometry. *Nature* **293**, 374–376.

14. Disteche, C. M., Kunkel, L. M., Lojewski, A., Orkin, S. H., Eisenhard, M., Sahar, E., Travis, B., and Latt, S. A. (1982). Isolation of mouse X-chromosome specific DNA from an X-enriched lambda phage library derived from flow sorted chromosomes. *Cytometry* **2**, 282–286.

15. Frischauf, A. M., Lehrach, H., Poustka, A., and Murray, N. (1983). Lambda replacement vectors carrying polylinker sequences. *J. Mol. Biol.* **170**, 827–842.

16. Griffith, J. K., Cram, L. S., Crawford, B. D., Jackson, P. J., Schilling, J., Schimke, R. T., Walters, R. A., Wilder, M. E., and Jett, J. H. (1984). Construction and analysis of DNA sequence libraries from flow sorted chromosomes: practical and theoretical considerations. *Nucl. Acids Res.* **12**, 4019–4034.

17. Grosveld, F. G., Dahl, H. M., De Boer, E., and Flavell, R. A. (1981). Isolation of β-globin related genes from a human cosmid library. *Gene* **13**, 227–237.

18. Gusella, J. F., Wexler, N. S., Conneally, P. M., Naylor, S. L., Anderson, M. A., Tanzi, R. E., Watkins, P. C., Ottina, K., Wallace, M. R., Sakagushi, A. Y., Yong, A. B., Shoulson, I., Bonilla, E., and Martin, J. B. (1983). A polymorphic DNA marker genetically linked to Huntingdon's disease. *Nature* **306**, 234–238.

19. Hall, A., Marshall, C. J., Spurr, N. K., and Weiss, R. A. (1983). Identification of transforming gene into human sarcoma cell lines as a new member of the ras gene family located on chromosome 1. *Nature* **303**, 396–400.

20. Harper, M. E., and Saunder, G. F. (1981). Localization of single copy DNA sequences on G-banded human chromosomes by *in situ* hybridization. *Chromosoma* **83**, 431–439.

21. Hughes, S. H., Stubblefield, E., Payvar, F., Engel, J. D., Dogson, J. B., Spector, D., Cordeel, B., Schimke, R. T., and Varmus, H. E. (1979). Gene localization by chromosome fractionation: globin genes are on at least two chromosomes and three estrogen-inducible genes are on three chromosomes. *Proc. Natl Acad. Sci. USA* **76**, 1348–1352.

22. Janssen, J. W. G., Collard, J. G., Tulp, A., Cox, D., Millington-Ward, A., and Pearson, P. (1986). Construction and analysis of an EMBL-3 phage library representing partially digested human chromosome 21 specific DNA inserts. *Cytometry* **7**, 411–417.

23. Janssen, J. W. G., Vernole, P., Boer, P. A. J. de., Oosterhuis, J. W., and Collard, J. G. (1986). Sublocalization of c-myb to 6q21-q23 and c-myb expression in a human teratocarcinoma with 6q rearrangements. *Cytogenet. Cell Genet.,* **41**, 129–135.

24. Kao, F., Jones, C., and Puck, T. T. (1976). Genetics of somatic mammalian cells: genetic immunologic and biochemical analysis with Chinese hamster cell hybrids containing selected human chromosomes. *Proc. Natl Acad. Sci. USA* **73**, 193–197.

25. Kit, S., Hazen, M., Otsuka, H., Quavi, H., Frkula, S., and Dubbs, D. (1981). The site of integration of the herpes simplex virus type 1 thymidine kinase gene in human cells transformed by an HSV-1 DNA fragment. *Int. J. Cancer* **28**, 767–776.

26. Krumlauf, R., Jeanpierre, M., and Young, B. D.

(1982). Construction and characterization of genomic libraries from specific human chromosomes. *Proc. Natl Acad. Sci. USA* **79**, 2971–2975.

27. Kunkel, L. M., Tantrahavi, U., Eisenhard, M., and Latt, S. A. (1982). Regional localization on the human X of DNA segments cloned from flow sorted chromosomes. *Nucl. Acids Res.* **10**, 1557–1578.

28. Lalande, M., Kunkel, L. M., Flint, A., and Latt, S. A. (1984). Development and use of metaphase chromosome flow sorting methodology to obtain recombinant phage libraries enriched for parts of the human X chromosome. *Cytometry* **5**, 101–107.

29. Langlois, R. G., Yu, L.-C., Gray, J. W., and Carrano, A. V. (1982). Quantitative karyotyping of human chromosome by dual beam flow cytometry. *Proc. Natl Acad. Sci. USA* **79**, 7876–7880.

30. Latt, S. A. (1975). Fluorescent probes of chromosome structure replication. *Can. J. Genet. Cytol.* **19**, 603–623.

31. Latt, S. A., Alt, F. W., Schreck, R. R., Kanda, N., and Baltimore, D. (1982). *In* "Gene amplification" (R. T. Schimke, ed.), pp. 283–289. Cold Spring Harbor Laboratory, Cold Spring Harbor.

32. Latt, S. A. and Wohlleb, J. C. (1975). Optical studies of the interaction of 33258 Hoechst with DNA, chromatin, and metaphase chromosomes. *Chromosoma* **25**, 297–316.

33. Lebo, R. V. (1982). Chromosome sorting and DNA sequence localization. *Cytometry* **3**, 145–154.

34. Lebo, R. V., and Bastian, A. M. (1982). Design and operation of a dual laser chromosome sorter. *Cytometry* **3**, 213–219.

35. Lebo, R. V., Carrano, A. V., Burkart-Schultz, K. J., Dozy, A. M., Yu, L.-C. and Kan, Y. W. (1979). Assignment of human β-, γ-globin and δ-globin genes to the short arm of chromosomes 11 by chromosome sorting and DNA restriction enzyme analysis. *Proc. Natl Acad. Sci. USA* **76**, 5804–5808.

36. Lebo, R. V., Gorin, F., Fletterick, R. J., Kao, F.-T., Cheung, M.-C., Bruce, B. D., and Kan, Y. W. (1984). High resolution chromosome sorting and DNA spot–blot analysis assign McArdle's syndrome to chromosome 11. *Science* **225**, 57–59.

37. Lebo, R. V., Kan, Y. W., Cheung, M.-C., Carrano, A. V., Yu, L.-C., Chang, J. C., Cordell, B., and Goodman, H. M. (1982). Assigning the polymorphic human insulin gene to the short arm of chromosome 11 by chromosome sorting. *Hum. Genet.* **60**, 10–15.

38. Lebo, R. V., Tolan, D. R., Bruce, B. D., Cheung, M.-C., and Kan, Y. W. (1985). Spot–blot analysis of sorted chromosomes assigns a fructose intolerance disease locus to chromosome 9. *Cytometry* **6**, 478–483.

39. McCutcheon, M. J., and Miller, R. G. (1982). Flexible sorting decision and droplet charging control electronic circuitry for flow cytometer-cell sorters. *Cytometry* **2**, 219–225.

40. Mendelsohn, J., Moore, D. E., and Salzman, N. P. (1968). Separation of isolated Chinese hamster chromosomes into three size-groups. *J. Mol. Biol.*, **32**, 101–108.

41. Miller, R. G. (1973). Separation of cells by velocity sedimentation. *In* "Biophysics Cell Biology" (R. H. Pain and B. J. Smith eds), pp. 87–112. Wiley, New York.

42. Miller, R. G. (1984). Separation of cells by velocity sedimentation. *In* "Methods in Enzymology"(G. Di Sabato, J. Langone, and H. Van Vunakis, eds), Vol. 108, pp. 64–87. Academic Press, New York.

43. Murray, J. M., Davies, K. E., Harper, P. S., Meredith, L., Mueller, C. R., and Williamson, R. (1982). Linkage relationship of a cloned DNA sequence on the short arm of the X chromosome to Duchenne muscular dystrophy. *Nature* **300**, 69–71.

44. O'Brien, S. J., Nash, W. G., Goodwin, J. L., Lowy, D. R., and Chang, E. (1983). Dispersion of the ras family of transforming genes to four different chromosomes in man. *Nature* **302**, 839–842.

45. Rohme, D., Fox, H., Herrman, B., Frischauf, A. M., Edstrom, J.E., Mains, P., Silver, L. M., and Lehrach, H.J. (1984). Molecular clones of the mouse t complex derived from microdissected metaphase chromosome. *Cell* **36**, 783–788.

46. Rubin, C. M., Houck, C. M., Deininger, P. L., Friedmann, T., and Schmid, C. W. (1980). Partial nucleotide sequence of the 300-nucleotide interspersed repeated human DNA sequences. *Nature* **284**, 372–374.

47. Ruddle, F. H., and Creagan, R. P. (1975). Parasexual approaches to the genetics of man. *Ann. Rev. Genet.* **9**, 407–486.

48. Schneider, E. L., and Salzman, N. P. (1970). Isolation and zonal fractionation of metaphase chromosomes from human diploid cells. *Science* **167**, 1141–1143.

49. Schroder, J., Nikinmaa, B., Kavathas, B., and Herzenberg, L. (1983). Fluorescence activated cell sorting of mouse–human hybrid cells aids in locating the gene for the Leu 7 (HNK-1) antigen to human chromosome 11. *Proc. Natl Acad. Sci. USA* **80**, 3421–3424.

50. Sillar, R., and Young, B. D. (1981). A new method for the preparation of metaphase chromosomes for flow analysis. *J. Histochem. Cytochem.* **29**, 74–78.

51. Stubblefield, E., and Oro, J. (1982). The isolation of specific chicken macrochromosomes by zonal centrifugation and flow sorting. *Cytometry* **2**, 273–281.

52. Stubblefield, E., and Wray, W. (1978). Isolation of specific human metaphase chromosomes. *Biochem. Biophys. Res. Commun.* **83**, 1404–1414.

53. Tulp, A., Barnhoorn, M., Collard, J. G., Kraan, W., Sluyser, M., Polak, F., and Aten, J. A. (1982). Separation of mammalian cells and cell organelles at low *g*-forces. *In* "Methodological Surveys" (E., Reid, G. Cook, and D. J. Morre, eds), Vol. II, pp. 158–170. Wiley, New York.

54. Tulp, A., and Collard, J. G. (1986). Methods for enrichment of cellular specimins for flow cytometry

and sorting. *In* "Flow cytometry and Sorting" (M. R. Melamed, T. Lindmo, and M. L. Mendelsohn, eds). Alan R. Liss, New York.

55. Tulp, A., Hart, A. A. M., Barnhoorn, M. G., and Stukart, M. J. (1986). Separation of cells and cell organelles by low gravity forces. *In* "Techniques in the Life Sciences, Cell Biology" (E. Kurstak, ed.), pp. 1–31. Elsevier, Ireland.

56. Tulp, A., Kooi, M. W., Kipp, J. B.A., Barnhoorn, M. G., and Polak, F. (1981). A separation chamber to sort cells, nuclei and chromosomes at moderate *g*-forces. II. Studies on velocity sedimentation and equilibrium density centrifugation of mammalian cells. *Anal. Biochem.* **117**, 354–365.

57. Tulp, A., De Leng, P., and Barnhoorn, M. (1980). A sedimentation chamber to sort cells, nuclei and chromosomes at ten times gravity. *Anal. Biochem.* **107**, 32–43.

58. Tunnacliffe, A., Parkar, M., Povey, S., Bengtsson, B. O., Stanley, K., Solomon, E., and Goodfellow, P. (1983). Integration of Ecogpt and SV40 early region sequences into human chromosome 17: a dominant selection system in whole cell and microcell human-mouse hybrids. *EMBO J.* **2**, 1577–1584.

59. van den Engh, G. J., Trask, B. J., Gray, J. W., Langlois, R. G. and Yu, L.-C. (1985). Preparation and bivariate analysis of suspensions of human chromosomes. *Cytometry* **6**, 92–100.

60. Young, B. D., Ferguson-Smith, M. A., Sillar, R., and Boyd, E. (1981). High resolution analysis of human peripheral lymphocyte chromosomes by flow cytometry. *Proc. Natl Acad. Sci. USA* **78**, 7727–7731.

Chromosome Purification by High-speed Sorting

DONALD C. PETERS

Biomedical Sciences Division
Lawrence Livermore National Laboratory
Livermore, CA 94550, USA

I. INTRODUCTION

The flow cytometer/cell sorter, as described in the previous chapter, can be used to identify and purify subpopulations of cells and organelles, such as chromosomes, with great accuracy. Most commercially available cell sorters can scan several thousand objects per second, selecting from these the objects to be purified. Unfortunately, this processing rate makes purification of bulk quantities of human chromosomes of a single type extremely time consuming. This severely limits the utility of chromosome sorting for gene mapping or for production of recombinant DNA libraries. For example, with a conventional cell sorter processing 2000 objects/sec, over 120 h of sorting time is required to purify 1 μg of DNA from the human Y chromosome (2). The time required to purify bulk quantities of chromosomes can be reduced almost ten-fold by the use of a high-speed (rather than conventional) cell sorter (4a,10).

The high-speed sorter differs from a conventional cell sorter mainly in an increased droplet production rate. Maintaining high-purity sorted populations requires that the probability of having more than one particle per drop must be low. Since the particles occur at random intervals, their occurrence can be treated as a Poisson process. The probability that the interval between two particles is larger than the droplet period is $\exp(-nt)$, where nt is the average number of particles during the period. This is also the probability that a desired particle will be sorted (sorting is aborted when more than one object would be included in a sort event). Therefore, it represents the sorting efficiency. Figure 1 shows the effect of input particle rate and droplet rate on sorting efficiency when three drops are sorted for each particle (a means to compensate for timing errors). Assuming a 75% sorting efficiency, the following comparison can be made. In conventional sorters, the droplet production rate is about 30 000–40 000/sec and the maximum particle processing rate is around 3000/sec. The high-speed sorter, with a droplet production rate of 220 000/sec, can process over 20 000 particles/sec.

The purpose of this chapter is to describe some of the theoretical, design and developmental aspects of the high-speed sorter that differentiate it from conventional cell sorters and to discuss its use in purifying human chromosomes for recombinant DNA libraries. I will refer to the high-speed sorter, which was designed and developed at the Lawrence Livermore National Laboratory, as the HiSS (*4a,10*).

II. SYSTEM DESIGN CONSIDERATIONS

A. FLUIDICS THEORY

The frequency of droplet production in any cell sorter is determined largely by its liquid jet velocity and diameter, and the magnitude of the surface disturbance induced on the jet by the action of a piezoelectric transducer. Rayleigh (*12*) showed that, for a given surface disturbance, droplets are produced most efficiently (i.e. jet breakup occurs most rapidly) when the frequency of droplet production, f, is

$$f = v/4.5\,d \qquad (1)$$

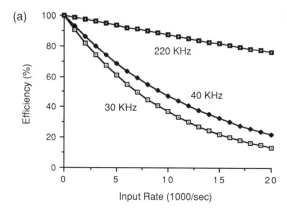

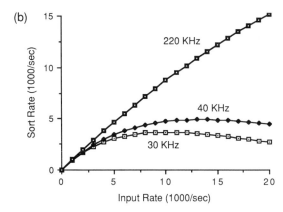

Fig. 1. Sorting efficiency comparison. Maintaining high-purity sorted populations requires that the probability of having more than one particle per drop must be low. Since the particles occur at random intervals, their occurrence can be treated as a Poisson process. The probability P that the interval between two particles is larger than the droplet period is $P = \exp(-nt)$, where nt is the average number of particles during the period. Thus, P is also the probability that a desired particle will be sorted. Therefore, it represents the sorting efficiency. (a) The efficiency when three drops are used to sort a particle is plotted versus the input particle rate for droplet rates of 30, 40 and 220 kHz. (b) The actual sort rate (efficiency times input rate) is plotted for the same conditions.

and that the maximum droplet frequency is

$$f = v/3.14\,d, \qquad (2)$$

where v is the jet velocity and d is the jet diameter. The magnitude of the surface disturbance induced on the jet is determined by the energy supplied by the piezoelectric transducer and by the acoustic transmission characteristics of the flow chamber and nozzle assembly. Webber (17), Kachel and Menke (6), Sweet (13), and Pinkel and Stovel (11) provide more information about the origin of these equations and about the general hydrodynamic behavior of sorters. The nozzle diameter is a compromise between a small size for increased droplet frequency and one large enough to produce little damage to cells or chromosomes and that, with some care, will rarely become plugged. An 80 μm diameter nozzle is used on the HiSS. Experiments carried out during the development of the HiSS indicated that stable droplet production could be achieved at droplet production frequencies up to 500 kHz by increasing the jet velocity to 100 m/sec (data not shown). However, the droplet production stability was marginal at 500 kHz and precise droplet charging was impossible (see below). Thus, somewhat arbitrarily, an intermediate velocity of 50 m/sec was selected for routine operation. The diameter of the jet emanating from the 80 μm diameter orifice is approximately 65 μm. With these data, Eq. (1) would predict a frequency of 170 kHz for minimum jet length and Eq. (2) a maximum frequency of 240 kHz. The HiSS operates at 220 kHz, close to the maximum, in order to maximize the particle processing rate.

According to Bernoulli's equation, the velocity of the sorter jet is related to the pressure according to the equation

$$v = \sqrt{2\,P/r},$$

where P is the water pressure and r is the density of the sheath liquid. Guided by this equation, the operating pressure on the sheath is adjusted to 200 psi to produce a 50 m/sec jet.

B. FLOW CHAMBER DESIGN

The design of the flow chamber and nozzle assembly is critical for stable droplet production (a requirement for accurate sorting). Figure 2(a) shows the flow chamber design. A key feature of this chamber is the length to diameter ratio of the orifice (shown in the expanded view) forming the jet in air. The stability of droplet formation increases significantly as this ratio is reduced, thereby minimizing the turbulence induced by the interaction between the flowing liquid and the orifice walls (4). The actual orifice is a synthetic ruby jewel through the center of which is a hemispherical depression centered on a 80 μm diameter hole. The side of the jewel opposite the hemispherical depression is polished to reduce the length of the 80 μm diameter region to about 40 μm. The jewel is epoxied, hemispherical depression down, into a cylindrical recess in the end of a square quartz flow chamber with a 250 μm square flow channel (11). Cytometric measurements are performed while the particles are in the square cross-section part of the chamber. This is advantageous because the high optical quality of this part of the chamber allows effective use of high-quality optical components for laser beam shaping and fluorescence collection, and because the jet disturbances induced by the piezoelectric transducer to force accurate droplet production do not affect the dual-beam flow cytometric measurements. In addition, the flow velocity in this region is only about one-tenth of that in the jet in air. The relatively low flow velocity allows the cells to remain in the measurement region longer so more accurate measurements of total particulate fluorescence are possible.

C. DROPLET CHARGING
AND DEFLECTION

The entire flow chamber is caused to vibrate at approximately 220 kHz by a piezoelectric crystal mounted axially on the top of the chamber. The pressure waves transmitted to the liquid jet result in the formation of sinusoidal variations of the jet diameter that grow exponentially with time and eventually cause the jet to break into droplets of equal size and spacing. The influence of the piezoelectric crystal insures that the length of the liquid jet is constant. When particles to be sorted reach the end of the liquid jet, a 150 V pulse is applied to

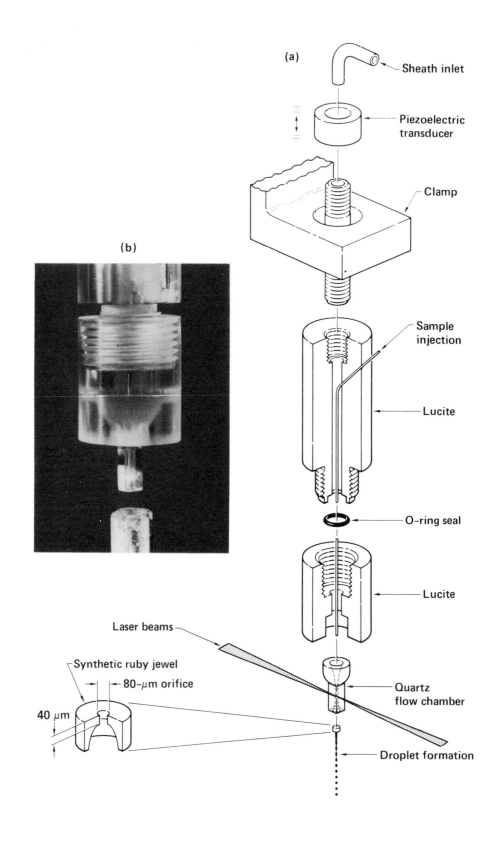

(a)

Sheath inlet

Piezoelectric transducer

Clamp

Sample injection

Lucite

O-ring seal

Lucite

Laser beams

Synthetic ruby jewel

80-μm orifice

40 μm

Quartz flow chamber

Droplet formation

(b)

the liquid jet for a time equal to that required for one or more droplets to break off. Each droplet that breaks off during this time has a charge proportional to its capacitance to ground and to the voltage at the end of the jet. The droplets then continue downward through an electric field. The charged droplets containing the particles of interest are deflected into a collection device. In the HiSS, two classes of particles can be sorted by applying a negative charge to droplets containing one class and a positive charge to droplets containing the other class.

Droplet charging and deflection in the HiSS is complicated by the shape of the jet at the break-off point and by the velocity of the drops during passage through the deflecting electric field. Figure 3 shows a picture of the liquid jet in the vicinity of the droplet break-off region. This region is approximately 2 cm downstream from the point where particles are measured and the transit time from the measurement region to the droplet break-off point is about $500\,\mu$sec. The most important differences between the HiSS and conventional sorters in this region are (1) the extended liquid neck between droplets in the HiSS (see arrow in Fig. 3) and (2) the higher jet velocity. The size of the neck region is sufficiently small (approximately $2\,\mu$m) to begin to interfere with the induction of charge on the separating droplets thereby interfering with proper droplet deflection during sorting. At higher jet velocity and droplet production frequencies, the neck diameter and extent decreases still further and eventually prevents proper droplet charging. The high jet velocity results in the charged droplets moving rapidly through the deflecting DC electric field used

to separate charged and uncharged droplets. To obtain sufficient droplet deflection the deflection plate length was increased from the conventional length of about 4 cm to approximately 18 cm. In addition, the potential difference between the plates is set at 10 kV to maximize the deflecting force on each charged droplet. The plate separation is greater than 7 mm to minimize arcing between the plates.

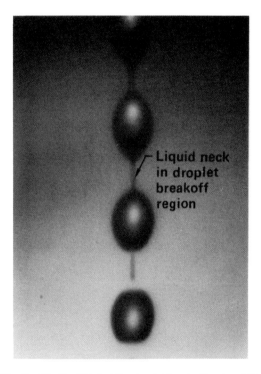

Fig. 3. The liquid jet at the droplet breakoff region. This photograph shows the long, thin neck that connects droplets prior to breakoff. The jet was illuminated by a pulsed light emitting diode (LED) driven at the droplet production frequency of 220 kHz.

Fig. 2. Diagram of the flow chamber. (a) Sheath liquid is injected axially through a tube and through a piezoelectric crystal mounted on top of the flow chamber. The liquid flows downward into a tapered quartz section where the sample is injected. Particles are entrained in the liquid and flow in single line through a $250\,\mu$m square quartz channel where they intersect the two laser beams, vertically separated by $75\,\mu$m. The fluid then exits through a $80\,\mu$m diameter orifice. The enlargement shows the orientation and shape of the orifice that is formed by a synthetic ruby jewel, polished to produce an orifice length to diameter ratio of 0.5 and inserted cone side down. (b) Shown here is a photograph of the lower portion of the flow chamber. Also shown is the upper portion of the grounding collar through which the liquid jet passes.

D. Overall System

The overall high-speed sorter system is illustrated in Fig. 4. The sorter is equipped with two argon ion lasers. To excite the dyes Hoechst 33258 (HO) and Chromomycin A3 (CA3) used for chromosome staining, one beam is adjusted to operate in the visible portion of the spectrum (458 nm), the other in the ultraviolet (351 and 363 nm). The visible output beam passes through a 300 mm focal length cylindrical lens, and is focused onto the sample stream in the quartz flow chamber with a 25 mm focal length spherical lens. The minor dimension of its elliptical cross-section is about 7 μm, minimizing the possibility of multiple objects passing through the beam at the same time. The major dimension of the eclipse is about 270 μm. The ultraviolet beam passes through 600 mm and 38 mm lenses, respectively. The minor

dimension of the ultraviolet beam is about 11 μm at focus. The major dimension of its elliptical cross-section is about 200 μm. The fluorescence emission and light scattered from particles passing through the two beams (which are separated by 75 μm) is collected by a 40×, 0.60 NA long working distance microscope objective lens. The light from the two laser beam intersection points is imaged through two pinholes, one above the other (one for each beam intersection point). The light from the lower laser beam intersection passes through the upper pinhole and goes straight to a detector assembly consisting of a beam splitter and a pair of photomultipliers with barrier filters. The light from the first laser beam intersection point goes through the lower pinhole and is reflected with a mirror to a second detector assembly. A lens in each of the detector assemblies forms an image of the back focal plane of the

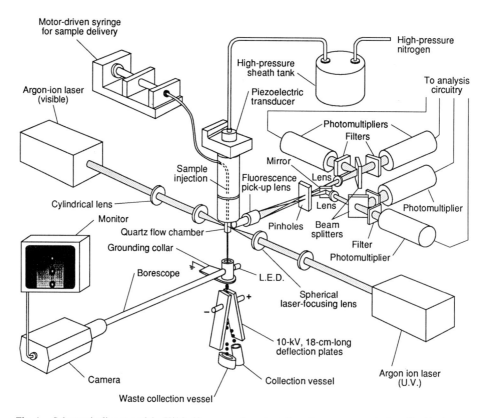

Fig. 4. Schematic diagram of the HiSS. Shown are the principle system components described in the text.

fluorescence pickup objective lens on the photomultiplier tube. This minimizes the movement of light on the photomultiplier tube's photocathode (which may have a position sensitive response) due to the movement of particles through the laser beam. When used to sort chromosomes stained with HO and CA3, the chromosomes cross the 458 nm beam first. Fluorescence from the 458 nm excitation passes through a 520 nm long-pass color filter and is measured as indication of CA3 stain. Fluorescence from ultraviolet excitation passes through a 425 nm long-pass filter and is measured as indication of HO stain.

The particles to be sorted are introduced onto the axis of the system flow chamber from a motor driven syringe. They are surrounded with a conductive sheath liquid flowing from a stainless-steel sheath container pressurized to 200 psi with dry nitrogen. The sheath stream carries the particles in the sample core through the square cross-section flow channel [see Fig. 2(A) and (B)] where they are illuminated with the two laser beams.

The droplet break-off region of the liquid jet is illuminated by a pulsed (200 nsec duration) light-emitting diode driven at the droplet frequency of 220 kHz. The image of the stroboscopically illuminated jet is collected by a lens system and projected into a video camera to permit visual monitoring of the jet length stability on a video monitor. A ground reference collar, seen at the bottom of Fig. 2(B), surrounds most of the liquid jet between the end of the flow chamber and the droplet breakoff point. It shields the break-off point from the electrostatic field of the deflection plates and increases the capacitive coupling of the jet to ground, thereby increasing the induced charge on drops separating from the jet during the application of the 150 V charging pulse.

E. ELECTRONICS

Figure 5 is a block diagram of the system electronics. The photomultiplier tubes convert the fluorescence intensity of each particle passing through the two laser beams into pulses that are amplified and actively integrated. The integrator is turned on and off by a threshold detector. The signal from one amplifier's threshold detector is designated as the system master signal and is assumed to be produced by every particle passing through the system. This signal is used to sense closely spaced objects and to inhibit sorting when an undesired object might contaminate a sort event. Signals from all active amplifiers are routed to window discriminators which are used to select particles to be sorted. Window discriminators, rather than a pulse height analyzer, are used to determine which events will be sorted in order to increase the rate at which objects can be processed. At the same time, the signals are routed to the analog-to-digital converter (ADC) inputs of a pulse height analyzer for digitization and accumulation into univariate or bivariate distributions. At present, the system can analyze and sort on a maximum of two signals. The settings of the discriminators used to select particles for sorting are read by a computer and are transferred to the analyzer where they are displayed as intensified regions superimposed on the accumulating bivariate distributions. Thus, the sorting windows selected by the discriminators are clearly visible on the bivariate distributions thereby facilitating accurate window setting. The computer also is used for setting the delay between particle detection and droplet charging and for selecting the number of droplets to be charged per sorting event.

III. PERFORMANCE EVALUATIONS

We have conducted a number of studies to determine the operating characteristics of the high-speed sorter. Specifically, we have determined its total fluorescence measurement resolution at several analysis and sorting rates, its sorting efficiency, and its physical effects on chromosomes, fixed cells and living cells.

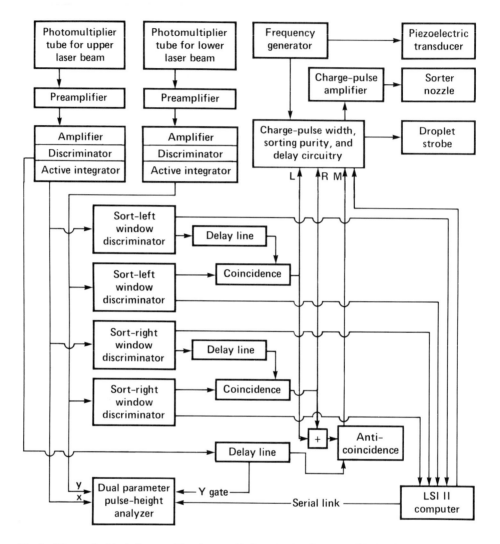

Fig. 5. Electronics block diagram. The photomultiplier tube signals are amplified and actively integrated (the integrator is turned on and off by a threshold detector). The output of a master discriminator, which detects all particles passing through the sorter, is used to recognize closely spaced objects and to inhibit sorting when an undesired object might be sorted simultaneously with a desired object. The discriminator signals derived from the upper, or first, laser beam are delayed to coincide with the lower beam signals. The coincidence time window of the dual-parameter pulse height analyzer (PHA) was extended to accommodate the time delayed signals, while a delayed gate controls the input for the lower beam signal. An anti-coincidence circuit inhibits the master discriminator signal if it was produced by a particle that falls within a sorting window (the master signal could occur in a different droplet period and negate the event). The computer monitors the threshold levels of the sorting window discriminators and displays them as intensified regions on the PHA. The computer also is used for setting the delay between particle detection and droplet charging and for selecting the number of droplets to be charged. The piezoelectric transducer (PZT) used for vibrating the flow chamber is driven with a square wave from the frequency generator. The droplet break-off region of the liquid jet is stroboscopically illuminated by a light emitting diode (LED) driven at the droplet frequency with a 200 ns-wide pulse.

A. System Resolution

The total fluorescence measurement resolution of the system was estimated using 1.8 μm diameter fluorescent microspheres (Polysciences, Warrington, PA). Univariate fluorescence distributions were measured for these microspheres at several analysis rates ranging from 2500/sec to 46 000/sec. The diameter of the sample core was kept constant during these experiments and the measurement rate was varied by changing the concentration of the microspheres. Thus, these data illustrate the effect of increased measurement rates on the system resolution. The coefficients of variation increased from 1.2 to 1.4% as the rate increased from 2500 to 46 000 microspheres/sec. The microspheres were excited at 458 nm (0.1 W) and fluorescence was collected through a colored glass filter (3-71; Corning Glass Works, Corning, NY), transmitting fluorescence at wavelengths longer than 520 nm, for these measurements.

The diameter of the sample core also affects the total fluorescence measurement resolution. The magnitude of this effect was estimated by analyzing fluorescence distributions for microsphere samples measured at several sample core diameters. The sample concentrations were adjusted during these studies to keep the measurement rate constant at about 2500/sec. The coefficients of variation increased from 1.1 to 1.2% as the sample stream diameter increased from 3.7 to 8.3 μm.

B. Recovery of Sorted Objects

The ultimate test of any sorter is whether objects of known characterics can be selected accurately for sorting and effectively recovered. These aspects of the HiSS were tested by counting the number of microspheres (1.8 μm diameter) collected on millipore filters after sorting known numbers (typically 20–50) directly onto the filters (0.8 μm pore size). We found the measured sorting efficiency to range from 95 to 100% for both one-drop and three-drop sorting (i.e. when the drop presumed to contain the microsphere and one droplet on either side were charged and deflected). We also evaluated the efficiency of the system by sorting microspheres onto microscope slides coated with a thin layer of grease. The slide was moved between each sort so that each deflected droplet (the drops of a three-drop sort appear to merge together) struck the slide in a new location. Each deflected drop left an impact crater in the grease thus allowing correlation of the sorting event (i.e. the appearance of a crater) with a proper sort (i.e. with the appearance of a fluorescent microsphere in the crater). We found, on average, the ratio of craters to sort pulses was greater than 0.95 and that the ratio of microspheres to craters was also greater than 0.95. We found it necessary, however, to insure good separation (7 mm) between the long deflection plates and the deflected stream to minimize the possibility of a charged droplet striking a deflection plate.

We also were concerned about the possible loss of sorted material due to splashing while sorting into a test tube. The splashing loss was determined by comparing the mass of liquid collected while sorting a fixed number of drops with that expected. Using a three-drop sort, the average number of sorted events needed to collect a gram of liquid was 391 000. The variation was from -7000 to $+8000$ over 16 experiments. The expected mass of a droplet in the HiSS was determined to be 8.5×10^{-7} g by collecting a volume of sheath from the liquid jet for a preset length of time and dividing its weight by the number of droplet periods during that time. Thus, 394 000 events should produce a gram of liquid with a three-drop sort. This value is in good agreement with the actual number per gram determined during sorting, indicating that splashing in the test tube is not a problem.

We also tested our ability to sort rare events in a concentrated mixture of unstained objects. Human erythrocytes fluorescently stained using a two-step antibody procedure with the primary antibody specific for hemoglobin (1) were mixed with unstained human erythrocytes at a ratio of 1 to 10^7 and processed through the HiSS

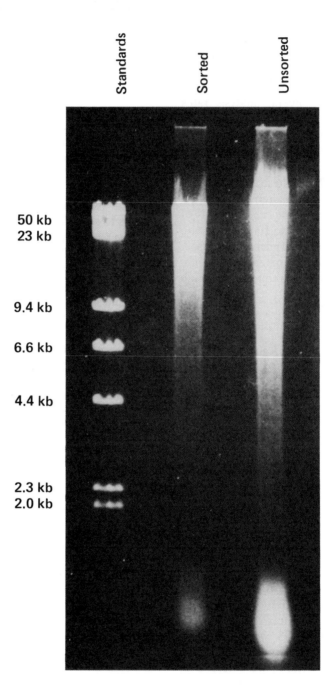

Fig. 6. Chromosome DNA size analysis. The sizes of the DNA fragments from isolated chromosomes before and after sorting were analyzed by agarose gel electrophoresis. The sorted chromosomes were frozen and stored at −70°C prior to fragment size analysis. A complete HindIII digest of lambda phage DNA was run as a size standard. The DNA fragments from the sorted chromosomes do not appear to be smaller than those from the unsorted chromosomes.

at a rate of 10^6/sec. In this mode of operation, the system did not "see" the nonfluorescent objects. One-hundred fluorescent cells were sorted onto hot microscope slides (which speeds evaporation of the sorted water in order to reduce the loss of cells by splashing) in each of seven different trials and counted microscopically. The fluorescent cell recovery efficiency was 88% ±8% for the seven trials. Thus, the system recovery efficiency was not significantly affected by the presence of unstained erythrocytes in most or all of the droplets.

C. EFFECTS OF SORTING

The size of DNA fragments from isolated chromosomes before and after sorting was analyzed by agarose gel electrophoresis. The sorted chromosomes were frozen and stored at −70°C prior to DNA size analysis. Sorted and unsorted chromosomes stained with propidium iodide were pelleted by centrifuging at 40 000g for 30 min at 4°C. The supernatant was removed and the chromatin was hydrolyzed with proteinase K overnight at 37°C. The proteins were then extracted with phenol and chloroform and the purified DNA was loaded onto a 0.4% agarose gel for electrophoresis. The gel photograph shown in Fig. 6 indicates that the sorted chromosomes did not have shorter fragments than the unsorted ones. The DNA sizes ranged above 50 kbp for both the sorted and unsorted chromosomes.

IV. EXPERIENCE PURIFYING HUMAN CHROMOSOMES

A. RESOLUTION

Chromosomes were isolated from the human fibroblast line 811 and the human/hamster hybrid cell line UV20HL21-27 (a gift of Dr L. H. Thompson) according to the procedure of van den Engh et al. (14,15), stained with HO and CA3 and measured with the sorter operating in the dual-beam mode. One beam was adjusted to the ultraviolet (0.3 W) to excite the HO and the other was adjusted to 458 nm (0.2 W) to excite the CA3. HO fluorescence was measured through a 425 nm long wavelength pass filter and CA3 fluorescence was measured through a 520 nm long wavelength pass filter (Corning 3-71). Figure 7 is a bivariate distribution showing the performance of the HiSS in resolving human chromosomes 9–22 at a measurement rate of about 15 000/sec. This performance is comparable to that obtained in conventional dual-beam cytometers and sorters (5,8). Chromosomes isolated from the human/hamster line (retaining human chromosomes 21, 8 and 4) were measured in the same way. Fig. 8 is a comparison of bivariate HO/CA3 distributions measured for these chromosomes at rates of 2000 and 20 000/sec in the HiSS. The resolution of the human chromosomes 8 and 21 remained approximately the same. The human chromosome 4 is off-scale and not shown. The high measurement rates in both of the above cases did not degrade the ability to sort the chromosomes of interest.

B. GENE LIBRARY PROJECT

The HiSS has been used by our laboratory for the sorting of individual human chromosomes for the National Gene Library Project (16). Since the machine operates about 10-fold faster than conventional sorters, it has greatly reduced the time required to sort enough chromosomes for cloning. The number of sorted chromosomes required to provide enough DNA for one cloning experiment was estimated at about 10^6. This number was based on previous experience of other laboratories (16), and has turned out so far to be a reasonable estimate (although, as cloning efficiency improves, fewer chromosomes will be required). Since not all cloning attempts are successful, a safety factor of four was built into the sorting requirements, so that 4×10^6 chromosomes of each type was established as the target. Sorting time requirements for autosomes from human diploid fibroblast cell lines could then be estimated at about 9 working days for the conventional speed sorters and half

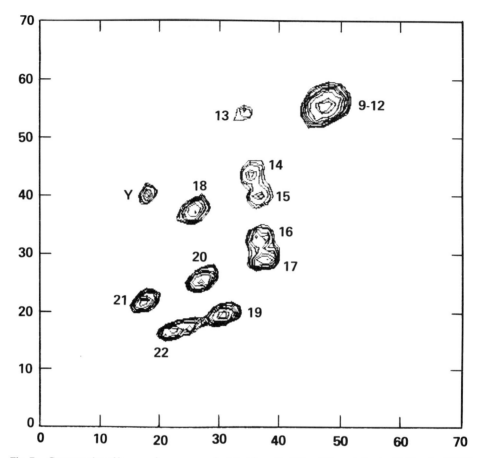

Fig. 7. Contour plot of human chromosomes isolated from the 811 cell line, stained with Hoechst 33258 ($2\,\mu$g/ml) and Chromomycin A3 ($40\,\mu$g/ml) and measured with the sorter operating in the dual-beam mode. The analysis rate was 15 000/sec. The chromosomes producing each peak are indicated on the plot.

a day for the HiSS. Experience over the past 1½ years has generally followed these initial estimates. The course of an actual sorting run depends greatly on the quality and quantity of the chromosome preparation and on ease of machine alignment, stability and freedom from nozzle plugs.

Experience on the HiSS for the Gene Library Project can be summarized as follows:

(1) The results of approximately 1 day of sorting for each of four chromosome types is shown in Table I. With current cloning techniques, enough chromosomes can be sorted for 2–5 cloning procedures for each chromosome type. The chromosomes 14 and 15 were combined because of too much histogram peak overlap for individual resolution.

(2) Average yield for the Lawrence Livermore National Laboratory dual-beam sorter (3) (a conventional speed sorter) is 0.3×10^6 chromosomes/day. Thus, the high-speed sorter is up to an order of magnitude faster.

(3) Overall efficiency, calculated as the ratio of the number of sorted chromosomes to the number of chromosomes of that type in the mitotic cells used in the isolation procedure, varies from 4–11%. Losses occur partly in the sorting procedure (efficiency = 10–30%) and partly in the isolation procedure (efficiency = 25–50%). Some of the sorting loss is due to the action of the purity assurance circuitry. A greater loss is due to chromosome settling and clumping during the time they are in

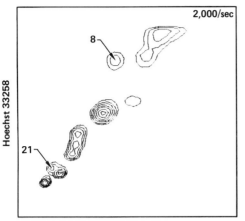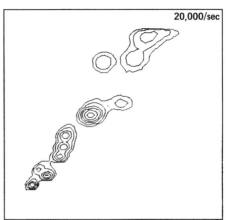

Fig. 8. Bivariate contour plots of chromosomes isolated from a human/Chinese hamster hybrid cell line retaining human chromosomes 21, 8 and 4, and stained with Hoechst 33258 and Chromomycin A3. Analysis was in dual-beam mode at rates of 2000 and 20 000/sec. The human chromosome 4 population is off-scale and does not appear in the plot. The peaks produced by human chromosomes 8 and 21 are indicated on the plot.

Table I

HISS OPERATING CHARACTERISTICS DURING THE SORTING OF FOUR CHROMOSOME TYPES

	Normal human fibroblasts (LLNL line 811)		Hybrid line (UV24HL5)	
Chromosome	20	14–15	2	X
Analysis rate (K/sec)	15	15	15	15
Sort rate (sec)	120	120	200	170
Sorter on time (h)	6	6	8.5	8.5
Yield (M)	1.3	2.6	5.3	4.4
Isolation efficiency (%)			60–70	60–70
Sorting efficiency (%)			17	11
Overall efficiency (%)			17	11

the sample injection syringe and the tubing leading to the flow chamber. This problem is compounded by the very high chromosome concentration, and low volume flow rate, needed to maintain the small diameter sample core required for high-resolution measurements. Means to reduce or eliminate these losses are currently being investigated, since their reduction will reduce the size of the mass culture and chromosome isolation required.

(4) Sorting reduces the chromosome concentration to 3.9×10^5/ml, about three orders of magnitude less than the value at the sorter input. The chromosome concentration after sorting is independent of the input concentration, and is simply the inverse of the volume of the three droplets deflected in a sorting event. This low concentration was increased by centrifugation at the beginning of the cloning procedure.

(5) Purity is high as long as the chromosomes to be separated can be cleanly resolved from other chromosomes or DNA-containing debris fragments. For example, sorting the human chromosomes in human/hamster hybrid cell lines is a problem when the histogram peak for the human chromosome is close to the hamster background. Purity is, in general, correlated with flow karyotype quality, i.e. impurity level is low for high-quality flow karyotypes.

V. CONCLUSION

The high-speed sorter described in this chapter has a significant advantage over conventional sorters. Particles can be processed and sorted at rates up to ten times faster than previously possible using an instrument that is no more difficult to operate than conventional sorters. This has been accomplished by constructing a system capable of operating at a pressure of 200 psi and at a droplet production frequency of 220 kHz. The HiSS seems to cause little or no damage to fixed cells or to chromosomes and thus seems well suited to the purification of cells and organelles for biochemical studies where large amounts of material are required. The HiSS also seems well suited to the purification of objects that occur at low frequency in a population. Thus, the HiSS seems especially suited to purification of bulk quantities of chromosomes of a single type for use in gene mapping (9) or in the production of recombinant DNA libraries (7).

ACKNOWLEDGEMENTS

Work performed under the auspices of the U.S. Department of Energy by the Lawrence Livermore National Laboratory under contract number W-7405-ENG-48.

REFERENCES

1. Bigbee, W. L., Branscomb, E. W., Weintraub, H. B. Papayannopoulou, T., and Stamatoyannopoulos, G. (1981). Cell sorter immunofluorescence detection of human erythrocytes labeled in suspension with antibodies specific for hemoglobin S and C. *J. Immunol. Meth.* **45**, 117–127.

2. Cremer, C., Gray, J. W., and Ropers, H. H. (1982). Flow cytometric characterization of a Chinese hamster X man hybrid cell line retaining the human Y chromosome. *Hum. Genet.,* **60**, 262–266.

3. Dean, P. N. (1980). Dual beam sorting at Livermore. *In* "Flow Cytometry" (O. D. Laerum, T. Lindmo and E. Thorud, eds), Vol. IV, pp. 41–44. Universitetsforlaget, Bergen.

4. Gray, J. W., Alger, T. W., and Lord, D. E. (1982). Fluidic assembly for an ultra-high-speed chromosome flow sorter. United States Patent 4, 361, 400.

4a. Gray, J., Dean, P., Fuscoe, J., Peters, D., Trask, B., van den Engh, G. and Van Dilla, M. (1987). High speed chromosome sorting. *Science* **238**, 323–329.

5. Gray, J. W., Langlois, R. G., Carrano, A. V., Burkhart-Schultz, K., and Van Dilla, M. A. (1979). High resolution chromosome analysis: one and two parameter flow cytometry. *Chromosoma* **73**, 9–27.

6. Kachel, V., and Menke, E. (1979). Hydrodynamic properties of flow cytometric instruments. *In* "Flow Cytometry and Sorting" (M. R. Melamed, P. F. Mullaney, and M. L. Mendelsohn, eds), pp. 41–60. Wiley, New York.

7. Krumlauf, R., Jeanpierre, M., and Young, B. D. (1982). Construction and characterization of genomic libraries from specific human chromosomes. *Proc. Natl Acad. Sci. USA* **79**, 2971–2975.

8. Langlois, R., Yu, L. -C., Gray, J. W. and Carrano, A. V. (1982). Quantitative karyotyping of human chromosomes by flow cytometry. *Proc. Natl Acad. Sci. USA* **79**, 7876–7880.

9. Lebo, R. V., Carrano, A. V., Burkhart-Schultz, K., Dozy, A. M., Yu, L.-C., and Kan, Y. W. (1979). Assignment of human alpha-, beta-, and gamma-globin genes to the short arm of chromosome 11 by chromosome sorting and DNA restriction enzyme analysis. *Proc. Natl Acad. Sci. USA* **76**, 5804–5808.

10. Peters, D., Branscomb, E., Dean, P., Merrill, T., Pinkel, D., Van Dilla, M., and Gray, J. W. (1985). The LLNL high-speed sorter: design features, operational characteristics, and biological utility. *Cytometry* **4**, 290–301.

11. Pinkel, D., and Stovel, R. (1985). Flow chambers and sample handling. *In* "Flow Cytometry: Instrumentation and Data Analysis" (M. A. Van Dilla, P. N. Dean, O. D. Laerum and M. R. Melamed, eds), pp. 77–127. Academic Press, New York.

12. Rayleigh, J. W. S. (1945). "The Theory of Sound" 2nd ed., Vol. II, Ch. XX. Dover Publications, New York.

13. Sweet, R. G. (1964). High frequency oscillography with electrostatically deflected ink jets. Stanford Electronics Laboratory Technical Report No. 1722–1.

14. van den Engh, G., Trask, B., Cram, S., and Bartholdi, M. (1984). Preparation of chromosome suspensions for flow cytometry. *Cytometry* **5**, 108–123.

15. van den Engh, G., Trask, B., Gray, J., Langlois, R., and Yu, L.-C. (1989). Preparation and bivariate analysis of suspensions of human of chromosomes. *Cytometry* **6**, 92–100.

16. Van Dilla, M. A., Deaven, L. L., Albright, K. L., Allen, N. A., Aubuchon, M. R., Bartholdi, M. F., Browne, N. C., Campbell, E. W., Carrano, A. V., Clark, L. M., Cram, L. S., Fuscoe, J. C., Gray, J. W., Hildebrand, C. E., Jackson, P. J., Jett, J. H., Longmire, J. L., Lozes, C. R., Luedemann, M. L., Martin, J. C., McNinch, J. S., Meincke, L. J., Mendelsohn, M. L., Meyne, J., Moyzis, R. K., Munk, A. C., Perlman, J., Peters, D. C., Silva, A. J., and Trask, B. J. (1986). Human chromosome-specific DNA libraries: construction and availability. *Biotechnology* **4**, 537–552.

17. Weber, C. (1931). Zum Zerfall eines Flussigeitsstrahles. *Z. Angewandte Math Mech.* **11**, 136.

Gene Mapping Strategies and Bivariate Flow Cytogenetics

ROGER V. LEBO

Department of Obstetrics, Gynecology, and Pediatrics
University of California, San Francisco, CA 94143, USA

I. INTRODUCTION

The human genome is composed of 22 pairs of chromosomes plus 2 sex chromosomes that replicate during each cell cycle so that duplicate chromatids can be segregated to each daughter cell at mitosis. Chromosomes carry several classes of genes and other DNA sequences. Some DNA sequences like the cloned insulin gene only hybridize to a unique genomic sequence found in the case of insulin on the short arm of chromosome 11 (36). A unique gene may belong to a family of homologous genes on two different chromosomes like the cellular-src-1 (c-src-1) and cellular-src-2 (c-src-2) oncogenes on chromosomes 20 and 1,

respectively (54), and the β-globin and α-globin gene clusters on chromosomes 11 and 16 (12,33,57). Both globin locations include unexpressed pseudo-genes as well as genes expressed in the embryo, fetus and adult arranged in order of ontogeny within each gene cluster (15,55). The homologous phosphoglycerokinase (16) and ferritin light-chain loci (41) are each found on three chromosomes and the aldolase loci are on four different chromosomes (59). The low-repetitive DNA sequences represent another DNA class that have been used in one instance to map chromosome deletions (31). Moderately repetitive DNA sequences like the 3.2 kbp *Hae*III Y-specific sequence can be chromosome specific (32), and satellite DNAs are repeated many times (47). Finally, the human alu sequence is so ubiquitous that it has been used to identify human DNA clones in a cloned DNA library constructed from somatic cell hybrids (30). Many abnormal gene loci cause human genetic disease. For instance, an abnormal aldolase A locus encodes an unstable enzyme that results in hemolytic anemia (59) and the α- and β-thalassemias are caused by abnormal or deleted globin genes (51).

Three different maps of these various sequences may be assembled for each species' karyotype (29): (1) the physical map identifies the actual location of each gene on the chromosome(s); (2) the genetic map gives the probability of any gene locus recombining with neighboring genes during gametogenesis; (3) the chiasma map is constructed by scoring chromatid association of homologous sister chromosomes during meiotic prophase. During the characterization of a cloned DNA segment, the primary sequence and a number of homologous sequences may be mapped to one or more of a subset of the 24 chromosome types. Then each sequence(s) may be sublocalized (1) physically to a morphological segment of a chromosome or (2) genetically with respect to other loci by the frequency of recombination with neighboring polymorphic gene sequences.

This paper will review the spot–blot gene mapping method and the developments that led to its current efficacy (38,43). Then, the other physical gene mapping methods will be compared, including somatic cell hybridization (56), in situ hybridization (6) and identification of a sex-linked gene by familial inheritance (49). DNA clones isolated from chromosome-specific recombinant DNA libraries are being studied to facilitate saturating the physical and genetic maps with polymorphic restriction enzyme fragments (13,42,50). These polymorphic fragments will enable improved diagnoses of linked genetic diseases. Comparative gene mapping in different species can reveal evolutionary relationships between different homologous sequences (25). Finally,

myc oncogene mapping will be discussed to illustrate the role mapping played in uncovering the molecular events in Burkitt's lymphoma (11).

II. FLOW SORTING IMPROVEMENTS

Flow analysis and sorting of human chromosomes (3,10,18,38) has been used for gene mapping (2,7,33,38,39,40), construction of chromosome-specific recombinant DNA libraries (13,42) and flow karyotyping (1,18,43). Sorting capability has evolved from sorting 12 human chromosome fractions with one laser (3) via successive improvements (18,35,62) to dual-laser sorting 23 identifiable fractions of 24 different chromosome types (38,44). This discussion will begin by reviewing improvements we have made in bivariate chromosome analysis and sorting that have made gene mapping by splot–blot analysis rapid and straightforward.

A. CHROMOSOME SUSPENSION PREPARATION

High-resolution chromosome sorting applies protocols from several disciplines. Cultured cells are partially synchronized to the G_0G_1 portion of the cell cycle by growing the cultures to stationary phase. Then, lymphocyte cultures are diluted with fresh medium and fibroblast cultures are passed so that the cells proceed through the cell cycle. Colcemid is added to arrest cells in metaphase when the largest portion of cells is about to divide. The time before and during the colcemid block is dependent on the cell cycle time of the culture, which is shorter for lymphocytes than for fibroblasts (39,40,44). We prepare all our chromosome suspensions in spermine buffer since preparations are superior, the chromosomal DNA remains more intact for subsequent large-insert chromosome library construction (42) and the sorted chromosomes can be identified (38). Single mitotic chromosomes are released from cultured fibroblasts or

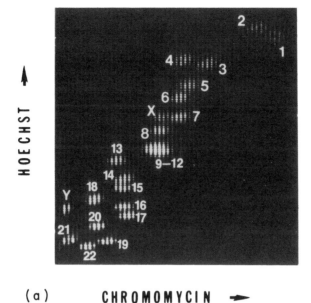

(a)   CHROMOMYCIN ➡

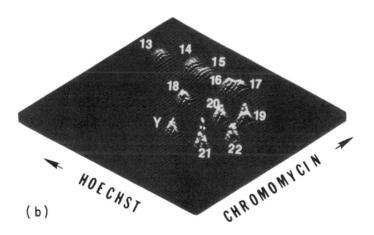

(b)

Fig. 1. (a) Flow histogram of normal male human chromosomes. A 64 × 128-channel three-dimensional histogram of flow-analyzed lymphocyte chromosomes viewed from above on a Nuclear Data electronic display. The two laser-excited fluorescence signals were measured separately and correspond to chromomycin A3 fluorescence from the first laser on the abscissa and to Hoechst 33258 fluorescence from the second laser on the ordinate. The light intensity for each dot in the three-dimensional histogram is linearly proportional to the number of chromosomes recorded in each channel. [Reprinted from *Science* **225**, 57–59. (1984). Copyright © 1984 by the AAAS.] (b) Three-dimensional histogram of smaller chromosomes. A 64–128-channel three-dimensional flow histogram of the smaller human lymphocyte chromosomes viewed from the side. The dual-laser fluorescence signals correspond to chromomycin A3 fluorescence on the abscissa and Hoechst 33258 fluorescence on the ordinate. Note that few chromosome clumps or fragments are recorded between the histogram peaks.

lymphocytes by vortexing or homogenizing [see Lebo and Bruce (44) for details]. During vortexing (39) or homogenizing (40) to segregate individual chromosomes from the mitotic nuclei, care is exercized not to over-extend the procedure and split the mitotic chromosomes in half to chromatids. This can present a problem since smaller chromosome peaks would then contain larger chromatids that are sorted together. The first sign that significant numbers of chromatids have been separated is the appearance of the 9–12 chromatid peak beneath the normal chromo-some 18 peak on the flow histogram. This chromatid peak is observed first not only because it is well separated from the normal chromosome peaks, but because four times the frequency of chromatids are recorded at this peak position.

B. Normal Chromosome Separations

Homogenized chromosome suspensions are stained with two complementary DNA-specific dyes (38) and analyzed and sorted by dual-laser excitation (35,45). The two fluorescent signals from each chromosome are collected and recorded separately (Fig. 1). All chromosomes stained with Hoechst 33258 plus chromomycin A3 may be resolved and sorted uniquely with the exception of chromosomes 9–12 (38). Another stain pair, DIPI plus chromomycin A3, was found to resolve chromosome 9 from chromosomes 10–12 (38) and chromosomes 9 and 12 from chromosomes 10 and 11 (44). In this fashion, 21 of the 24 human chromosome types were resolved uniquely. Improved chromosome suspensions and optics have permitted sorting most chromosome fractions to a purity of 98–99% (35,45).

The identify of the sorted chromosomes is determined by quinacrine banding patterns on slides prepared with modified Leif buckets (44,46). Chromosomes are sorted into Leif buckets at room temperature, fixed immedi-ately, centrifuged onto slides, fixed again

overnight and aged 1 week to improve banding (44). Then the chromosomes are stained with quinacrine and studied with a high-power optical microscope at 1500× (15× eyepieces with a 100× objective) to identify the condensed chromosome bands (44). Although only a fraction of the sorted chromosomes can be identified, these have been found to repre-sent a random sample of the entire sorted fraction. Spot–blot gene mapping data with chromosome-specific probes have confirmed the banding results.

C. Optical Bench

Chromosome suspensions are prepared in Falcon 2063 polypropylene tubes and introduced into the sorter through a siliconized, glass-lined flow restricter and then floppy siliconized tubing (35). These precautions permit us to deliver an average of 15% of the suspended lymphocyte chromosomes prepared from the cultures to the focused laser beams for analysis and sorting. In addition to gene mapping applications, this improved recovery is particularly important to chromosome library construction (13,42).

Improved optical bench design increases the chromosome sorting resolution to separate similar chromosomes with increased purity. High-power laser beams delivered from 18 W argon ion lasers were chosen to optimally excite chromosomes, which are small cellular organelles. In order to deliver as much of this light as possible, we configure and coat all subsequent optics to optimally reflect or pass the critical wavelengths for chromosome analysis (35,45). The beams are expanded immediately upon leaving the laser to decrease the light flux on each subsequent surface and extend the life of the optical coatings from hours to months. Then the beams are focused 150–200 μm apart onto the sample stream with independent optics (35). The most recent triple beam system can be focused optimally for any experimental combination of one to three beams and one to five signals in less than 20 min

(45). We have also added a rapid sheath switch-over to permit switching sterile cell and chromosome sheath buffers in 90 sec without resterilizing (45). The samples exit a Becton–Dickinson 80 μm ceramic nozzle tip that resists the high-intensity heat delivered when the beam being focused strikes the nozzle (Fig. 2). Our first nozzle tip lasted 3 years. In contrast, plastic nozzle tips are destroyed in seconds. The sheath and sample lines are oxidized with sodium hypochlorite (bleach) and then washed with 70% ethanol and water (44). This ensures that nonspecific autoradiographic spots will not appear in sorted chromosome locations on spot–blot filters.

The sheath exiting the nozzle tip is collected by a vacuum tube attached to a peristaltic pump to transport the unsorted droplets to a collection reservoir (Fig. 2). The waste reservoir liquid is decontaminated with an iodine-based detergent or autoclaved before disposal, consistent with P-2 biosafety requirements for animal cells. The desired chromosome fractions are deflected to either side onto a single 2.5 cm diameter nitrocellulose filter disk (Fig. 2). To avoid confusion, pairs of chromosome peaks are sorted left and right according to the relative positions on the flow histogram. The filter disk is placed on a scintered glass filter holder (Millipore) mounted in a wooden block. Building vacuum

Fig. 2. Sorting spot–blot filter construction. The chromosome suspension is directed through the focused laser beams, and the chromosome-containing droplets of interest are deflected onto one of two spots of a nitrocellulose filter applied to a scintered glass millipore filter holder under vacuum. The remaining chromosomes in undeflected drops are collected by the central vacuum tube. Thirty-thousand chromosomes of each type were deflected to either side directly onto the nitrocellulose filter on spots delineated by the circles following an initial 1 sec test sort. The chromosomes are sorted onto the same spot with 30 000 chromosomes being sorted onto a pile observed under 160× fluorescence microscopy to be about 100 chromosomes deep. The minute size of this chromosome pile is not related to the spot around the chromosomes wet by the sheat fluid which varies in diameter with filter lot number and the vacuum applied. Chromosome spots are circled for reference for subsequent comparison with autoradiographic spot positions following hybridization to radiolabeled gene probes (SA = 2×10^8–3×10^9 cpm/μg). The chromosomal DNA is denatured directly on the filters, hybridized to ^{32}P-labeled gene probe, washed and autoradiographed. Gene-specific signal is visible in the gene-carrying chromosome spot (Fig. 6). An entire filter set of 11 filters can be sorted in 5 h. Once a filter set has been prepared it may be hybridized and rehybridized about eight times to locate additional cloned genes quickly. Only the positive filters are denatured to extend filter life. [Repringed from *Cytometry* **6**, 478 (1985)].

($\sim$500 mmHg) is applied to the millipore filter unit and a nitrocellulose filter added. A 1 sec test pulse is added to deflect enough droplets in order to circle the spots with a VWR labmarker pen (San Francisco, CA) for reference when studying the autoradiographs. (Eight other inks tested hybridized nonspecifically to radio-labeled gene probes). Then two fractions of 30 000 chromosomes each are sorted onto each spot in about 15 min. Sorting chromosomes directly onto nitrocellulose filter spots maximizes DNA yield and gene-specific signal and avoids introducing contaminating DNA sequences during DNA extraction. [See Lebo and Bruce (44) for further protocol details.]

D. PREDICTING ABNORMAL CHROMOSOME PEAK POSITIONS

Usually, chromosome deletions, translocations or insertions are sufficiently different in length to be sorted separately from the normal remaining homologous chromosomes (40,43). Even the identity of these rearrangements can be determined from flow histograms according to the position and frequency of events in abnormal peaks even when the karyotype is unknown (43). Predicting the positions of characterized derivative chromosomes in the three-dimensional flow histogram has expedited subchromosomal gene localization on numerous occasions. For example, cell line GM201 carries a translocation between chromosomes 1 and 17 that has been sorted and tested to sublocalize the c-src-2 oncogene (Fig. 3). The positions of these chromosome rearrangements can be illustrated with sliced chromosome idiograms. Chromosome idiograms are drawn as a composite of the average chromosome banding pattern observed under the microscope by many laboratories (24,53). Idiograms of

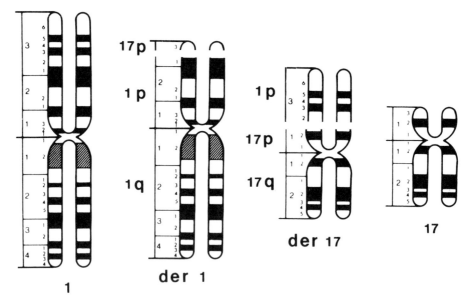

Fig. 3. Idiograms of a translocation involving chromosomes 1 and 17. The normal chromosome 1 is illustrated on the left and the normal chromosome 17 on the right. A reciprocal translocation has occurred between one chromosome 17 and one chromosome 1 to give two intermediate size chromosomes in cell line GM 201. The terminal portion of the chromosome 1 short arm has been exchanged for the tip of the short arm of 17. Based on idiogram size and banding patterns, the four chromosomes may be expected to be resolved by flow analysis of chromosomes stained with Hoechst 33258 plus chromomycin A3. Derivative chromosome 17 with the 17 centromere has gained more Giemsa-light idiogram band bmaterial than Giemsa-dark Hoechst-33258 positive band material. Thus, the derivative chromosome 17 is expected to move farther in the chromomycin histogram than in the Hoechst 33258 histogram direction. [Figure reprinted from Meth. Enzymol. **151**, 292–313 (1987)].

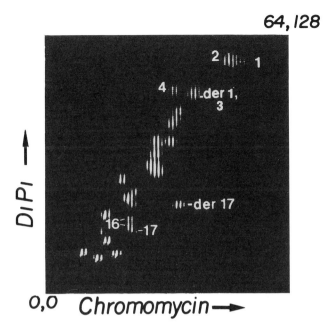

Fig. 4. DIPI–chromomycin flow analysis of translocations 1 and 17. Chromomycin A3 fluorescence increases illustrated linearly on the abscissa and DIPI fluorescence on the ordinate. The normal chromosomes 1 and 17 peak positions are indicated along with the intrmediate size derivative chromosome 17 (der 17) peak and the derivative chromosome 1 found in the normal chromosome 3 peak. The peaks of the largest chromosomes are exposed longer for clarity in this histogram only. Chromosome rearrangements have been predicted with good agreement by abnormal peak position and frequency in a blind study of 11 cell lines with 30 chromosome abnormalities (*43*). [Figure reprinted from *Meth. Enzymol.* **151**, 292–313 (1987)].

Giemsa-banded chromosomes represent the Giemsa-positive bands in black and the Giemsa-negative bands in white (Fig. 3). Quinacrine, Hoechst 33258 and DIPI bands are similar to Giemsa-positive bands and chromomycin A3 and reverse- or R-bands are similar to Giemsa-negative bands. Measuring the lengths of Giemsa-positive and negative idiogram bands with the exceptions required for each stain pair difference predicts the position of the chromosome peak in a three-dimensional flow idiogram (*43*). This prediction has been most useful in choosing cell lines with derivative chromosomes that can be sorted and tested for subchromosomal localization. Many such cell lines are available from a central repository (*23*).

Reciprocal translocations between two different chromosomes usually result in two new derivative chromosomes without a loss in genetic material. Individuals who carry a balanced *de novo* reciprocal translocation without a microscopically detectable deleted segment are normal 95% of the time. Reciprocal translocations from normal subjects are quite useful for gene mapping by chromosome sorting since the two derivative chromosomes resulting from the translocation are usually different in size from the normal homologous chromosomes and all four can be sorted from each other. Such is the case for the chromosomes in cell line GM201 (Figs 4 and 5). Other useful derivative chromosomes include chromosomes with deletions that are smaller than the normal chromosome and chromosomes with insertions that are larger.

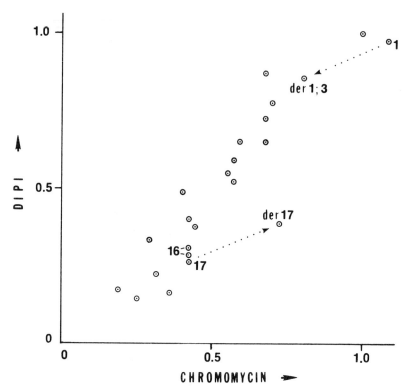

Fig. 5. Positions of peaks for translocation 1;17. The graph illustrates the peak positions of the derivative chromosomes in the bivariate flow karyotype relative to the chromosome 2 peak position defined as coordinate (1, 1). Note that the derivative chromosomes move in approximately equal and opposite parallel directions as expected for a reciprocal chromosome translocation with no change in total DNA content. The slight deviation from this observation is most likely caused by the variable heterochromatic regions immediately beneath the centromere of the two chromosomes 1. Measuring chromosomes to correctly predict peak positions was reported previously (*43*). [Figure reprinted from *Metho. Enzymol.* **151**, 292–313. (1987)]

III. GENE MAPPING

A. CHROMOSOME SORTING

Recently, spot–blot analysis has permitted rapid assignment of cloned genes to individual chromosomes by direct hybridization of radio-labeled gene probe to sorted chromosomal DNA (*38,39*). Similarly, genes have been localized to subchromosomal regions by spot–blot analysis of different size-rearranged chromosomes (*40,44*). Unique gene-carrying restriction fragments located on more than one chromosome can be mapped rapidly by spot–blot analysis followed by miniaturized restriction enzyme analysis of additional sorted DNA from the gene-carrying chromosomes identified by spot–blot analysis (*41*).

a. *Spot–Blot Analysis*

Genes or low-copy DNA sequences may be assigned to whole normal chromosomes that can be sorted uniquely onto spots of nitrocellulose filter paper. A panel can be sorted from chromosome suspensions prepared from 200 ml lymphocyte suspension culture. When 30 000 chromosomes of each type are sorted two fractions at a time onto filters disks, an entire panel consisting of each separable chromosome fraction can be sorted in less than

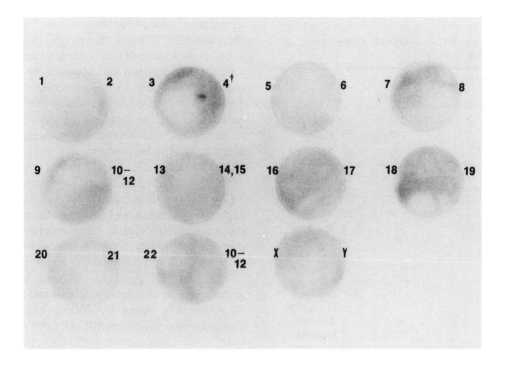

Fig. 6. Cloned albumin gene mapped to chromosome 4. An entire spot–blot filter panel from normal male cell line GM130 (Fig. 1) stained and sorted with DIPI plus chromomycin A3 (*38*) was hybridized to the radiolabeled putative albumin gene clone from Dr Nancy Cooke. The chromosome 4 spot hybridized uniquely to this probe as expected from previous gene mapping results (*20*). This result provided additional evidence that this separately isolated albumin gene clone was actually the correct clone.

5 h. After an entire panel of chromosomes has been sorted, the chromosomal DNA is denatured onto a spot less than 1 mm in diameter and hybridized to radiolabeled gene probe. The cloned albumin gene hybridized to chromosome 4 as expected (Fig. 6).

Using this spot–blot technique we have tested 13 genes which mapped to the chromosome expected, based on assignments by other laboratories (Table I). In one other instance a cloned probe did not map to the chromosome expected. In this case DNA sequencing revealed that the putative gene clone had an incorrect DNA sequence and another clone had to be chosen. Thus the results of mapping by spot–blot analysis can serve as initial proof that a putative gene clone which has been chosen truly represents a previously mapped gene, or

that substantial clone sequencing and characterization efforts would best be reserved for another putative clone.

We then used this method to determine the location of 40 previously unmapped cloned genes (Table II). This list of new gene assignments further emphasizes the rapidity with which genes may be localized and sublocalized by spot–blot analysis. Nineteen of these gene assignments have since been confirmed by another gene mapping method. We made one error. A degraded probe hybridized nonspecifically to reveal a more intense signal on the spot with chromosomes 9–12 which has four times more total DNA. Alternatively, we may have tested the wrong probe. We then tested translocations involving chromosomes 9–12 with a newly labeled probe in a different

Table I

AGREEMENT WITH ASSIGNED GENE
LOCATIONS

Gene	Chromosome location	Probe source
Amylase	1^a	B. Wise
Albumin	4^a	N. Cooke
HLA-DR-α	6	S. Weismann
Clotting Factor XIII	6^a	L. Weisberg
Epidermal growth factor receptor	7	G. Merlino and M. Rosenfeld
Thyroglobulin	8^a	G. Vassart
Clotting Factor IX	X	K. H. Choo
LDH-X	11	M. Law and Y.-F. Lau
Insulin	11p15.1 → pter	H. Goodman
β-Globin	11p15.1 → pter	T. Maniatis
γ-Interferon	14	P. Gray and D. Goeddel
ζ-Globin	16	T. Maniatis
α-Globin	16	T. Maniatis

^{a} Used to help confirm identity of clone.

shipment and found a 4;11 translocated chromosome carried the gene. Since we had failed to test a normal control chromosome 4 simultaneously, we incorrectly concluded the gene was on chromosome 11. In fact, the gene is on the chromosome 4 fragment in the 4;11 translocated chromosome. Further studies with the only gene that was localized to different chromosomes by spot–blot and *in situ* hybridization confirmed the spot–blot assignment. Some of these genes are located on a single chromosome and are not related to other known gene loci like growth hormone releasing factor which has been assigned to chromosome 20 (*48*). Most mapped genes belong to a family of homologous genes like the homologous human sequence to the drosophila homeo box on the long arm of chromosome 17 (*25*). Testing the spot–blots with separate probes to each locus under high-stringency hybridization conditions usually differentiates between the identical cloned DNA sequence locus and homologous loci.

Mapping cloned genes to two separate chromosomes is most meaningful when the gene clones have been characterized. For instance, we mapped the homologous c-src-1 oncogene uniquely to chromosome 20 (*54*). This gene sequence is expressed in every tissue tested by Northern analysis. Another c-src-2 sequence has been isolated and sequenced and found to be 90% homologous to c-src-1 in three of the 3′ exons. This c-src-2 probe hybridized uniquely under high stringency to chromosome 1 (*54*) and has subsequently been shown to be expressed in placenta (J. M. Bishop, personal communication). These data emphasize the complementary interaction between molecular clone analysis and chromosome localization.

The spot–blot protocol not only localizes genes to whole chromosomes, but can be used to sublocalize genes by testing rearranged chromosomes. As discussed, cell lines carrying rearranged chromosomes involving segments of a chromosome to which the gene was assigned are chosen so that the different size derivative chromosome(s) can be sorted from the normal chromosome and tested. If the sorted derivative chromosome carries the gene of interest, the gene must be on the chromosome segment carried by the derivative chromosome. Since we had localized the c-src-2 oncogene to chromosome 1 (Table II), we chose a cell line with segments of chromosome 1 carried by different size chromosomes. Since the c-src-2 oncogene had previously hybridized only to chromosome 1 on duplicate spot–blot filter sets, we sorted and tested the two derivative chromosome fractions. Spot–blot analysis of these two fractions revealed derivative 17 DNA hybridized to c-src-2 gene probe but not derivative 1 DNA (Fig. 7). This result indicates that the c-src-2 locus is on the terminal portion of the short arm of chromosome 1 (1p32 → 1pter).

b. Miniaturized Restriction Enzyme Analysis

Restriction enzyme analysis may be used to test isolated chromosomal DNAs to determine which of several unique restriction enzyme

Table II

NEW GENE ASSIGNMENTS BY SPOT–BLOT ANALYSIS

Gene	Chromosome location	Probe source
FcγRII, FcγRIII	1q21.2 → q25.1[a]	K. Moore and G. Peltz
Human tissue factor	1q32 → 1pter	E. Sadler
c-src-2 oncogene	1p32 → 1pter[a]	J. M. Bishop and R. Parker
Na$^+$–K$^+$ATPase	1[a]	R. Blostein and Y. W. Kan
RBC 4.1 protein	1p32 → 1pter	J. Conboy and Y. W. Kan
Fetal alkaline phosphatase	2[a]	W. Kam and Y. W. Kan
Apoliprotein B	2[a]	S. Deeb
HLA-DRγ	5[a]	H. Erlich
Glucocorticoid receptor	5, 16[a]	R. M. Evans
Phosphoglycerokinase	6, X, 19[a]	S. Gartler
Multiple drug resistance	7	I. Pastan
Erythropoietin	7[a]	J. Powell
Tissue-type plasminogen activator inhibitor	7	D. Ginsberg
Tissue-type plasminogen activator	8[a]	G. Opdenakker
Aldolase B	9	D. Tolan
Pseudoaldolase	10q11 → pter	D. Tolan and E. Penhoet
Interleukin-2 receptor	10	W. Greene
Myophosphorylase	11q13 → pter[a]	F. Gorin and R. Fletterick
Parathyroid hormone	11p15.1 → pter	P. O'Connell and R. White
ADJ-762	11p15.1 → pter[a]	R. White
LDH-A	11p15.1 → pter	G. Bruns
Calcitonin	11p15.1 → pter[a]	M. Rosenfeld
Brain phosphofructokinase	12[a]	P. L. Ashley and D. R. Cox
Von Willebrand factor	12[a], 22	F. Chehab and Y. W. Kan
Procollagen (α1 type IV)	13[a], 8	C. Boyd and Prokoff
Her's Disease	14	R. Fletterick
β-Spectrin	14	J. Prchal and Y. W. Kan
Aldolasc A	16	D. Tolan and E. Penhoet
Aldolase C	17	D. Tolan and E. Penhoet
Thyroid hormone receptor nerve	17	R. Evans
Glycoprotein II-B	17	P. Bray and Y. W. Kan
Homeo Box	17q11 → qter[a]	R. Tjian and G. Martin
GM colony stimulating factor	18	J. Powell
Placental tissue plasminogen activator inhibitor	18	E. Sadler
Insulin receptor	19[a]	W. Rutter and Y. W. Kan
Ferritin light chain	19, 20, X[a]	J. Drysdale and D. Boyd
Growth hormone releasing factor	20	K. E. Mayo and R. M. Evans
c-src-1 oncogene	20[a]	J. M. Bishop and R. Parker
Human HCK	20	N. Quintrell and M. Bishop
Prion	20	Yu-Cheng Liao
G-6-PD	X, 17[a]	A. Yoshida

[a] Confirmed by another method.

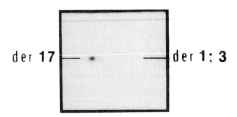

der 17 ──── der 1; 3

Fig. 7. c-src-2 oncogene probe hybridization. c-src-2 oncogene probe (44) hybridized to derivative chromosome 17 DNA to localize the gene to the terminal portion of chromosome 1 in the same region as RBC 4.1 protein (9). [Figure reprinted from *Meth. Enzymol.* **151**, 292–313 (1987).]

fragments are carried by each chromosome type (36,41). Our initial restriction enzyme analysis studies of sorted human chromosomes mapped the β-globin gene cluster to the short arm of human chromosome 11 by standard size restriction enzyme analysis of chromosomes sorted with one laser (33). This protocol has two major drawbacks. First, about 7 days of sorting time were required to obtain two fractions for analysis. Second, once a gene was found in any fraction with 2–4 chromosomes, large quantities of cells with translocations had to be prepared and sorted to test each chromosome in the positive chromosome peak. As previously mentioned, high-resolution chromosome sorting now separates nearly all chromosome types uniquely so that translocations from only chromosomes 10 and 11 are now necessary to map genes. In addition, restriction enzyme analysis of sorted chromosomal DNA has been miniaturized so that considerably smaller quantities of sorted chromosomes are used for analysis. Together, these improvements have been used to map unique or low-copy gene sequences to individual chromosomes or segments thereof. For instance, the five BglII ferritin light-chain restriction enzyme fragments were assigned to a few chromosomes by spot–blot analysis (41). In order to locate which of the five fragment lengths are located on each chromosome, we tested sorted chromosomal DNA by miniaturized restriction enzyme analysis. This revealed that two unique

restriction fragment lengths are carried on chromosome 19, two on chromosome 20 and one on chromosome X (Fig. 8). Analysis of messenger RNAs from different tissues will be required to determine whether these different fragments carry tissue-specific alleles or pseudogenes. The ferritin-L results emphasize the importance of further molecular characterization and restriction enzyme analysis to characterize chromosomal DNA when homologous gene sequences are found on several chromosomes. One further way to test tissue-specific expression of a unique restriction fragment is to study gene expression in hybrids

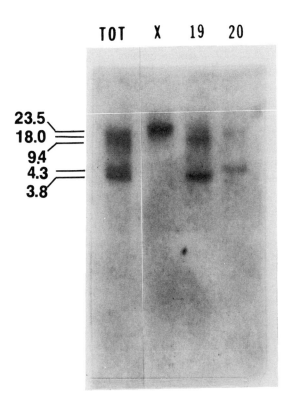

TOT   X   19   20

23.5
18.0
9.4
4.3
3.8

Fig. 8. Miniature restriction enzyme analysis of sorted chromosomes. In the left lane, restriction enzyme analysis of total BglII-digested human DNA (TOT) reveals all five restriction fragment lengths. BglII-digested sorted X, 19 and 20 chromosomal DNA hybridized to ferritin-L probe reveals each of the five restriction fragment lengths are carried by unique chromosomes. [Reprinted from *Hum. Genet.* **71**, 325–328 (1985).]

of different somatic cell types. In this case of the ferritin-L gene, somatic cell hybrid mapping indicated the expressed gene sequence is on chromosome 19 (4,61).

B. OTHER GENE MAPPING METHODS

Somatic cell hybridization might also be viewed as a chromosome segregation system like chromosome sorting (56). Somatic cell hybrids are constructed with cells from two different species by fusion. Hybrids are chosen by requiring selectable characteristics of both parent cell types for growth in culture. Then the hybrids preferentially lose chromosomes from one parent cell line depending upon the parental tissue of origin, both parental species and permanence of both parental cell lines. Hybrid clones each with a few remaining chromosomes from one parent are tested for gene product from both parental species. A gene is assigned to a chromosome according to multiple correlations between retained chromosomes and the presence of that species' gene product. Although constructing hybrids, waiting for chromosome loss and karyotyping the hybrids is time consuming, several laboratories have constructed hybrid panels which may be tested readily for concordance between chromosome type and gene product. A few chromosomes carry genes that can be selected for in culture. Chromosomes without selectable markers are lost randomly and the chromosome content of hybrids can change during clonal expansion. Thus, these unstable hybrids must be karyotyped regularly to maintain useful hybrid panels. Panels of some derivative chromosomes have also been constructed in somatic cell hybrids. Sub-localizing genes on chromosomes for which derivative chromosome panels have not been constructed is still time consuming. Nevertheless, this method remains the only procedure to localize genes for which the gene product can be assayed but the gene has not been cloned.

More recently, chromosome-specific cell surface antigens have been mapped to unique chromosomes by expression in somatic cell hybrids and selection by the fluorescence-activated cell sorter (27,28). The cell sorter was used to sort and clone hybrid cells which were positive for several specific cell surface antigens. Then these hybrids were karyotyped. In this fashion, one surface antigen S11 was mapped to the terminal portion of the long arm of the human X chromosome between the human hypoxanthine phosphoribosyl transferase and glucose-6-phosphate dehydrogenase loci (28).

In situ hybridization of tritium-labeled or biotinylated, cloned gene probes has been used to subchromosomally localize single-copy genes, beginning with the insulin gene (19). Cells enriched in mitotic cells by colcemid block are swollen in hypotonic KC1, fixed and spread on slides. The chromosomal DNA is then denatured, prehybridized and hybridized to radiolabeled gene probe. Excess, unbound radiolabeled probe is washed away, the slides dipped in emulsion and exposed several days to give sufficient signal after developing the photographic emulsion. In one instance, *in situ* hybridization localized a homologous gene sequence but not the primary gene sequence (17). Because of occasional inconsistencies, convention has been that a gene assignment is provisional unless made by more than one method. Two complementary methods would be to localize a gene to a whole chromosome by spot–blot analysis and then to proceed with *in situ* mapping, since only the chromosome of interest need be scored for the subchromosomal localization.

In addition to the three gene mapping methods discussed so far, human X-linked genetic diseases and polymorphic DNA fragments and genes have all been mapped to the X-chromosome by characteristic segregation in pedigrees. Since a male has one X and one Y chromosome and a female has two X chromosomes, a father transmits an X chromosome only to his daughters while a mother transmits an X chromosome to both daughters and sons. Thus absence of male-to-male transmission characterizes X-linked markers (*34*). Another method used to map X-linked restriction fragments is to compare restriction enzyme analysis of male (XY), female (XX) and abnormal (XXXX) cells simultaneously with autosomal and X-linked probes. X-linked probes reveal bands that are more intense than autosomal probes on XXXX cell lines compared to XY male cell lines. This protocol has been used to characterize an X chromosome library (*50*).

Y-linked genes are transmitted from father to son since only males have Y chromosomes. Y-linked probes can be differentiated by hybridization to total male but total female DNA by restriction enzyme analysis. Furthermore, Y-linked probes can be sub-chromosomally mapped by hybridization to total DNA from cell lines with different size deletions of the Y chromosome.

IV. COMPARISON OF PHYSICAL, GENETIC AND CHIASMA MAPS

If the cytogenetically scored chiasma in meiotic prophase chromosomes actually represent physical recombination, then the chiasma map indicates that more frequent meiotic recombination generally occurs near the ends of the chromosome arms (telomeres) than near the centromeres. This predicts that the genetic map distance (or recombination frequency) between genes located near the telomeres should be greater than predicted by the physical map distance. Physical and genetic map distances compared on the X chromosome are consistent with more chiasma near the telomeres than the centromeres (*21*). Each chiasma representing one crossover event that gives rise to two recombinants and two nonrecombinants and is taken to correspond to a genetic distance of 0.5 morgan or 50 centimorgans. With an average 60 chiasma scored per meiosis in the total human genome (*5,22*), this frequency gives a total genetic map distance of 3000 centimorgans. With a total of 2.87×10^9 bp per haploid genome, 1 centimorgan or a 1% crossover frequency

represents about 1000 kbp (or 1 Mbp). Thus, a genetic map distance between 2 and 3 centimorgans (2% crossover frequency) would represent about 2000 kbp of DNA. If the X chromosome data is typical (*21*), then two loci that are 2 centimorgans apart and physically located near the telomere are actually separated by less than 2000 kbp, while two loci 2 centimorgans apart near the centromere are more than 2000 kbp apart. This is important to investigators searching for a disease locus of unknown function beginning at a closely linked polymorphic DNA fragment. Still, we know "hot spots" of recombination, like the one 5' to the insulin gene (*37*), exist along the chromosome and may change the estimates in subchromosomal regions. Investigators moving toward disease loci from a closely linked DNA polymorphism can look for additional polymorphic markers on the same 100–500 kbp DNA fragment separated by pulse field electrophoresis (*58*). Others may "hop" with libraries constructed to facilitate moving more rapidly than with 15 kbp or 40 kbp inserts in phage or cosmid vectors, respectively (*8*). Others looking for fragments in the Duchenne's locus screened small chromosome deletions in total male DNA with many probes enriched for that subchromosomal region (*50*).

V. EVOLUTIONARY IMPLICATIONS OF GENE LOCATION

Comparing families of linked genes in organisms with vastly differing karyotypes has revealed which gene linkage groups have been conserved. In the case of the two homeo box gene loci on human chromosome 17, each map to a homologous mouse chromosome segment, indicating that each locus has been maintained through at least 50×10^6 years of evolution and has not moved to either location recently (*25*). This suggests that the homeotic box sequence serves some normal function in segmented organisms.

In another study the homologous aldolase genes were mapped to four different chromosomes. These four genes appear to have arisen from a single ancestral gene that duplicated twice at the time the chordate and vertebrate DNAs duplicated (59). The three active genes each code for a tissue-specific tetrameric isozyme that catalyzes the conversion of the six-carbon fructose diphosphate to two three-carbon chains. We have mapped the four aldolase genes to chromosomes 9, 10, 16 and 17. Aldolase B (chromosome 9) is the predominant enzyme in liver (39), aldolase C (chromosome 17) in brain and aldolase A (chromosome 16) in erythrocytes. Deficiency of aldolase B results in hereditary fructose intolerance and aldolase A deficiency causes hemolytic anemia. Mapping and sequencing these genes showed data consistent with a conservative evolution of the banded human chromosome complement. These four genes were mapped to two similarly banded chromosome pairs considered to be generated by one or two duplications of the entire genome at the time the aldolase isozymes diverged. [See Tolan *et al.* (59) for a thorough explanation.]

VI. DISCUSSION

Chromosomes sorted by bivariate flow cytogenetics have been quite useful in gene mapping. First, spot–blot analysis locates unique or low-copy genes to normal chromosomes. Secondly, spot–blot analysis of translocated or deleted chromosomes sublocalizes genes to chromosome segments. Thirdly, miniaturized restriction enzyme analysis of sorted chromosomes assigns unique restriction fragments to specific chromosomes. Originally, mitotic human chromosomes were sorted on the basis of total Hoechst 33258 fluorescence with laser excitation (33). Two-million chromosomes of each type were sorted and tested by restriction enzyme analysis. Since more than one chromosome type was sorted in a given fraction, further sorts of different size chromosomes carrying segments of the chromosomes in the gene-carrying peak were required. Discovery of the stain pairs Hoechst

33258 plus chromomycin A3 (18) and DIPI plus chromomycin A3 (38), which each enhance different chromosome characteristics, permitted separation of 22 of the 24 human chromosome types. Refinements in chromosome preparation substantially decreased tissue culture requirements (38), and along with improvements in sorter design (35,45) increased sorted chromosome purity. Sorting chromosomes directly onto nitrocellulose filters obviated the need to extract the DNA and increased the speed with which a cloned gene could be mapped since no more than 30 000 chromosomes were required to detect a gene-positive signal (38).

The human genome has been mapped more slowly and with considerably more effort than the *Drosphila* (fruit fly) since the generation time is longer, the number of offspring is smaller, the matings are uncontrolled, and the number of chromosomal linkage groups is several times larger. Since characterization of the human genome has been more relevant to human genetic disease, more sophisticated techniques were developed to characterize it. Gene mapping has been done successfully with somatic cell hybrids, *in situ* hybridization and now by spot–blot analysis of sorted human chromosomes. Polymorphic restriction fragments that were first used to prenatally diagnose a human genetic disease (26) can now generate many more informative chromosomal sites to test linkage between two gene loci (37). Human chromosome-specific recombinant DNA libraries (13,42,60) facilitate isolating many chromosome-specific fragments to saturate the human map by linkage analysis of polymorphic restriction enzyme fragments. The classical genetics of the laboratory fruit fly has provided the basis of understanding of linkage, but its application to human pedigrees are now applied with sophisticated computer programs that calculate recombination frequences and LOD scores (52) of multiple gene loci and multiple restriction enzymes that can be used to define ever larger numbers of polymorphic loci. The human genome no longer consists of 46 unbanded chromosomes, but rather banded chromosome karyotypes with up to 2000 bands per haploid genome (*14,63*). Indeed, human genetics has come a long way and the tools are now available to advance more rapidly than ever. Undoubtedly these tools will contribute to further understanding gene action in health and disease. For example, some of the first evidence that chromosome translocations can cause cancer was suggested when oncogenes and immunoglobin genes were mapped to specific chromosome bands often involved in cancer-specific translocations. Surely many more candidate genes will be identified and confirmed as the entire human genome is mapped and sequenced.

REFERENCES

1. Bartholdi, M. F., Ray, F. A., Cram, L. S., and Kramer, P. M. (1984). Flow karyology of serially cultured Chinese hamster cell lineages. *Cytometry* **5**, 534–538.
2. Bernheim, A., Metezeu, P., Guellaen, G., Fellous, M., Goldberg, M. E. and Berger, R. (1983). Direct hybridization of sorted human chromosomes: Localization of the Y chromosomes on the flow karyotype. *Proc. Natl Acad. Sci. USA* **80**, 7571–7575.
3. Carrano, A. V., Gray, J. W., Langlois, R. G., Burkhardt-Schultz, K. J. and Van Dilla, M. A. (1979). Measurement and purification of human chromosomes by flow cytometry and sorting. *Proc. Natl Acad. Sci. USA* **76**, 1382–1384.
4. Caskey, J. H., Jones, C., Miller, Y. E. and Seligman, P. A. (1983). Human ferritin gene is assigned to chromosome 19. *Proc. Natl Acad. Sci. USA* **80**, 482–486.
5. Cavalli-Sforza, L. L. and Bodmer, W. F. (1971). "*The Genetics of Human Populations*", p. 883. Freeman, San Francisco.
6. Chandler, M. E. and Saunders, G. F. (1981). Localization of single copy DNA sequences on G-banded human chromosomes by *in situ* hybridization. *Chromosoma* **83**, 431–439.
7. Collard, J. G., de Boer, P. A. J., Janssen, J. W. G., Schijven, J. F., and Oosterhuis, J. W. (1985). Gene mapping by chromosome spot hybridization. *Cytometry* **6**, 179–185.
8. Collins, F. S., Drumm, M. L., Cole, J. L. and Weissman, S. M. (1986). Generation of "chromosome hopping" libraries for fine mapping the human genome. *Clin. Res.* **34**, 448A.
9. Conboy, J. G., Mohandas, N., Wang, C., Chernia, G. T., Shohet, S. B., and Kan, Y. W. (1985). Molecular cloning and characterization of the gene coding for red cell membrane skeletal protein 4.1. *Blood* **66**, 31a.
10. Davies, K. E., Young, B. D., Elles, R. G., Hill, M. E. and Williamson, R. (1981). Cloning of a representative genomic library of the human X chromosome after sorting by flow cytometry. *Nature* **293**, 374–376.
11. Davis, M., Malcom, S. and Rabbitts, T. H. (1984). Chromosome translocation can occur on either side of the c-myc oncogene in Burkitt's lymphoma cells. *Nature* **308**, 286–288.
12. Deisseroth, A., Nienhuis, A., Turner, P., Velez, R., Anderson, W. F., Lawrence, J., Creagan, R. and Kucherlapati, R. (1977). Localization of the human α-globin structural gene to chromosome 16 in somatic cell hybrids by molecular hybridization assay. *Cell* **12**, 205–218.
13. Disteche, C. M., Kunkel, L. M., Lojewski, A., Orkin, S. H., Eisenhard, M., Sahar, E., Travis, B., and Latt, S. A. (1982). Isolation of mouse X-chromosome specific DNA from an X-enriched lambda phage library derived from flow sorted chromosomes. *Cytometry* **2**, 282–286.
14. Francke, U. and Oliver, N. (1978). Quantitative analysis of high-resolution trypsin–Giemsa bands on human prometaphase chromosomes. *Hum. Genet.* **45**, 137–165.
15. Fritsch, E. F., Lawn, R. M., and Maniatis, T. (1980). Molecular cloning and characterization of the human β-like globin gene cluster. *Cell* **19**, 959–972.
16. Gartler, S. M., Riley, D. E., Lebo, R. V., Cheung, M. C., Eddy, R. L. and Shows, T. B. (1986). Mapping of a human autosomal phosphoglycerate kinase sequence to chromosome 19. *Somat. Cell Molec. Genet.* **12**, 395–401.
17. Goode, M. E., Ledbetter, D. H., and Daiger, S. P. (1984). Failure of *in situ* hybridization to detect the primary copy of ωH3. *Am. J. Hum. Genet.* **36**, 138S.
18. Gray, J. W., Langlois, R. G., Carranno, A. V., Burkhart-Schultz, K., and Van Dilla, M. A. (1979). High resolution chromosome analysis: one and two parameter flow cytometry. *Chromosoma* **73**, 9–27.
19. Harper, M. E., Ullrich, A., and Saunders, G. F. (1981). Localization of the human insulin gene to the distal end of the short arm of chromosome 11. *Proc. Natl. Acad. Sci. USA* **78**, 4458–4460.
20. Harper, M. E., and Saunders, G. F. (1981). Chromosomal localization of human insulin gene, placental lactogen-growth hormone genes, and other single copy genes by in situ hybridization. *Am. J. Hum. Genet.* **33**, 105A.
21. Hartley, D. A., Davies, K. E., Drayna, D., White, R. L. and Williamson, R. (1984). A cytological map of the human X chromosome-evidence for non-random recombination. *Nucl. Acid. Res.* **12**, 5277–5285.
22. Hulten, M. (1974). Chiasma distribution at diakinesis in the normal human male. *Hereditas* **76**, 55–78.
23. Human Genetic Mutant Cell Repository, Coriell

Institute for Medical Research, Copewood St, Camden, NJ.

24. ISCN (1981). An international system for human cytogenetic nomenclature high-resolution banding. *Cytogenet. Cell Genet.* **31**, 5–21.

25. Joyner, A. L., Lebo, R. V., Kan, Y. W., Tjian, R., Cox, D. R., and Martin, G. R. (1985). Comparative chromosome mapping of a human homeo box region: conservation of linkage in humans and mice. *Nature* **314**, 173–175.

26. Karmarck, M. E., Peters, P. M., Hemler, M. E., Strominger, J. L., and Ruddle, F. H. (1984). Genetic and biochemical characterization of human lymphocyte cell surface antigens: the A-1A5 and A-3A4 determinants. *Anal. Cytol.* **X**, D29.

27. Kamarck, M. E., Roberts, M. P., Ruddle, F. H., and Hart, J. T. (1985). Genetic analysis of the human S11 surface antigen. *Anal. Cytol.* **XI**, 128.

28. Kan, Y. W., and Dozy, A. M. (1978). Polymorphism of DNA sequence adjacent to human β-globin structural gene: relationship to sickle mutation. *Proc. Natl Acad. Sci. USA* **75**, 5631–5635.

29. Keats, B. J. B., Morton, N. E., Rao, D. C., and Williams, W. R. (1979). "A Source Book for Linkage in Man", p. 197. Johns Hopkins University Press, Baltimore.

30. Korenberg, J. R., Croyle, M. L., and Cox, D. R. (1985). Isolation and characterization of DNA sequences unique to human chromosome 21. *Am. J. Hum. Genet.* **37**, A162.

31. Law, M. L., Tung, L., Graw, S., and Van Keuren, M. (1984). Identification of chromosomal deletions in somatic cell hybrids by cloned repetitive sequences. *Am. J. Hum. Genet.* **36**, 203S.

32. Lau, Y.-F., Dozy, A. M., Huang, J. C., and Kan, Y. W. (1984). A rapid screening test for antenatal sex determination. *Lancet* **i**, 14–16.

33. Lebo, R. V., Carrano, A. V., Burkhart-Schultz, K. J., Dozy, A. M., Yu, L.-C., and Kan, Y. W. (1979). Assignment of human β-, γ- and δ-globin genes to the short arm of chromosome 11 by chromosome sorting and DNA restriction enzyme analysis. *Proc. Natl Acad. Sci. USA* **76**, 5804–5808.

34. Lebo, R. V. (1982). Chromosome sorting and DNA sequence localization: a review. *Cytometry* **3**, 145–154.

35. Lebo, R. V. and Bastian, A. M. (1982). Design and operation of a dual laser chromosome sorter. *Cytometry* **3**, 213–219.

36. Lebo, R. V., Kan, Y. W., Cheung, M.-C., Carrano, A. V., Yu, L.-C., Chang, J. C., Cordell, B., and Goodman, H. M. (1982). Assigning the polymorphic human insulin gene to the short arm of chromosome 11 by chromosome sorting. *Hum. Genet.* **60**, 10–16.

37. Lebo, R. V., Chakravarti, A., Buetow, K. H., Cann, H., Cheung, M.-C., Cordell, B. and Goodman, H. (1983). Chiasma within and between the human insulin and β-globin gene loci. *Proc. Natl Acad. Sci. USA* **80**, 4808–4812.

38. Lebo, R. V., Gorin, F., Fletterick, R. J., Kao, F. -T., Cheung, M. -C., Bruce, B. D., and Kan, Y. W. (1984). High-resolution chromosome sorting and DNA spot–blot analysis assign McArdle's syndrome to chromosome 11. *Science* **225**, 57–59.

39. Lebo, R. V., Tolan, D. R., Bruce, B. D., Cheung, M.-C., Penhoet, E. D., and Kan, Y. W. (1985). Spot–blot analysis of sorted chromosomes assigns the aldolase B gene to chromosome 9. *Cytometry* **6**, 478–483.

40. Lebo, R. V., Cheung, M.-C., Bruce, B. D., Riccardi, V. M., Kao, F.-T., and Kan, Y. W. (1985). Mapping parathyroid hormone, β-globin, insulin and LDH-A genes within the human chromosome 11 short arm by spot blotting sorted chromosomes. *Hum. Genet.* **69**, 316–320.

41. Lebo, R. V., Kan, Y. W., Cheung, M.-C., Jain, S. K., and Drysdale, J. (1985). Human ferritin light chain gene sequences mapped to several sorted chromosomes. *Hum. Genet.* **71**, 325–328.

42. Lebo, R., Anderson, L., Lau, C., Flandermeyer, R., Kan, Y. W. (1986). Flow sorting analysis of normal and abnormal human genomes. *Cold Spring Harbor Symp.* **51**, 169–176.

43. Lebo, R. V., Golbus, M. S., and Cheung, M.-C. (1986). Detecting abnormal human chromosome constitutions by dual laser flow cytogenetics. *Am. J. Med. Genet.* **25**, 519–529.

44. Lebo, R. V. and Bruce, B. D. (1987). Gene mapping with sorted chromosomes. *Meth. Enzymol.* **151**, 292–313.

45. Lebo, R. V., Bruce, B. D., Dazin, P. F., and Payan, D. G. (1987). Design and operation of a versatile triple laser cell and chromosome sorter. *Cytometry* **8**, 71–82.

46. Leif, R. C., Easter, J. R., Warters, R. L., Thomas, R. A., Dunlap, L. A., and Austin, M. F. (1971). Quantitative technique for the preparation of glutaraldehyde-fixed cells for the light and scanning electron microscope. *J. Histochem. Cytochem.* **19**, 203–215.

47. Macaya, G., Thiery, J.-P., and Bernardi, G. (1977). DNA sequences in man. In "Molecular Structure of Human Chromosomes", (J. J. Yunis, ed.), pp. 35–38. Academic Press, New York.

48. Mayo, K. E., Cerelli, G., Lebo, R. V., Bruce, B. D., Rosenfeld, M. G., and Evans, R. M. (1985). Structure, sequence, and chromosomal assignment of the gene encoding human growth hormone releasing factor. *Proc. Natl Acad. Sci. USA* **82**, 63–67.

49. McKusick, V. A. (1983). "*Mendelian Inheritance in Man*" 6th ed. Johns Hopkins University Press, Baltimore.

50. Monaco, A. P., Neve, R. L., Colletti-Feener, C., Bertelson, C. J., Kurnit, D. M., and Konkel, L. M. (1986). Isolation of candidate cDNAs for portions of the Duchenne muscular dystrophy gene. *Nature* **323**, 646–650.

51. Orkin, S. H., and Kazazian, H. H. (1984). The

mutation and polymorphism of the human β-globin gene and its surrounding DNA. *Ann. Rev. Genet.* **18**, 131–172.

52. Ott, J. (1974). Estimation of the recombination fraction in human pedigrees: Efficient computation of the likelihood for human linkage studies. *Am. J. Hum. Genet.*, **26**, 588–597.

53. Paris Conference (1972). Standardization in human cytogenetics. National Foundation-March of Dimes. Birth Defects (1971): Original Article Series VIII, Vol. 7.

54. Parker, R. C., Mardon, G., Lebo, R. V., Varmus, H. E., and Bishop, J. M. (1985). Isolation of duplicated human c-src genes located on chromosomes 1 and 20. *Mol. Cell Biol.* **5**, 831–838.

55. Proudfoot, N. J., Gil, A. and Maniatis, T. (1982). The structure of the human zeta-globin gene and a closely linked, nearly identical pseudogene. *Cell.* **31**, 553–563.

56. Ruddle, F. H., and Creagan, R. P. (1981). Parasexual approaches to the genetics of man. *Ann. Rev. Genet.* **9**, 407–486.

57. Sanders-Haigh, L., Anderson, W. F., and Francke, U. (1980). The β-globin gene is on the short arm of human chromosome 11. *Nature* **283**, 683–686.

58. Schwartz, D. C., and Cantor, C. R. (1984). Separation of yeast chromosome-sized DNAs by pulsed field gradient gel electrophoresis. *Cell.* **37**, 67–75.

59. Tolan, D. R., Niclas, J., Bruce, B. D., and Lebo, R. (1987). Evolutionary implications of the human aldolase A, B, C and pseudogene chromosome locations. *Am. J. Hum. Genet.* **41**, 240–257.

60. Deaven, L. L., Van Dilla, M. A., Bartholdi, M. F., Carrano, A. V., Cram, L. S., Fuscoe, J. C., Gray, J. W., Hildebrand, C. E., Moyzis, R. K. and Perlman, J. (1986). Construction of human chromosome-specific DNA libraries from flow sorted chromosomes. *Cold Spring Harbor Symp.* **51**, 159–167.

61. Worwood, M., Brook, J. D., Cragg, S. J., Hellkuhl, B., Jones, B. M., Perera, P., Roberts, S. H. and Shaw, D. J. (1985). Assignment of human ferritin genes to chromosomes 11 and 19q13.3→19qter. *Hum. Genet.* **69**, 371–374.

62. Young, B. D., Ferguson-Smith, M. A., Sillar, R, and Boyd, E. (1982). High resolution analysis of human peripheral lymphocyte chromosomes by flow cytometry. *Proc. Natl Acad. Sci. USA* **78**, 7727–7731.

63. Yunis, J. J., and Chandler, M. E. (1978). High-resolution chromosome analysis in clinical medicine. *Prog. Clin. Pathol.*, **7**, 267–88.

Metaphase Chromosome Flow Sorting and Cloning; Rationale, Approaches and Applications

S. A. LATT,**† M. LALANDE,* A. FLINT,* P. HARRIS,*
U. MULLER,* T. DONLON,* U. TANTRAVAHI,* G. BRUNS,*
D. KURNIT,* R. NEVE,* and L. KUNKEL*

Mental Retardation Center, Division of Genetics,
The Children's Hospital and Departments of
Pediatrics* and Genetics,** Harvard Medical School, MA 02115, U.S.A.

I. INTRODUCTION

A. OVERALL DESCRIPTION AND RATIONALE

Metaphase chromosome flow sorting permits one to enrich for the DNA (or protein) of a given chromosome and thereby focus more efficiently on molecular studies of that chromosome. Cells are trapped at metaphase, metaphase chromosomes are isolated and the chromosomes are stained by one or more fluorescent dyes. The fluorescence emitted by these dyes guides subsequent chromosome

† Deceased.

FLOW CYTOGENETICS
ISBN 0-12-296110-2

sorting. The net result can be up to a 10–50× purification of a given human chromosome, depending on chromosome size, typically in amounts up to a few micrograms (as DNA). In practice, absolute purity of sorted chromosomes is not achieved, e.g. because of chromosome fragmentation and aggregation, but even 80–90% purity constitutes an appreciable enrichment that is useful for most purposes, such as gene mapping or the construction of chromosome-enriched recombinant DNA libraries (*99*).

The significance of enriching 10–50× for a given chromosome can be appreciated in the context of the approximate size of the human haploid genome (3×10^9 bp) (*81*), the goal of obtaining a polymorphic DNA marker near a disease of interest and the number of cloned DNA fragments (100–1000) that a given laboratory can screen within a reasonable time. Without chromosome sorting, analysis of 1000 random DNA fragments will lead to an average fragment spacing of 3×10^6 bp and hence an average closest distance between an arbitrary point in the genome, such as a disease to which linkage is sought, of the order of 10^6 bp. For linkage purposes, 10^6 bp corresponds on the average to approximately 1% genetic recombination (*71*), i.e. fairly tight linkage. However, only a small fraction of arbitrary DNA fragments are sufficiently polymorphic (heterozygosity exceeding 0.3–0.4) to be of significant use in linkage analysis (*1,7,34, 68,92*).

Presorting of metaphase chromosomes to a 10–50× enrichment prior to recombinant library construction permits the same level of subsequent effort to localize a useful polymorphic DNA segment within 0.1–1% recombination units of an arbitrary gene whose chromosomal location is known. One can enhance this situation further, either by increasing the number of laboratories screening a given library, such as made possible by the Lawrence Livermore–Los Alamos National Laboratory Chromosome Sorting Project(*89*; see Chapter 16). The end result is the ability to identify, within a reasonable time, multiple DNA fragments closely linked to an arbitrary

gene, and thus is useful to follow the inheritance of this gene from one generation to the next, even though the gene itself might not yet have been cloned or understood at a functional level.

Another important goal is the assignment of multiple DNA fragments to a small subregion of a metaphase chromosome, e.g. 1.5–3×10^6 bp, essentially the limit of resolution at a cytological level (*100*). This latter potential has already been realized by several investigators (e.g. *16,20,23,42,90*). Such probes, once mapped, can provide an alternative, using quantitative DNA blotting, to high-resolution chromosome banding for detecting small, chromosome-specific microdeletions.

B. Basic Requirements

The required amount and condition of sorted chromosomes depends greatly on the use intended for the sorted material. Five-million of an average-sized chromosome contain nearly 750 ng DNA of which at least 250–300 ng should be recovered. This amount of DNA is sufficient when digested (e.g. with Eco RI or Hind III) and ligated to DNA, identically digested, from a bacteriophage such as Charon 21A, to construct a recombinant library containing inserts averaging 4–5 kbp in size. When highly efficient packaging extracts are used, the complexity of such libraries can be in the 100 000+ plaque forming units (pfu) range (e.g. *42*). This number of phage should then contain, on the average, several copies of any segments in the total digest that are present in this chromosome. If a recombinant library containing large DNA inserts, isolated by size fractionation, is desired, then the DNA requirements could increase approximately ten-fold. At a complexity approaching 10^5 pfu, this could mean isolating 10^6 or more chromosomes (*26,41*).

If chromosomes are to be used for gene mapping rather than recombinant library construction, then the needs vary further. Dot–blot analysis done by sorting chromosomes directly onto nitrocellulose disks requires no more than 50 000 chromosomes (*11,55*, see Chapter 14) but cannot yield restriction fragis subject to false positive signals, e.g. for technical reasons or from pseudogenes. If the sorted material is to be used for DNA isolation, restriction enzyme digestion, gel electrophoresis and blotting, then more chromosomes are typically employed (31,53,54,85). Finally, if chromosomal proteins are to be examined (87,94,95), μg amounts of chromosomal proteins may be desired and tens of millions of chromosomes would then be required.

For these applications, needed chromosome amounts are not limited by current instrumentation. For example, on a FACS IV, at an analysis rate of 1000–2000 chromosomes/ sec, and sorting a chromosome present at a fraction of 1/23, one can obtain approximately 10^6 chromosomes/day (e.g. 42). Collection of tens of millions of chromosomes extends sorting time to weeks. This can be countered either by chromosome size enrichment, by velocity sedimentation (10), or by the use of a high-pressure chromosome sorter.

The condition of the sorted material is very important and is an important consideration when selecting among described chromosome isolation procedures. In one frequently used protocol (80), divalent cations which activate nucleases are chelated by EDTA, while oligoamines further protect the DNA. Isolation of DNA from such chromosomes with a size in excess of 50 kbp is routine. This size is large enough for phage libraries which accommodate DNA inserts up to the 20 kbp range and perhaps sufficient for cosmid libraries (e.g. 27,62), which can accomodate DNA inserts of 40 kbp or more.

II. SPECIFIC CHROMOSOME ISOLATION

A. Univariate (One-fluorochrome) Metaphase Chromosome Flow Sorting

Initial flow cytometric chromosome identification and isolation for DNA cloning employed a single DNA-binding fluorochrome (13,17,18,21,37,38,39,42,43) (e.g. Fig. 1).

Ethidium bromide was among the most convenient dyes because of its appreciable absorptivity at the 488 and 514 nm high-intensity output lines of argon ion lasers and its high fluorescence quantum yield when bound to double-stranded nucleic acids (52,58). However, ethidium bromide can also fluoresce brightly when bound to double-stranded RNA, which can adsorb to metaphase chromosomes. Chromomycin A3 is more DNA specific, with a GC-binding preference [see Latt and Langlois (52) and Chapter 5 for a detailed review of these and other dyes], but it has only modest absorptivity near the low-intensity output 457 nm argon ion laser line and a low fluorescence quantum yield in its DNA–magnesium complex (48). Hence, while useful as a single dye with whole cells, it too is suboptimal as a single stain for chromosomes. The bisbenzimidazole dye Hoechst 33258 (or its structurally related derivatives), has a much higher absorptivity and quantum yield than the aforementioned dyes, while exhibiting a nearly absolute DNA-binding specificity (47). It also binds preferentially to AT base pair clusters, distorting slightly the linearity between fluorescence intensity and DNA content. Its major drawback is its need for near-ultraviolet excitation. However, this can now readily be accomplished with the 351–364 nm doublet of 4–6 W argon ion lasers.

A significant amount of chromosome sorting and cloning has been accomplished using chromosomes stained with only one dye and instruments with a single laser.

(1) Structurally abnormal mouse X chromosomes (containing inserted mouse chromosome 7, i.e. the Cattanach X) were among the first X chromosomes sorted preparatively (18). This abnormal X chromosome could be resolved by flow cytometry using either ethidium bromide or Hoechst 33258 (17) staining. The latter yielded somewhat better separation from the next largest mouse chromosome (the 1). Hoechst 33258 was thus used in preparative sorting of the Cattanach X for subsequent cloning (18).

(2) Human X chromosomes were sorted preparatively using either ethidium bromide

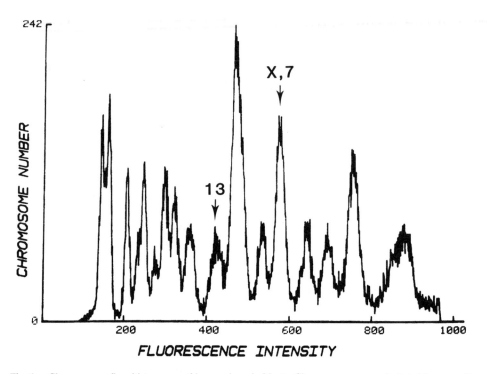

Fig. 1. Chromosome flow histograms of human lymphoblasts. Chromosomes were isolated from a cell line with a 49,XXXXY karyotype, stained with Hoechst 33258 dye, and analyzed on a FACS II flow system at a flow rate of 3000 chromosomes/sec. The ordinate is the chromosome number, and the abscissa is a linear (1024 channel) scale of fluorescence intensity. The labeled fluorescence peaks contain chromosome 13 and chromosomes 7 and X, respectively (*43*).

(13) or Hoechst 33258 (*38*) as stains. Both libraries have yielded useful X-specific probes (e.g. *1,32*). An example of an X-specific probe from the Kunkel *et al.* (*38*) library, which is approximately 60–65% pure, is shown in Fig. 2. This figure also illustrates the type of confirmatory screening necessary for DNA probes derived from "chromosome-enriched" libraries. A related study utilized Hoechst 33258 as the stain in preparative sorting of a dicentric human X [dic(X)(q24)] (*43*).

(3) Soon after the early X chromosome preparative sorting came experiments, typically using ethidium bromide, to isolate other chromosomes, such as chromosomes 21 or 22 (*37*) and 6 (*6*). Another study succeeded in isolating human chromosome 13 (*42*), using Hoechst 33258 as a stain (Fig. 1). Crucial to such one-color chromosome sorting is the choice

of a cell line which, especially in the case of the acrocentric chromosomes, possessed polymorphic variation causing the chromosome of choice to fall into a zone of fluorescence well separated from the others (e.g. *30*).

(4) Human Y chromosome sorting, exploiting polymorphic variations to obtain an optimally sized Y chromosome, succeeded with Hoechst 33258 as a stain (*21*).

B. Two-color Flow Sorting

The combined use of Hoechst 33258 plus chromomycin A3 has proved to be the workhorse of most modern preparative chromosome sorting. The AT specificity of Hoechst 33258, plus its AT base pair cluster-dependent energy transfer to the GC-specific

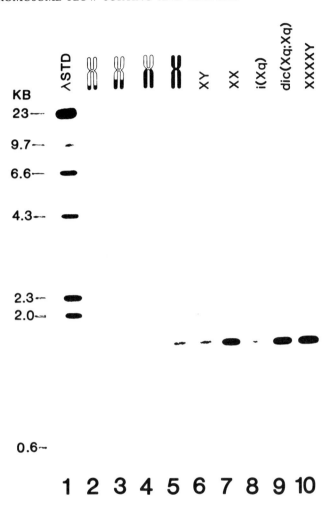

Fig. 2. DNA blot localizing a probe from the Kunkel *et al.* (*38*) X-enriched library to Xp. All lanes have equal amounts of DNA, cut with Hind III. Lane 1 represents Hind-III-digested ^{32}P-labeled phage lambda standards. Lanes 2–5 are hybrids retaining segments of the X shaded as shown. Presence of Xp (lane 5) is needed for a signal (at about 1.8 kbp). Lanes 6–10 are from human lymphoblast cells showing a dosage of the blotting signal with the number of X chromosomes per genome (lanes 6, 7, 10) and the number of X short arms [the i(Xq)-containing cells in lane 8 have one copy of Xp while the dic(X)(q24) cells in lane 9 have three copies of Xp].

chromomycin A3 (*74,75*) results in an "off-diagonal" position for chromosomes with a base composition bias (*25,39,45*). AT-rich chromosomes such as the Y chromosome are deflected off the "equifluorescent diagonal" in one direction (*64,65*), while GC-rich chromosomes such as chromosome 19 are deflected in the other direction (Fig. 3). Of equal importance, with small chromosomes, is the dispersal (but not elimination) of the debris background, so that even small derivative chromosomes [e.g. inv dup (15)] (*20,44*) can be resolved. Nonspecific contamination of sorted chromosomes occurs both from particles with subthreshold fluorescence and from debris underlying chromosome "peaks" delineated

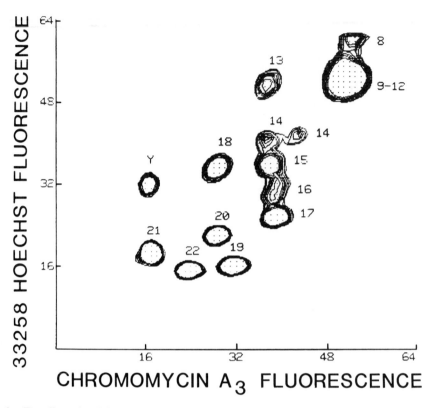

Fig. 3. Two-dimensional flow sorting of the smaller human metaphase chromosomes from the same 49,XXXXY lymphoblast line as used in Fig. 1. The chromosomes were stained with Hoechst 33258 and chromomycin A3 (*25,39,65*) with Hoechst 33258 fluorescence plotted on the ordinate and chromomycin A3 fluorescence on the abscissa. A FACS IV was used. Contour lines start at 150 chromosomes and range up to 300 chromosomes. Hence, the peaks of the small chromosomes are clearly resolved, although subchromosomal debris is not apparent.

by contour lines that start well above zero fluorescence. Two-dimensional sorts, using appropriate cell lines, can also resolve large chromosomes, e.g. chromosome 1 from 2, and should be very useful in separating human chromosomes in hybrid cells from the rodent chromosome background.

Additional multistain procedures can selectively alter the fluorescence of human chromosomes 1, 9, 15, 16 and Y either by adding a third (nonfluorescent) dye, such as netropsin or distamycin (*44,63*), or by substituting phenylindole derivatives (e.g. DAPI or DIPI) for the bisbenzimidazole dye Hoechst 33258 (*55*). In no case has the chromosome 9–12 cluster been resolved

completely. The use of polymorphic variations, chromosome abnormalities or judicious choice of chromosome-reduced human–rodent hybrid cells (*35*) is needed to isolate the members of this chromosome set.

III. TYPES OF RECOMBINANT LIBRARIES

A. INSERT TYPE AND SIZE

The past few years have witnessed an ever growing spectrum of cloning vectors and screening strategies for libraries constructed from the DNA of flow-sorted chromosomes that can be used. In general, libraries have been constructed from DNA which had been digested to completion with a given restriction enzyme, e.g. Hind III. Hence these are comprised of small inserts and contain most but not all of the DNA of a given chromosome. These are best for gaining a molecular "foothold" on the chromosome. The small fragments so derived can then be used directly or as starting points for molecular walks with larger insert, partial digest libraries. These large inserts can themselves be constructed from the DNA of flow-sorted and partially purified chromosomes or from total genomic DNA, subjected to incomplete enzyme digestion and size fractionation. Either large-insert-accepting phage or cosmids can be used, the latter typically requiring more DNA.

B. Cloning Vector

The choice of vector is intimately linked with the size of insert used. The phage Charon 21A (49,93) can accommodate 8–9 kbp DNA fragments inserted into its unique Hind III or EcoRI sites. Lambda gt wes (86), which accepts slightly larger inserts, has also been used for complete EcoRI digests (13,18).

Clearly, plasmids (5) would accommodate either of the above, although screening of the resulting library might be more laborious. One exception is the miniplasmid set (PiVX or derivatives) introduced by Seed (79). In this case, one can insert fragments (66,67), including small Hind III–EcoRI fragments of sorted chromosome DNA (U. Tantravahi et al., unpublished work) and then screen a second phage library, such as that of Maniatis et al. (61).

Other strategies for cloning small DNA inserts have not been completely exploited. For example, based on experience with MboI total digest fragments in phenol-enhanced reassociation technique (PERT) libraries (40), it is evident that small, renatured DNA fragments with two MboI termini, subcloned, e.g. into the BamHI site of a high-yield plasmid (e.g. PUC18) would yield inserts, few of

which contained confounding chromosome nonspecific repeats.

One could also exploit the lambda gt10– lambda gt11 pair (33,97,98), both accepting EcoRI inserts. The former phage gives a high yield per infected bacterium and utilizes cloning into the "imm" gene to permit efficient selection of recombinants as clear plaques on a turbid plaque background. Inserts could then be transferred in bulk to phage lambda gt11, which couples inserts to the beta galactosidase gene so that, on the average, one in six coding sequences will be "in frame" and in the correct orientation to yield a hybrid protein which might then be screened with an appropriate antibody.

At the opposite extreme are large-insert libraries, typically prepared from partially digested or sheared DNA and constructed to represent most of a chromosome. Some success with this has been realized with an MboI partial digest of human X chromosome DNA cloned in lambda phage Charon 30 (41, 73) and a partial EcoRI digest of hamster chromosome 2 DNA cloned in Charon 4A (26,61). Problems with some phage include the absence of "chi" sequences, e.g., in Charon 30, making amplification of phage uneven [depending on "chi" (83) contained in inserts] and an unexplained inability of some phage to grow on recombination-proficient hosts (96). Cosmid libraries, with larger inserts, may present even larger insert preparation and cloning stability problems. Also, large-insert libraries are poor as initial sources of chromosome-specific DNA, since they invariably contain chromosome-nonspecific repeat DNA which must be removed by insert fragmentation or otherwise "blocked-out" before screening. However, such large-insert libraries can be very useful for genomic walking or for rapidly obtaining the genomic representation of a cDNA clone.

C. Bacterial Host

The inability of phage (small or large insert) to propagate efficiently on standard bacteria can be overcome by utilizing CES200 recombination modified bacteria. Wyman *et al.* (*96*) showed that recBC⁻ sbcB⁻ bacteria favored the propagation of phage with inserts containing inverted repeats. This was exploited by Donlon *et al.* (*20*) in isolating DNA inserts from the proximal long arm of human chromosome 15. These inserts included two that appear to be deleted in one chromosome 15 of a Prader–Willi patient, providing a potentially useful diagnostic tool and perhaps some insight into the cytological instability of this region of human chromosome 15. Other recombination altered hosts might favor propagation of phage with different types of inserts.

IV. LIBRARY SCREENING STRATEGIES

Detailed library screening strategies obviously depend on the type of insert and the ultimate goal for which the library was constructed. Small inserts are generally used as molecular linkage markers, or for detecting chromosome deletions, while large inserts have the additional potential for use in delineating or even isolating complete genes.

A. DIRECT

Phage with inserts carrying chromosome nonspecific repeat sequences can be detected, e.g. by Benton Davis (*4*) and/or Southern (*82*) hybridization. Inserts lacking repeats, or subclones of such inserts lacking nonspecific repeats are then relatively easy to map. Alternatively, some amount of repeat background can be blanked-out by adding an excess of cleaved total or cloned AluI (*76*) fragment DNA, which competes with repeats for target DNA (*59,78*). Once conditions for repeat-insensitive screening are established, one can then employ large DNA inserts for mapping with hybrid cell panels. Recent use of chromosome-sorted dot–blots (*11,55*, see Chapter 14) for probe mapping represents an alternate means of probe mapping.

In the case of X chromosomes, one can also utilize DNA from cells with one, two, three, four or even five X chromosomes to check X-localization by hybridization dosage. As a modification of this strategy, one can use cells with partial duplications and deficiencies for subchromosomal regions, e.g. i(Xq) or dic(X)(q24) for dosage blotting (*38,84*) (Fig. 2). For example, probes localized to Xp will hybridize weakly to 46,X,i(Xq) (one copy/genome) DNA but strongly to 46,X,dic(X)(q24) DNA (three copies per genome). Conversely, probes from Xq24 → Xqter will show strong hybridization with 46,X,i(Xq) DNA but weak hybridization with 46,X,dic(X) (q24) DNA.

The Y chromosome segments of course can be detected simply by hybridization to male but not female DNA, although the non-Y homology of certain sequences must also be considered. Higher-resolution mapping of chromosomes can then use somatic cell hybrids, diploid cells with small duplications or deletions, or metaphase chromosomes and *in situ* hybridization.

Since the total number of DNA fragments now obtainable vastly exceeds the number of available chromosome aberrations, uncertainties in precise positioning of the latter can in principle be overcome by recursive use of probes and cell lines until internally consistent chromosome break-point and ordering of probes is achieved (*69*).

B. SEQUENTIAL

One could utilize a large-insert library, against which small-chromosome-specific genomic DNA segments or cDNA clones are then used as probes. If the large-insert libraries contain overlapping clones via partial digestion, total digestion with different enzymes, or inclusion of junctions between other clones, then genomic walking can follow. For example, the ability of long-range walking via jump libraries (*12*), which aim to clone ends of large DNA fragments, or pulse gel electrophoresis (*8,77*), which can separate very large DNA fragments, permits probes to be linked onto individual or small numbers of overlapping DNA fragments of a size resolvable at a cytological level. This "bridge" ensures an ultimate probe map of the human genome upon which cytological aberrations can be ordered and from which extensive or even complete regional DNA sequencing can be initiated.

C. Cross-Screening

Screening by recombination is illustrated by the PiVX miniplasmid method (79), described earlier, in which the miniplasmid carries a needed (suppressor) gene, to override a translation termination signal, and a small insert that recognizes the existence and multiplicity of sequences in a phage library by homologous recombination between inserts. Only phage which have recombined with the miniplasmid, via insert homology, will now grow in a suppressor negative host. Ideally, this library would itself be a chromosome-enriched library, e.g. so that the two libraries have a small subchromosomal intersection. However, for reasons not totally clear, some so-called amber (suppressable, non-sense) mutation vectors, e.g. Charon 21A, have a high non-amber background (D. Kurnit, personal communication) which, at least for now, require elaborate maneuvers to circumvent. In contrast, the original Maniatis Charon 4A partial digest total representation large-insert human library (61) has a low background (66,67). Technical improvements in this approach will stimulate more widespread use.

V. CURRENT AND PROJECTED APPLICATIONS

A. General

(1) Chromosome-specific DNA segments are useful adjuncts to cytology for analyzing chromosome composition and rearrangement. For example, DNA segments detecting (via reduced hybridization intensity) and differentiating small deletions of chromosome 13 [in retinoblastoma, (36)] and chromosome 15 [in the Prader–Willi syndrome (PWS) (20, 56, 57)] have been isolated from flow-sorted chromosome-derived recombinant libraries. Structural aberrations of the human X chromosome have been characterized by X-specific probes and aberrations in turn used for fine mapping of subsequent X probes (69,70,84).

(2) Those chromosome-library-derived DNA probes that can be shown to recognize restriction fragment length polymorphisms (RFLPs) (7) can be used in molecular linkage studies. For example, such probes, mapping to Xp and linked to Duchenne muscular dystrophy (DMD), were instrumental in early work localizing the DMD gene to Xp21 (1,2,14) and following its segregation in pedigrees. These probes are now being used in studies of the fragile X (e.g. 70). A multitude of other such applications will soon follow. Polymorphic probes from chromosome 13, derived primarily from cell hybrids (9), have been instrumental in demonstrating the mitotic recombination and/or gene conversion leading to homozygosity for the retinoblastoma gene in retinoblastomas (9) and osteosarcoma (29), a finding soon extended by many others to other neoplasms involving different chromosomes.

It may be true that the single most efficient way to create a molecular linkage grid for the entire genome is to identify hundreds of informative RFLPs and *then* map them (D. Botstein, Plenary lecture, HGM8, 1985, Helsinki). However, chromosome-specific libraries permit studies on localized regions of the genome in individual laboratories, and they may be useful for analysis of gaps in the genome resistant to the above global approach.

B. Specific Biological Problems

The problems approachable using DNA from flow-sorted libraries are multiple. X probes can, if recognizing expressed sequences, shed some light on X-inactivation (24), a process that appears to be more complicated the more it is studied. Y probes, especially those present in XX males (*15,22,28,65,72,91*) and absent in XY females, e.g. with visible deletions (*19,60*), should facilitate isolation of a male sex determinant. Chromosome-specific probes can help characterize cancer-associated chromosome rearrangements, detect and ultimately delineate the chromosomal breakpoints of deletions associated with neoplasias or developmental defects, and, by virtue of their content (e.g. inverted repeats) provide insight about genomic fluidity in a particular region, e.g. 15q11.2 (*20*). DNA amplification can also be studied using DNA probes from flow-sorted chromosomes and novel features relating to DNA relocation, long-range splicing and rearrangements associated with amplification of oncogenes [reviewed by Latt *et al.* (*50,51*)].

C. FUTURE DIRECTIONS

The future should see even more important applications of chromosome-library-derived probes. For example, one should be able to quantitate (and with RFLPs trace ancestry of) human autosomes, e.g. 13, 18 and 21, which are involved in common human trisomies. Using new DNA hybridization techniques (*88*, see Chapter 17), and perhaps 100 or so probes per chromosome to provide sufficient sensitivity, flow cytometric detection of trisomy might be possible. Also, insight into the mechanisms of chromosome rearrangements, e.g. isochromosome formation, should follow once enough pericentromeric RFLPs are available.

With chromosome sorting scaled up by high-speed instruments (see Chapter 13), large-insert chromosome-specific libraries for gene structure and chromosome walking (*3*) should become plentiful, as should chromosomes for protein analysis, including that possibly relevant to understanding X inactivation. Were $10–100 \mu g$ amounts of sorted chromosome DNA available, then DNA transfection experiments, e.g. for gene isolation, could

be made much more efficient, were the chromosomal location of the gene known.

Still, the most obvious prediction of the use of multiple chromosome-specific probes isolated from flow-sorted chromosome-enriched libraries is that the genome will become saturated by accurately mapped, then ordered, DNA fragments. Assuming that many of these fragments can be used either directly or via locus expansion to find RFLPs, then molecular linkage to human diseases of both simple and complex inheritance modes should be possible, with resultant manifold clinical applications. Such efforts should eventually result in a detailed characterization of the entire human genome at a molecular, if not sequence level. From this should emerge new global insights about structure–function relationships of the human genome as a whole.

ACKNOWLEDGEMENTS

This work has been supported by grants from the National Institutes of Health (GM33579, HD18658, HD20118, and HD00687).

REFERENCES

1. Aldridge, J., Kunkel, L., Bruns, G., Tantravahi, U., Lalande, M., Brewster, T., Moreau, E., Wilson, M., Bromley, W., Roderick, T., and Latt, S. A. (1984). A strategy to reveal high frequency RFLPs along the human X chromosome. *Am. J. Hum. Genet.* **36**, 546–564.
2. Bakker, E., Hofker, M. H., Goor, N., Mandel, J. L., Wrogemann, K., Davies, K. E., Kunkel, L. M., Willard, H. G., *et al.* (*1985*). Prenatal diagnosis and carrier detection of Duchenne muscular dystrophy with closely linked RFLPs. *Lancet* i, 655–658.
3. Bender, W., Arkham, M., Karch, F., Beachy, P. A., Pfeifer, M., Spierer, P., Lewis, E. B., and Hogness, D. (1983). Molecular genetics of the bithorax complex in drosophila melanogaster. *Science* **221**, 23–29.
4. Benton, W. D., and Davis, R. W. (1977). Screening lambda gt recombinant clones by hybridization to single plaques in situ. *Science* **196**, 180–182.
5. Bolivar, I., Rodriguez, R. C., Green, P. J., Betlach, M. C., Heyneker, H. L., Boyer, H. W., Crosa, J. H., and Falkow, S. (1977). Construction and

characterization of new cloning vehicles. II. A multipurpose cloning system. *Gene* **2**, 95–100.

6. Boncinelli, E., Goyns, M. H., Scotto, L., Simeone, A., Harris, P., Kellow, J. E., and Young, B. D. (1984). Isolation of a series of HLA Class I clones from a human chromosome 6 genomic library. *Mol. Biol. Med.* **2**, 1–14.

7. Botstein, D., White, R. L., Skolnick, M., and Davis, R. W. (1980). Construction of a genetic linkage map in man using restriction fragment length polymorphisms. *Am. J. Hum. Genet.* **32**, 314–441.

8. Carle, G. F., Frank, M., and Olson, M. V. (1986). Electrophoretic separations of large DNA molecules by periodic inversion of the electric field. *Science* **232**, 65–68.

9. Cavenee, W., Leach, R. J., Mohandas, T., Pearson, P., and White, R. L. (1984). Isolation and regional localization of DNA segments revealing polymorphic loci from human chromosome 13. *Am. J. Hum. Genet.* **36**, 10–26.

10. Collard, J. G., Philippus, E., Tulp, A., Lebo, R. V., and Gray, J. W. (1984). Separation and analysis of human chromosomes by combined velocity sedimentation and flow sorting applying single- and dual-laser flow cytometry. *Cytometry* **5**, 9–19.

11. Collard, J. G., de Boer, P. A. J., Janssen, J. W. G., Schijven, J. F., and de Jong, B. (1985). Gene mapping by chromosome spot hybridization. *Cytometry* **6**, 179–185.

12. Collins, F., and Weissman, S. (1984). Directional cloning of DNA fragments at a large distance from an initial probe. A circularization method. *Proc. Natl Acad. Sci. USA* **81**, 6812–6816.

13. Davies, K. E., Young, B. D., Elles, R. G., Hill, M. E., and Williamson, R. (1981). Cloning of a representative genomic library of the human X-chromosome after sorting by flow cytometry. *Nature* **293**, 374–376.

14. Davies, K. E., Pearson, P. L., Harper, P. S., Murray, J. M., O'Brien, T., Sarfarazi, M., and Williamson, R. (1983). Linkage analysis of two cloned DNA sequences flanking the Duchenne muscular dystrophy locus on the short arm of the human X chromosome. *Nucleic Acids Res.* **11**, 2303–2312.

15. de la Chapelle, A., Tippett, P. A., Wetterstrand, G., and Page, D. (1984). Genetic evidence of X–Y interchange in a human XX male. *Nature* **307**, 170–171.

16. de Martinville, B., Kunkel, L. M., Bruns, G., Morle, F., Koenig, M., Mandel, J. L., Horwich, A., Latt, S. A., Gusella, J. F., Housman, D., and Francke, U. (1985). Localization of DNA sequences in region Xp21 of the human X chromosome: search for molecular markers close to the Duchenne muscular dystrophy locus. *Am. J. Hum. Genet.* **37**, 235–249.

17. Disteche, C. M., Carrano, A. V., Ashworth, L. K., Burkhart-Schultz, K., and Latt, S. A. (1981). Flow sorting of the Cattanach X chromosome in an active or inactive state. *Cytogenet. Cell Genet.* **29**, 189–197.

18. Disteche, C. M., Kunkel, L. M., Lojewski, A., Orkin, S. H., Eisenhard, M., Sahar, E., Travis, B., and Latt, S. A. (1982). Isolation of mouse X-chromosome specific DNA from an X-enriched lambda phage library derived from flow sorted chromosomes. *Cytometry* **2**, 282–286.

19. Disteche, C. M., Casanova, M., Saal, H., Friedman, C., Sybert, V., Graham, J., Thuline, H., Page, D., and Fellous, M. (1986). Small deletions of the short arm of the Y chromosome in 46, XY females. *Proc. Natl Acad. Sci. USA* **83**, 7841–7844.

20. Donlon, T. A., Lalande, M., Wyman, A., Bruns, G., and Latt, S. A. (1986). Isolation of molecular probes associated with the chromosome 15 instability of the Prader-Willi syndrome. *Proc. Natl Acad. Sci. USA* **83**, 4408–4412.

21. Fantes, J. D., Green, D., and Cooke, H. J. (1983). Purifying human Y chromosomes by flow cytometry and sorting. *Cytometry* **4**, 88–91.

22. Ferguson-Smith, M. A. (1966). X–Y chromosome interchange in the aetiology of true hermaphroditism and of XX Klinefelter's syndrome. *Lancet* **ii**, 475–476.

23. Francke, U., Ochs, H. D., de Martinville, B., Giacalone, J., Lindgren, V., Disteche, C., Pagon, R. A., Hofker, M. H., van Ommen, G.-J. B., Pearson, P. L., and Wedgewood, R. J. (1985). Minor Xp21 chromosome deletion in a male associated with expression of Duchenne muscular dystrophy, chronic granulomatous disease, retinitis pigmentosa and McLeod syndrome. *Am. J. Hum. Genet.* **37**, 250–267.

24. Gartler, S., and Riggs, A. (1983). Mammalian X chromosome inactivation. *Ann. Rev. Genet.*, **17**, 155–190.

25. Gray, J. W., Langlois, R. G., Carrano, A. V., Burkhart-Schultz, K., and Van Dilla, M. A. (1979). One and two parameter flow cytometry. *Chromosoma* **73**: 9–27.

26. Griffith, J. K., Cram, L. S., Crawford, B. D., Jackson, P. J., Schilling, J., Schimke, R. T., Walters, R. A., Wilder, M. E., and Jett, J. H. (1984). Construction and analysis of DNA sequence libraries from flow-sorted chromosomes: practical and theoretical considerations. *Nucleic Acids Res.* **12**, 4019–4034.

27. Grosveld, F. G., Lund, T., Murray, E. J., Mellor, A. L., Dahl, H. H. M., and Flavell, R. A. (1982). The construction of cosmid libraries which can be used to transform eukaryotic cells. *Nucleic Acids Res.* **10**, 6715–6732.

28. Guellaen, G., Casanova, M., Bishop, C., Geldwerth, D., Andre, G., Fellous, M., and

Weissenbach, J. (1984). Human XX males with Y single-copy DNA fragments. *Nature* **307**, 172–173.

29. Hansen, M. F., Koufos, A., Gallie, B., Phillips, R. A., Fodstad, O., Brogger, A., Gedde-Dahl, T., and Cavenee, W. K. (1985). Osteosarcoma and retinoblastoma: a shared chromosomal mechanism revealing recessive predisposition. *Proc. Natl Acad. Sci. USA* **82**, 6216–6220.

30. Harris, P., Boyd, E., and Ferguson-Smith, M. A. (1985). Optimising human chromosome separation for the production of chromosome-specific DNA libraries by flow sorting. *Hum. Genet.* **70**, 59–65.

31. Harris, P., Morton, C. C., Guglielmi, P., Li, F., Kelly, K., and Latt, S. A. (1986). Mapping by chromosome sorting of several gene probes, including c-myc, to the derivative chromosomes of a 3;8 translocation associated with familial renal cancer. *Cytometry* **7**, 589–594.

32. Hofker, M., Wapenaar, M., Goor, N., Bakker, E., Van Ommen, O., and Pearson, P. (1985). Isolation of probes detecting restriction fragment length polymorphisms from X chromosome-specific libraries: potential use for diagnosis of Duchenne muscular dystrophy. *Hum. Genet.* **70**, 148–156.

33. Huynh, T. V., Young, R. A., and Davis, R. W. (1985). Construction and screening of cDNA libraries in lambda gt10 and lambda gt11. *In* "DNA Cloning" (D. M. Glover, ed.), Vol. I, pp. 49–78. IRL Press, Oxford.

34. Jeffreys, A. J., Wilson, V., and Thien, S. L. (1985). Hypervariable "minisatellite" regions in human DNA. *Nature* **314**, 67–73.

35. Kao, F. T., Jones, C., and Puck, T. T. (1976). Genetics of somatic mammalian cells: genetic, immunologic, and biochemical analysis with Chinese hamster cell hybrids containing selected human chromosomes. *Proc. Natl Acad. Sci. USA* **73**, 193–197.

36. Knudson, A. G., Meadows, A. T., Nichols, W. W., and Hill, R. (1976). Chromosomal deletion and retinoblastoma. *New Eng. J. Med.* **295**, 1120–1123.

37. Krumlauf, R., Jeanpierre, M., and Young, B. D. (1982). Construction and characterization of genomic libraries from specific human chromosomes. *Proc. Natl Acad. Sci USA* **79**, 2971–2975.

38. Kunkel, L. M., Tantravahi, U., Eisenhard, M., and Latt, S. A. (1982). Regional localization on the human X of DNA segments cloned from flow sorted chromosomes. *Nucleic Acids Res.* **10**, 1557–1578.

39. Lalande, M., Schreck, R. R., Hoffman, R., and Latt, S. A. (1985). Identification of inverted duplicated # 15 chromosomes using bivariate flow cytometric analysis. *Cytometry* **6**, 1–6.

40. Kunkel, L. M., Monaco, A. P., Middlesworth, W., Ochs, H. D., and Latt, S. A. (1985). Specific cloning of DNA fragments absent from the DNA of a male patient with an X chromosome deletion. *Proc. Natl Acad. Sci. USA* **82**, 4778–4782.

41. Kunkel, L. M., Lalande, M., Monaco, A. P., Flint, A., Middlesworth, W., and Latt, S. A. (1985). Construction of a human X-chromosome-enriched phage library which facilitates analysis of specific loci. *Gene* **33**, 251–258.

42. Lalande, M., Dryja, T. P., Schreck, R. R., Shipley, J., Flint, A., and Latt, S. A. (1984). Isolation of human chromosome 13-specific DNA sequences cloned from flow sorted chromosomes and potentially linked to the retinoblastoma locus. *Cancer Genet. Cytogenet.* **13**, 283–295.

43. Lalande, M., Kunkel, L. M., Flint, A., and Latt, S. A. (1984). Development and use of metaphase chromosome flow sorting methodology to obtain recombinant phage libraries enriched for parts of the human X chromosome. *Cytometry* **5**, 101–107.

44. Lalande, M., Donlon, T., Petersen, R. A., Liberfarb, R., Manter, S., and Latt, S. A. (1986). Molecular detection and differentiation of deletions in band 13q14 in human retinoblastoma. *Cancer Genet. Cytogenet.* **23**, 151–157.

45. Langlois, R. G., and Jensen, R. H. (1979). Interactions between pairs of DNA-specific fluorescent stains bound to mammalian cells. *J. Histochem. Cytochem.* **27**, 72–79.

46. Langlois, R. G., Yu, L.-C., Gray, J. W., and Carrano, A. V. (1982). Quantitative karyotyping of human chromosomes by dual beam flow cytometry. *Proc. Natl Acad. Sci. USA* **79**, 7876–7880.

47. Latt, S. A., and Wohlleb, J. C. (1975). Optical studies of the interaction of 33258 Hoechst with DNA, chromatin, and metaphase chromosomes. *Chromosoma* **52**, 297.

48. Latt, S. A. (1977). Fluorescent probes of chromosome structure and replication. *Can. J. Genet. Cytol.,* **19**, 603–623.

49. Latt, S. A., Goff, S. P., Tabin, C. J., Paskind, M., Wang, Y. Y.-J., and Baltimore, D. (1983). Cloning and analysis of reverse transcript P160 genomes of abelson murine leukemia virus. *J. Virol.* **45**, 1195–1199.

50. Latt, S. A., Shiloh, Y., Sakai, K., Brodeur, G., Donlon, T., Korf, B., Shipley, J., Bruns, G., Heartlein, M., Kanda, N., Kohl, N., Alt, F., and Seeger, R. (1986). Novel DNA rearrangement phenomena associated with DNA amplification in human neuroblastomas and neuroblastoma cell lines. *In* "Genetic Toxicology of Environmental Chemicals" (C. Ramel, B. Lambert, and J. Magnusson, eds), Part A, pp. 601–612. Alan R. Liss, New York.

51. Latt, S. A., Shiloh, Y., Sakai, K., Rose, E., Brodeur, G., Donlon, T., Korf, B., Kanda, N., Heartlein, M., Kang, J., Stroh, H., Harris, P., Bruns, G., and Seeger, R. (1986). DNA rearrangement, relocation, and amplification in neuroblastoma cell lines and primary tumors. "Cellular and Molecular Biology of Tumors and Potential Clinical Applications,,. (In press).

52. Latt, S. A., and Langlois, R. G. (1986). Fluorescent probes of DNA synthesis and microstructure. In "Flow Cytometry" (M. Melamed ed.) 2nd edn. (In press).

53. Lebo, R. V., Carrano, A. V., Burkhart-Schultz, K., Dozy, A. M., Yu, L.-C., and Kan, Y. W. (1979). Assignment of human beta, gamma, and delta globin genes to the short arm of chromosome 11 by chromosome sorting and DNA restriction enzyme analysis. *Proc. Natl Acad. Sci. USA* **76**, 5804–5808.

54. Lebo, R. V. (1982). Chromosome sorting and DNA sequence organization. *Cytometry* **3**, 145–154.

55. Lebo, R. V., Gorin, F., Fletterick, A. J., Kao, F. T., Cheung, M. C., Bruce, B. D., and Kan, Y. W. (1984). High resolution chromosome sorting and DNA spot–blot analysis assign McArdle's syndrome to chromosome 11. *Science* **255**, 57–59.

56. Ledbetter, D. H., Riccardi, V. M., Airhart, S. D., Strobel, R. J., Keenan, S. B., and Crawford, J. D. (1981). Deletions of chromosome 15 as a cause of the Prader-Willi syndrome. *New Eng. J. Med.* **304**, 325–329.

57. Ledbetter, D. H., Mascarello, J. T., Riccardi, V. M., Harper, V. D., Airhart, S. D., and Strobel, R. J. (1982). Chromosome 15 abnormalities and the Prader-Willi syndrome. A follow-up report of 40 cases. *Am. J. Hum. Genet.* **34**, 278–285.

58. Le-Pecq, J. P., and Paoletti, C. (1967). A fluorescent complex between ethidium bromide and nucleic acids. *J. Mol. Biol.* **27**, 87–106.

59. Litt, M., and White, R. L. (1985). A highly polymorphic locus in human DNA revealed by cosmid-derived probes. *Proc. Natl Acad. Sci. USA* **82**, 6206–6210.

60. Magenis, R. E., Tochen, M. L., Holahan, T., Carey, L., Allen, L., and Brown, M. G. (1984). Turner syndrome resulting from partial deletion of Y chromosome short arm: localization of male determinants. *J. Pediatr.* **105**, 916–919.

61. Maniatis, T., Hardison, R. C., Lacy, E., Lauer, J., O'Connell, C., Quon, D., Sim, D. K., and Efstratiadis, A. (1978). The isolation of structural genes from libraries of eukaryotic DNA. *Cell* **15**, 687–701.

62. Maniatis, T., Fritsch, E. F., and Sambrook, J. (1982). "Molecular Cloning. A Laboratory Manual". Cold Spring Harbor Laboratory, Cold Spring Harbor, NY.

63. Meyne, J., Bartholdi, M. F., Travis, G., and Cram, L. S. (1984). Counterstaining human chromosomes for flow karyology. *Cytometry* **5**, 580–583.

64. Muller, U., Lalande, M., Donlon, T., and Latt, S. A. (1986). Moderately repeated DNA sequences specific for the short arm of the human Y chromosome are present in XX males and reduced in copy number in an XY female. *Nucleic Acids Res.* **14**, 1325–1340.

65. Muller, U., Lalande, M., Disteche, C. M., and Latt, S. A. (1986). Construction, analysis, and application to 46,XY gonadal dysgenesis, of a recombinant phage DNA library from flow sorted human Y chromosomes. *Cytometry* **7**, 418–424.

66. Neve, R., Bruns, G. A. P., Dryja, T. P., and Kurnit, D. M. (1983). Retrieval of human DNA from rodent-human genomic libraries by a recombination process. *Gene* **23**, 343–354.

67. Neve, R. L., and Kurnit, D. M. (1983). Comparison of sequence repetitiveness of human cDNA and genomic DNA using the miniplasmid vector, piVX. *Gene* **23**, 355–367.

68. Oberle, I., Drayna, D., Camerino, G., White, R., and Mandel, J. L. (1985). The telomeric region of the human X chromosome long arm: presence of a highly polymorphic DNA marker and analysis of recombination frequency. *Proc. Natl Acad. Sci. USA* **82**, 2824–2828.

69. Oberle, I., Camerino, G., Kloepfer, C., Moisan, J. P., Grzeschik, K. H., Hellkuhl, B., Hors-Cayla, M. C., van Cong, N., Weil, D., and Mandel, J. L. (1986). Characterization of a set of X-linked sequences and of a panel of somatic cell hybrids useful for the regional mapping of the human X chromosome. *Hum. Genet.* **72**, 43–49.

70. Oberle, I., Heilig, R., Moisan, J. P., Kloepfer, C., Mattei, M. G., Mattei, J. F., Bove, J., Froster-Iskenius, U., Jacobs, P. A., Lathrop, G. M., Lalouel, J. M., and Mandel, J. L. (1986). Genetic analysis of the fragile-X mental retardation syndrome with two flanking polymorphic DNA markers. *Proc. Natl Acad. Sci. USA* **83**, 1016–1020.

71. Ott, J. (1985). "Analysis of Human Genetic Linkage." Johns Hopkins University Press, Baltimore, MD.

72. Page, D. C., de la Chapelle, A., and Weissenbach, J. (1985). Chromosome Y specific DNA in related human XX males. *Nature* **315**, 224–226.

73. Rimm, D. L., Horness, D., Kucera, J., and Blattner, F. R. (1980). Construction of coliphage lambda Charon vectors with BamHI cloning sites. *Gene* **12**, 301–309.

74. Sahar, E., and Latt, S. A. (1978). Enhancement of banding patterns in human metaphase chromosomes by energy transfer. *Proc. Natl Acad. Sci. USA* **75**, 5650–5654.

75. Sahar, E., and Latt, S. A. (1980). Energy transfer and binding competition between dyes used to enhance staining differentiation in metaphase chromosomes. *Chromosoma* **79**, 1–28.

76. Schmid, D. W., and Jelinek, W. R. (1982). The Alu family of dispersed repetitive sequences. *Science* **216**, 1065–1070.

77. Schwartz, D. C., and Cantor, C. R. (1984). Separation of yeast chromosome-sized DNAs by pulsed field gradient gel electrophoresis. *Cell* **37**, 67–75.

78. Sealy, P. G., Whittaker, P. A., and Southern, E. M.

(1985). Removal of repeated sequences from hybridization probes. *Nucleic Acids Res.* **13**, 1905–1922.

79. Seed, B. (1983). Purification of genomic sequences from bacteriophage libraries by recombination and selection *in vivo. Nucleic Acids Res.* **11**, 2427–2445.

80. Sillar, R., and Young, B. D. (1981). A new method for the preparation of metaphase chromosomes for flow analysis. *J. Histochem. Cytochem.* **29**, 74–78.

81. Sober, H. A. (1970). "Handbook of Biochemistry: Selected Data for Molecular Biology," 2nd edn. Chemical Rubber Co., Boca Raton.

82. Southern, E. M. (1975). Detection of specific sequences among DNA fragments separated by gel electrophoresis. *J. Mol. Biol.* **98**, 503–517.

83. Stahl, F. W. (1979). Special sites in generalized recombination. *Ann. Rev. Genet.* **13**, 7–24.

84. Tantravahi, U., Kirschner, D. A., Beauregard, L., Page, L., Kunkel, L., and Latt, S. A. (1983). Cytological and molecular analysis of 46,XXq-cells to identify a DNA probe for a putative human X chromosome inactivation center. *Hum. Genet.* **64**, 33–38.

85. Taub, R., Kelly, K., Battey, J., Latt, S. A., Lenoir, G. M., Tantravahi, U., Tu, Z., and Leder, P. (1984). A novel alteration in the structure of an activated c-myc gene in a variant t(2/8) Burkitt lymphoma. *Cell* **37**, 511–520.

86. Tiemeier, D., Enquist, L., and Leder, P. (1976). Improved derivative of a phage lambda EK2 vector for cloning recombinant DNA. *Nature* **263**, 526–527.

87. Trask, B., van den Engh, G., Gray, J., Vanderlaan, M., and Turner, B. (1984). Immunofluorescent detection of histone 2B on metaphase chromosomes using flow cytometry. *Chromosoma* **90**, 295–302.

88. Trask, B., van den Engh, G., Landegant, J., Jansen in de Wal, N., and van der Ploeg, M. (1985). Detection of DNA sequences in nuclei in suspension by *in situ* hybridization and dual beam flow cytometry. *Science* **230**, 1401–1403.

89. Van Dilla, M. A., and Deaven, L. L., Albright, K. L. *et al.* (1986). Human chromosome-specific DNA libraries: construction and availability. *Biotechnol.* **4**, 537–552.

90. van Heynigen, V., Boyd, P. A., Seawright, A.,

Fletcher, J. M., Fantes, J. A., Buckton, K. E., Spowart, G., Porteous, D. J., Hill, R. E., Newton, M. S., and Hastie, N. D. (1985). Molecular analysis of chromosome 11 deletions in Aniridia-Wilms tumor syndrome. *Proc. Natl Acad. Sci. USA* **82**, 8592–8596.

91. Vergnaud, G., Page, D. C., Simmler, M.-C., Brown, L., Rouyer, F., Noel, B., Botstein, D., de la Chapelle, A., and Weissenbach, J. (1986). A deletion map of the human Y chromosome based on DNA hybridization. *Am. J. Hum. Genet.* **38**, 109–124.

92. Willard, H. F., Skolnick, M. H., Pearson, P. L., and Mandel, J. L. (1985). Report of the Committee on Human Gene Mapping by Recombinant DNA Techniques. Human Gene Mapping 8. *Cytogenet. Cell Genet.* **40**, 360–489.

93. Williams, B. G., and Blattner, F. R. (1980). Bacteriophage Lambda Vectors for DNA Cloning. *In* "Genetic Engineering" (J. Setlow and A. Hollaender, eds), Vol. 2, pp. 201–281. Plenum Press, New York.

94. Wray, W., and Stubblefield, E. (1970). A new method for the rapid isolation of chromosomes, mitotic apparatus, or nuclei from mammalian fibroblasts at near neutral. *Exp. Cell Res.* **59**, 469–478.

95. Wray, W., and Wray, V. P. (1980). Proteins from metaphase chromosomes treated with fluorochromes. *Cytometry* **1**, 18–20.

96. Wyman, A., Wolfe, L., and Botstein, D. (1985). Propagation of some human vectors requires mutant *Escherichia Coli* hosts. *Proc. Natl Acad. Sci. USA* **82**, 2880–2884.

97. Young, R. A., and Davis, R. W. (1983). Efficient isolation of genes by using antibody probes. *Proc. Natl. Acad. Sci. USA* **80**, 1194–1198.

98. Young, R. A., and Davis, R. W. (1983). Yeast RNA polymerase II genes: isolation with antibody probes. *Science* **222**, 778–782.

99. Young, B. D. (1984). Chromosome analysis by flow cytometry: a review. *Bas. Appl. Histochem.* **28**, 9–19.

100. Yunis, J. J. (1981). Mid-prophase human chromosomes. The attainment of 2000 bands. *Hum. Genet.* **56**, 293–298.

The National Laboratory Gene Library Project

M. A. VAN DILLA, N. A. ALLEN, A. V. CARRANO,
M. CHRISTENSEN, P. DE JONG, P. N. DEAN, J. C. FUSCOE*,
J. W. GRAY, C. R. LOZES, J. S. McNINCH,
M. L. MENDELSOHN, J. MULLIKIN, L. PEDERSON,
J. PERLMAN*, D. C. PETERS, A. J. SILVA, B. J. TRASK
and G. J. VAN DEN ENGH

Lawrence Livermore National Laboratory
Biomedical Sciences Division
Livermore, CA 94550, USA

L. L. DEAVEN, K. L. ALBRIGHT, M. F. BARTHOLDI,
N. C. BROWN, E. W. CAMPBELL, L. M. CLARK, L. S. CRAM,
J. J. FAWCETT, C. E. HILDEBRAND, P. J. JACKSON, J. H. JETT,
S. KOLLA, J. L. LONGMIRE, M. L. LUEDEMANN, J. MEYNE,
L. J. MEINCKE, R. K. MOYZIS, and A. C. MUNK

Los Alamos National Laboratory
Life Sciences Division
Los Alamos, NM 87545, USA

I. INTRODUCTION

The National Laboratory Gene Library Project was initiated in 1983 by the Department of Energy through its Office of Health and Environmental Research, as a contribution to the understanding of the human genome in general, and of the mutagenic and carcinogenic effects of energy-related environmental pollutants in particular. The project is a joint effort of the Lawrence Livermore National Laboratory (LLNL) and the Los Alamos National Laboratory (LANL). The first goal

* Present address: Center for Environmental Health, U-40, The University of Connecticut, Storrs, CT 06268, USA.

(phase 1) was the construction of small-insert (0–9 kbp) gene libraries for each of the 24 human chromosomal types. The DNA for each chromosome type is purified by fluorescence-activated cell sorting. The libraries are important to the medical genetics community for use in the study and diagnosis of genetic diseases by the restriction-fragment-length polymorphism method, and serve many other genetic research needs as well. The second goal of the project (phase 2) is the construction of large-insert libraries in phage (insert size 9–23 kbp) and cosmid (insert size 35–45 kbp) vectors. These libraries are better suited to the study of gene function, structure and chromosome organization. This chapter will review phase 1, which is completed, and then touch briefly on phase 2, now underway.

The specific strategy of phase 1 is outlined in the schematic diagram of Fig. 1, using the first library constructed from sorted chromosome 18 as an example. The human DNA source is cultured cells, either normal diploid human fibroblasts (as for the chromosome 18 library in Fig. 1), human lymphoblastoid cells, or human–hamster hybrids having a reduced number of human chromosomes. Metaphase chromosomes are isolated, stained with DNA-specific fluorescent dyes, and processed through a flow sorter to yield enough material for cloning (nominally 4×10^6 chromosomes, enough for several clonings). The chromosomal DNA is then extracted, purified and digested to completion with restriction enzyme. The vector DNA is cut with the same restriction enzyme and the arms are dephosphorylated with calf alkaline phosphatase to inhibit re-ligation. The chromosomal DNA fragments and the vector arms are mixed so that complementary (sticky) ends can anneal and join covalently in the ligation reaction. The resulting recombinant DNA molecules are packaged *in vitro* into infective virions using complementary packaging extracts from two mutant lysogenic bacterial strains, allowed to infect a lawn of *Escherichia coli* cells (strain LE392) and to multiply; the resulting phage lysate is the amplified chromosome 18 gene library.

Both laboratories have used the same bacteriophage lambda vector, Charon 21A. This vector has also been used to construct chromosome-specific libraries by others (20,21). Charon 21A is an insertion vector with a single restriction site for Hind III and another for Eco R1. It accepts inserts of size up to 9.1 kbp. Each laboratory has cloned all 24 human chromosomal types, LLNL using Hind III and LANL using Eco R1. In this way, it is likely that sequences larger than 9.1 kbp not cut with one restriction enzyme will be cut with the other and hence be cloned.

Libraries are available to the general scientific research community, both national and international. Close to 1200 library aliquots were sent to about 300 laboratories worldwide from LLNL and LANL through February 1986. The repository and distribution functions were transferred on that date to the American Type Culture Collection (ATCC), Rockville, MD; this repository is supported by the Division of Research Resources, NIH. The ATCC stores library stocks from LLNL and LANL frozen in liquid nitrogen, handles distributions, and receives and assembles characterization information from users into an information database available to the general scientific community. The total library distribution now exceeds 2000 worldwide.

II. METHODS

The methods used for the construction of the phase 1 libraries are described in general in Chapter 15 and specifically in previous publications (8,9,12,13,29), and so will only be outlined here. The process consists of chromosome purification and DNA cloning. Chromosome purification involves collection of cells in mitosis, mechanical shearing to release the chromosomes into a buffer designed to maintain chromosome integrity and high DNA molecular weight, staining with the DNA-specific dyes Hoechst 33258 (HO) and chromomycin A3 (CA3), and purifying by sorting according to HO and CA3 fluorescence

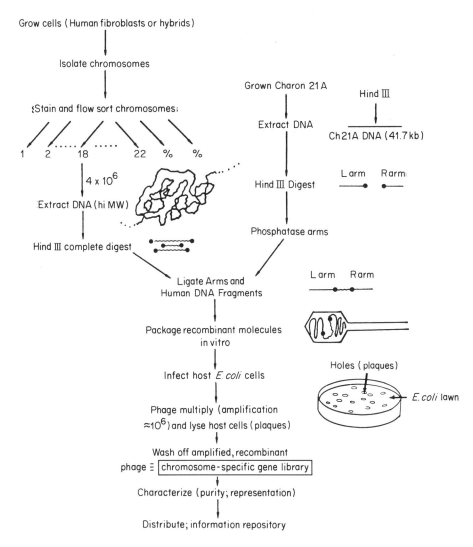

Fig. 1. The steps in constructing a chromosome-specific gene library using the Hind III site of the bacteriophage vector Charon 21A and flow sorted chromosome 18 are outlined. This procedure was used for all phase 1 library construction for the National Laboratory Gene Library Project; LLNL used the Hind III cloning site of Charon 21A and LANL used the Eco R1 cloning site of Charon 21A.

intensity. The DNA is then isolated from about 10^6 chromosomes and cloned into the vector.

The first steps in this process are critical. Judicious choice of the cell line from which chromosomes are isolated significantly improves the purity with which the target chromosome(s) can be sorted. Human diploid fibroblast or lymphoblastoid cells yield chromosome suspensions from which chromo-

somes 13, 18–22 and Y can be sorted at high purity. Chromosomes 14–17 can be sorted from cell lines in which one or more of these chromosomes carry advantageous polymorphisms so that the target chromosome(s) is well resolved from other chromosome types. For example, human lymphoblastoid line GM131 (Figs 2 and 3) would be unfavorable for sorting chromosomes 16 and 17 due to excessive

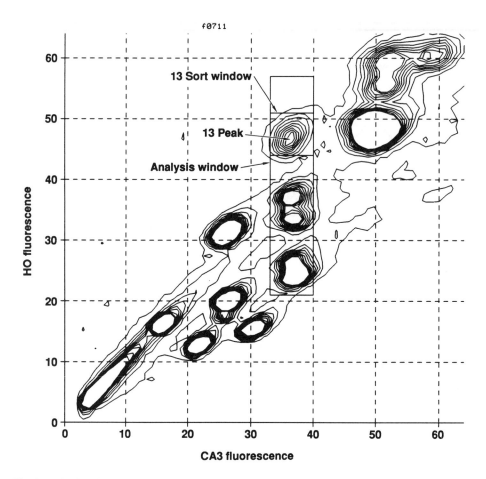

f0711

Fig. 2. The flow karyotype accumulated while sorting chromosome 13 from the apparently normal lymphoblastoid line GM131 is quantitatively analyzed by a univariate least squares best fitting program. The actual sort window is shown, along with an analysis window which is used only for calculation purposes. All the data in this analysis window is transformed into a univariate distribution of HO fluorescence, as shown in Fig. 3.

overlap of their peaks. Another line, human skin fibroblast line HSF7, affords much better discrimination between chromosomes 16 and 17 when stained with DAPI and CA3 instead of with the normal HO and CA3 protocol. Both DAPI and HO have a binding preference for AT-rich DNA. However, the fluorescence staining of heterochromatic DNA is often higher with DAPI staining than with HO staining (22, see Chapter 5). Since chromosome 16 has a large polymorphic heterochromatic centromere, the separation between chromo-

somes 16 and 17 peaks could be increased by DAPI plus CA3 staining for line HSF7 (but not for GM131). The larger chromosomes (1–12) from human cells are not as well resolved as the smaller ones. In addition, clumps of smaller chromosomes (which seem unavoidable) underlie the peaks for the larger chromosomes, reducing sort purity. Hence human–hamster hybrid lines containing a reduced number of human chromosomes are used as sources of the larger human chromosomes. The hybrid lines are selected so that the human chromosomes

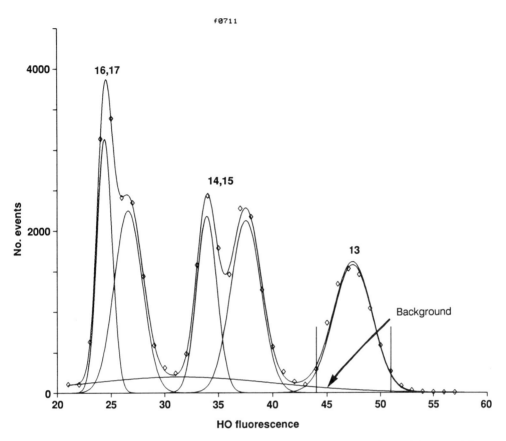

Fig. 3. The univariate transform of the data in the analysis window of Fig. 2 is fitted by a set of five univariate Gaussians representing chromosomes 13–17 and a low, broad Gaussian to represent underlying background. The sort window (channels 44–51) is shown. The six Gaussians, and their sum, are plotted along with the actual data points. The fit is seen to be very good. The result is an estimate of 95% sort purity; the impurity level of 5% is entirely due to background.

are well resolved from the hamster chromosomes. The price paid for this advantage is the karyotypic instability of many hybrid lines, so that the chromosome of interest is sometimes lost in the mass culture that precedes chromosome isolation. This problem can be overcome if the human chromosomes complement a genetic deficiency in the original hamster parent line, so that only cells containing the human chromosome correcting this deficiency will proliferate. An example is the human–hamster hybrid cell line J1 (*15*), in which human chromosome 11 seems to complement a genetic deficiency in the mutant CHO

parent line. An alternative approach is to integrate a selectable marker into the human chromosomes, so that only hybrid cells carrying marked human chromosomes can grow in the selection medium. This has been accomplished for human–mouse hybrid lines in which a human chromosome carries the bacterial Eco*gpt* gene (*4,27*). In general, human–mouse lines are not useful for flow sorting, since the human chromosomes are not well resolved from the mouse chromosomes. However, it is theoretically possible to transfer the human chromosomes from these hybrid lines to a hamster background, or develop new hybrid fusion with human cells. An advantage of rodent–human hybrid lines is that any contamination of the sorted chromosomes is most likely to be rodent, since cell lines are selected so that the number of human chromosomal types is small, and sometimes only one. Thus, library purity for the sorted human chromosomal types will be high. The rodent sequences can be easily screened when the library is used. In addition, only one homolog of the human chromosome of interest is present in hybrid lines, facilitating physical mapping (clone ordering) of that chromosome.

Chromosome isolation is a key step in the whole library construction procedure, since the quality of the chromosome preparation determines sorting quality, and hence ultimate library quality. The $MgSO_4$ and the polyamine chromosome isolation methods are both used in this project, the former at LLNL and the latter at LANL. Both are reviewed elsewhere in this book (see Chaper 4). Each isolation method has its advantages and disadvantages, and both are capable of producing chromosome preparations that yield high-resolution flow karyotypes with low debris. For the large-insert libraries of phase 2, high-molecular-weight DNA (over 100–150 kbp) is necessary. Here the polyamine method is advantageous, since it maintains very high molecular weight, probably as a result of inhibition of nuclease activity.

Flow sorting in general and its value to chromosome purification has been described in detail elsewhere (*5,19,28*; see Chapters 2 and 13) and needs only brief discussion here. Both conventional and high-speed sorters are used in this work, and both produce sufficient chromosomes (about 4×10^6) for phase 1 cloning. This quantity is sorted in a day by the high-speed sorter, while several days are required at conventional sorting speeds. This limits the amount of DNA that may be produced in a reasonable time to the order of a few micrograms.

The phase 1 cloning procedure (*12*) was designed to take advantage of prior experience of other laboratories, so that a complete set of libraries could be produced in minimum time —

a consideration deemed very important by the project's advisory committee and the medical genetics community. Hence the sorted DNA was subjected to complete digestion, and the insertion phage vector Charon 21A was used. The very small amount of starting DNA is a challenge to the cloning process, and so a minigel electrophoresis procedure was worked out (*12*) to check each step of the procedure (extraction and purification of the sorted chromosomal DNA, digestion to completion with Hind III or Eco R1 and ligation of vector arms with chromosomal DNA fragments). Many libraries were constructed in a single cloning attempt, while some libraries required two or more attempts. The amount of starting DNA varied considerably, but generally was in the 50–500 ng range.

Library characterization was recognized at the outset to be an important, but very labor intensive, aspect of the project. Unfortunately, resources adequate for the characterization process were not available. As a result, we have relied heavily on feedback from the user community. Information now available on the libraries both from our own studies and from the user community is described below.

Information on the properties of the libraries has been obtained in three ways: (1) quantitative analysis of the flow karyotypes recorded during the sorting process, (2) fluorescent *in situ* hybridization analysis of the sorted chromosomes and (3) molecular analysis of the libraries themselves.

Sort purity was estimated by mathematical analysis of the flow karyotypes generated during sorting. This analysis estimates the magnitude of the contamination in the sort window due to the neighboring peaks and the underlying debris continuum. Flow karyotypes are analyzed by bivariate least squares best-fitting programs, or by a simpler substitute procedure which takes advantage of univariate fitting methods (*7*). An example of the latter is shown in Figs 2 and 3, in which chromosomes 13 and 19 were sorted from GM 131, a lymphoblastoid line which is karyotypically normal. This is a good preparation with good

Table I

CHARACTERIZATION DATA ON LLNL LIBRARIES

Library	Cell source	Impurity (%) by quantitative flow karyotype analysis	Hamster impurity (%)		Unique sequence clones mapping to sorted chromosomes (%)	Average insert size (kbp)
			In situ fluor. hybridization on sorted chromosomes	Plaque hybridization		
LL*01*NS01	Hybrid	No data		28 (a)		
LL*01*NS02	Hybrid	21 (bkg)	17 (r)	26 (a)		
LL*02*NS01	Hybrid	25 (bkg)	12 (o)	16 (a)		
LL*03*NS01	Hybrid	16 (bkg)	18 (r)	18 (a)		
LL*04*NS01	Hybrid	No data		20 (b)	224/224 = 100% (b)	2.0
			17 (o)	17 (a)		
LL*04*NS02	Hybrid	5	5 (o)	5 (a)		
LL*05*NS01	Hybrid	38		39 (a)		
LL*06*NS01	Hybrid	No data	19 (r)	23 (a)		
LL*07*NS01	Human	12 (bkg)	N/A	N/A		
LL*08*NS02	Hybrid	No data	15 (r)	16 (a)		
				10 (c)	3/4 = 75% (c)	
LL*09*NS01	Hybrid	58	17 (r)	38 (a)	3/6 = 50% (i)	
LL*10*NS01	Hybrid	28 (mostly bkg)	38 (r)	38 (a)		
				33 (j)	152/238 = 64% (j)	
LL*11*NS01	Hybrid	No data	35 (r)	52 (a)		
				60 (d)	116/183 = 63% (d)	
LL*12*NS01	Hybrid	40 (bkg)		34 (a)		
LL*13*NS02	Human	5 (bkg)	N/A	N/A		
LL*14*NS01	Human	21 (bkg)	N/A	N/A		
LL*15*NS01	Human	21 (bkg)	N/A	N/A		
LL*45*NS01	Human	No data	N/A	N/A	3/4 = 75% (e)	3.5
LL*16*NS02	Human	No data	N/A	N/A	6/18 = 33% (f)	3.9
			N/A	N/A	11/47 = 23% (g)	
LL*16*NS03	Human	4 (bkg); 1 (#15)	N/A	N/A		
LL*17*NS01	Human	No data	N/A	N/A	2/3 = 67% (l)	
			N/A	N/A	10/20 = 50% (e)	2.0
LL*17*NS02	Human	2 (bkg); 2 (#16)	N/A	N/A		
LL*18*NS01	Human	No data	N/A	N/A	2/4 = 50% (m)	
			N/A	N/A	3/3 = 100% (a)	1.9
LL*19*NS01	Human	10 (bkg)	N/A	N/A		
LL*20*NS01	Human	No data	N/A	N/A	5/6 = 83% (a)	3.1
			N/A	N/A	5/8 = 63% (h)	
LL*21*NS02	Human	30 (bkg)	N/A	N/A	21/28 = 75% (a)	2.0
			N/A	N/A	15/20 = 75% (n)	
LL*22*NS01	Human	25	N/A	N/A	15/39 = 38% (p)	
			N/A	N/A	32/55 = 58% (q)	
LL*0X*NS01	Hybrid	24	16 (o)	37 (a)		
LL*0Y*NS01	Human	12 (bkg)	N/A	N/A		

(a)   Fuscoe *et al.* (1986, 1987).
(b)   Gilliam *et al.* (1986).
(c)   Wood (1986).
(d)   Davis and Shows (1987).
(e)   Van Tuinen (1986).
(f)   Hyland and Sutherland (1986).

(g)   Reeders (1986).
(h)   Deed (1987).
(i)   Smith (1987).
(j)   Bruns (1987).
(l)   Barker (1987).
(m)   Frelat (1987).

(n)   Antonarakis (1987).
(o)   van Dekken (1986).
(p)   Rouleau, MGH.
(q)   Budarf, Phila.
(r)   Fuscoe *et al.* (1988).

resolution and fairly low background. However, these are qualitative statements, and it is important to obtain quantitative estimates of sort purity. This was accomplished for the chromosome 13 sort by choosing a rectangular region of the bivariate flow karyotype containing the chromosome 13 sort window and the chromosomes 13–17 peaks. The data within this region were transformed into a univariate distribution of HO fluorescence, and fitted by iterative least squares methods with a set of five

univariate Gaussian functions which model the five chromosomal peaks and a sixth broad Gaussian to model background. The resulting excellent fit shows that the chromosome 13 sort window contains no contamination from its nearest neighbor, the chromosome 14 peak, since the Gaussian function representing it does not extend into the sort region. There is a small amount of background in the sort region, as indicated by the area under the broad background Gaussian function in the sort

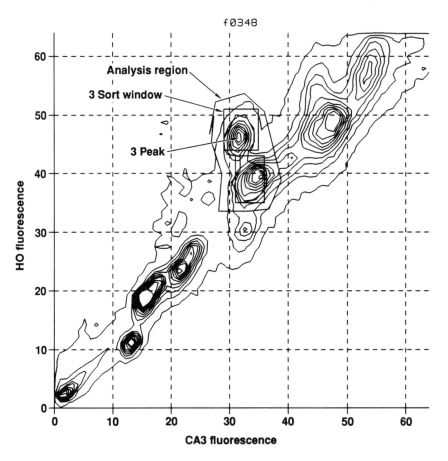

Fig. 4. The flow karyotype accumulated while sorting chromosome 3 from the human–hamster hybrid line 314-1b (containing chromosome 3 as the only human material) cannot be analyzed by the univariate method illustrated in Figs 2 and 3. Artifactual overlap of the chromosome 3 peak and the nearest hamster peak would produce an erroneous result. For cases like this, a bivariate fitting program was developed. The actual sort window around the chromosome 3 peak is shown, along with a polygonal analysis region which defines the region of the flow karyotype involved in the computation. Two bivariate Gaussians represent the chromosome 3 peak and the nearby hamster peak, and a bivariate second degree polynomial represents background.

region. This background impurity is 5%, and probably consists of fragments of large chromosomes or clumps of small ones. A similar analysis procedure for the chromosome 19 sort shows that the contamination from both the nearest neighboring peak chromosome (20) and background is under 1–2%. Thus, both chromosomes 13 and 19 sorting purity is high at this point in the total sorting period. Examination of Figs 2 and 3 shows that the peaks for chromosomes 14 and 15 overlap appreciably, and lie in a region of higher background. As a result, the contamination level increased about 10% (Table I). However, the considerable overlap of the peaks for chromosomes 16 and 17 rules out sorting; another more favorable line was used (HSF7; discussed above) which allowed sorting with only 5% contamination (Table I). All the smallest chromosomes (18–22) are favorable for sorting from this chromosome preparation, with the proviso that the amount of background underlying the chromosome 21 peak calls for either a better chromosome preparation from this line or another human line (in actuality, the normal human fibroblast line LLL811 was used). The goal was 90–95% sort purity, which could be achieved for the small chromosomes, except for chromosome 21.

The univariate approximation does not work for many flow karyotypes. An example is shown in Fig. 4, a flow karyotype accumulated while sorting chromosome 3 from the human–hamster hybrid line 314–1b, containing only the human chromosome 3. Here, transformation of the region of the chromosome 3 peak and the neighboring hamster peak to univariate distributions in either HO or CA3 fluorescence results in excessive peak overlap. Hence, a recently developed bivariate fitting program (7) was employed which models each of the peaks as a bivariate Gaussian and the background as a bivariate second degree polynomial. The calculated impurity in the chromosome 3 sort window is 9%. all due to background. Figure 5 shows isometric plots of data, the fitted function and the background function. Fluorescence *in situ* hybridization (Chapter 17) has also

been used to assess the purity of sorted chromosomes. The extent of contamination of hamster chromosomes in a sort of human chromosomes from a human–hamster hybrid can be assessed by measuring the fraction of chromosomes that hybridize to whole genomic hamster DNA. In addition, the purity of a sort from a human chromosome preparation can be assessed by the fraction of chromosomes which hybridize to a probe specific for the sorted chromosome. In practice, a sample of the sorted material is fixed, spun down on a microscope slide and hybridized with biotinylated probe DNA (14,25). Treatment with fluoresceinated avidin allows visualization of the hybridized chromosomes in the fluorescence microscope by virtue of their green fluorescence. Counterstaining with propidium iodide labels all chromosomes and makes them visible via their red fluorescence in the fluorescence microscope with the same exciting wavelength. The fraction of green-fluorescing chromosomes is the impurity level, using whole genomic hamster DNA as the probe. This procedure has been carried out for many of our libraries, and the results are shown in Tables I and II. Note that this assay method has an error associated with it. During examination of the slides in the fluorescence microscope, fluorescent objects of approximately chromosome size but of poor morphology are usually seen, and these are difficult to characterize and score. This phenomenon varies among sorts, and introduces some uncertainty in the results.

Molecular analysis of the libraries, the third method of analysis, is the most direct and important method of library characterization, but also the most difficult and time consuming. The information generated here (24) and made available to us by users is summarized in Tables I and II. Hamster contamination of libraries constructed from human–hamster hybrid cells was measured by a hybridization method similar in principle to that used for sorted chromosomes, except that DNA from phage plaques resulting from growth of the library on a lawn of the host cells (*E. coli* strain LE392) is

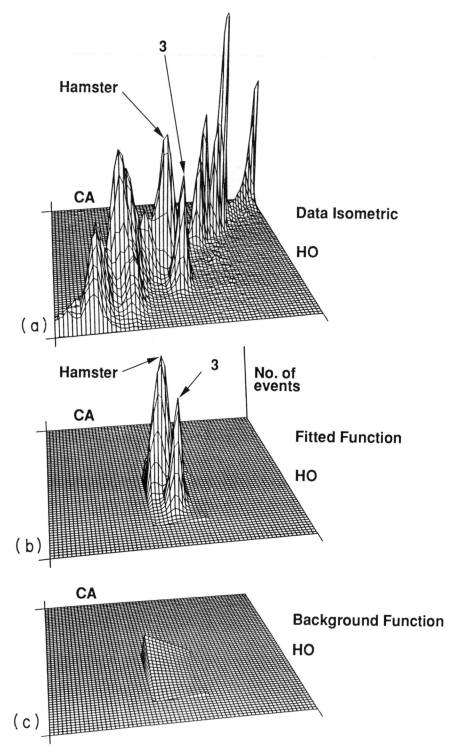

Fig. 5. Isometric plots of the original bivariate flow karyotype (a), the fitted function (b), and the background function (c) are shown. The computed functions seem to be a good fit to the original data. Predicted sort purity is 91%; the impurity level of 9% is due to underlying background.

Table II

CHARACTERIZATION DATA ON LANL LIBRARIES

Library	Cell source	Hamster impurity (%)		Unique sequence clones mapping to sorted chromosomes (%)	Average insert size (kbp)
		In situ fluor. hybridization on sorted chromosomes	Plaque hybridization		
LA01NS01	Hybrid	9 (a)	26 (a)		
LA02NS01	Hybrid	20 (a)	39 (a)		
LA03NS01	Hybrid	8 (a)	23 (a)		
LA04NS01	Hybrid	3 (a)	11 (a)		
LA05NS01	Hybrid	4 (a)	25 (a)		
LA06NS01	Hybrid	6 (a)	55 (a)		
LA07NS01	Hybrid			18/19 = 95% (b)	
			20–30 (c)	53/55 = 96% (c)	
		21 (a)	48 (a)		
LA08NS04	Hybrid	16 (a)	33 (a)		
LA10NS01	Hybrid	14 (a)	53 (a)		
LA11NS01	Hybrid		40–50 (d)	6/6 = 100% (d)	
LA11NS02	Hybrid	18 (a)	27 (d)	6/6 = 100% (i)	
LA12NS01	Hybrid	5 (a)	21 (a)		
LA15NS02	Hybrid	3 (a)	22 (a)		
LA16NS02	Human			4/7 = 57% (e)	2.3
LA16NS03	Human	5		4/4 = 100% (f)	4.1
		5		8/8 = 100% (g)	3.3
LA0XNS01	Hybrid	4 (a)	25 (a)	12/25 = 48% (h)	

(a)   Hildebrand et al. (1986).
(b)   Scambler (1986).
(c)   Barker (1986).
(d)   Shows (1986).
(e)   Hildebrand and Stallings (1986).
(f)   Keith (1987).
(g)   Reeders (1987).
(h)   Donis-Keller (1987).
(i)   Lewis (1986).

the target of the genomic hybridization probe—which is labeled with ^{32}P and detected autoradiographically. In the case of the LLNL chromosomes 18 and 20 libraries, the inserts of individual clones were characterized with regard to size, whether they contain repetitive or strictly unique sequences, and whether the unique sequences map to the correct chromosome. This more extensive characterization involves the following steps:

(1) Isolation and characterization of individual clones.

(2) Preparation of DNA from a panel of human–hamster hybrid cell lines containing overlapping sets of chromosomes which allow chromosomal assignment of the cloned single-copy sequence.

(3) Digestion of mapping DNAs, electrophoresis and transfer to nitrocellulose filters.

(4) Hybridization of single-copy clones with the DNA panel filters to verify the expected chromosomal assignment.

Similar procedures were used by several user groups who have kindly supplied us with their findings which are summarized in Tables I and II.

III. CHARACTERISTICS OF PHASE 1 LIBRARIES

The chromosome mapping data for most of the libraries show that the unique sequences map to the desired chromosome with high frequency, but a few (LL*11*NS01, LL*16*NS02, LL*17*NS01; see footnote to Table III for an explanation of this notation) show higher contamination from other chromosomes than anticipated. The impurities in the LLNL chromosomes 16 and 17 libraries probably result from a poor chromosome preparation from human fibroblast line LLL761 made in the early days of the project before the $MgSO_4$ method became standard. These two libraries have been recently recloned (LL*16*NS03; LL*17*NS02) from chromosomes sorted from a human fibroblast line (HSF7) with optimal resolution of the chromosomes 16 and 17 peaks. Quantitative flow karyotype analysis indicates 95–96% purity. The LLNL chromosome 11 library was constructed from a Chinese hamster–human hybrid line UV20HL4, containing nine other human chromosomes in addition to chromosome 11. The chromosome 11 peak lies close to a large hamster peak, probably accounting for the observed hamster contamination. The contamination by other human chromosomes is probably due to the fact that the hybrid line contains many human chromosomes (1, 4–6, 14–16, 19, 21). Thus, it is possible that fragments of large chromosomes and clumps of small chromosomes by chance fell in the sort window. We intend to use a more advantageous hybrid cell line for phase 2 cloning of chromosome 11, such as J1, which is both stable karyotypically and contains chromosome 11 as the only human material.

Some of the hamster contamination in libraries from hybrid lines, such as 17–20% in LL*04*NS01, was surprising at first—but further examination showed the same problem as with the LL*11*NS01, i.e. the presence of a large, nearby hamster ridge. Since the library was small (chromosome equivalent = 0.8), it was recloned (LL*04*NS02) from the same hybrid line (UV20 HL21-27, containing human chromosomes 4, 8 and 21), starting with

chromosomes from a better preparation and using a carefully placed sort window. Quantitative flow karyotype analysis indicated a 96% sort purity, confirmed by estimates from plaque hybridization and fluorescence *in situ* hybridization of sorted chromosomes.

The three methods (quantititive flow karyotype analysis, fluorescence *in situ* hybridization and plaque hybridization) for estimating hamster contamination of the LLNL libraries generally agree. In the two libraries, where we have results for all three methods (LL*03*NS01, LL*10*NS01), the agreement is excellent for the LL*03*NS01 (hamster impurity = 18–19%) and good for LL*10*NS01 (hamster impurity = 28–38%). In the case of LL*21*NS02, three independent measurements of purity agree extremely well. Quantitative flow karyotype analysis yielded 70% purity, whereas unique sequence mapping at two different laboratories (LLNL and Johns Hopkins) both yielded 75%. The two hybridization methods yield similar estimates of hamster purity in seven out of 11 libraries. There are discrepancies up to a factor of two for the other four libraries, with the plaque hybridization result consistently larger. This suggests a systematic error, the nature of which is not clear. Although some uncertainty exists in the various purity estimates, it is clear that these data, taken together, provide a good indication of library purity and the kinds and amounts of impurity they contain.

A comparison of fluorescence *in situ* hybridization data with plaque hybridization data for the libraries prepared from human–hamster hybrids at LANL is found in Table II. Again, the apparent level of impurity of each library as determined by plaque hybridization is greater than would be expected from estimates of sort purity, and the reasons are not clear. Two possible contributing factors should be mentioned. One is the general background underlying the peaks in the flow karyotypes. This background probably represents chromosomal fragments, which may not be recognized and scored in the *in situ* analysis, but which would contribute to the plaque hybridization result. The second possible

Table III

TWENTY-EIGHT CHROMOSOME-SPECIFIC DNA LIBRARIES (CHARON 21A; HIND III)

Library ID No.[a]	Independent recombinants	Frequency of non-recombinants	Chromosome equivalents[b]		Chromosome source[c]	Starting chromosomes ($\times 10^6$)	Starting DNA (ng)
LL01NS01	8.3×10^4	0.04	2.1		UV24HL10–12	0.5	270
LL01NS02	1.3×10^6	0.01	32	(20)	UV24HL6	0.5	270
LL02NS01	6.6×10^5	0.08	16	(5)	UV24HL5	0.5	270
LL03NS01	1.6×10^5	0.10	4.8		314–1b	0.5	220
LL04NS01	2.3×10^4	0.02	0.8		UV20HL21–27	1.0	400
LL04NS02	5×10^5	0.27	10		UV20HL21–27	0.5	210
LL05NS01	3.4×10^6	0.26	113	(30)	640–12	0.5	200
LL06NS01	7.6×10^5	0.06	27	(20)	UV20HL15–33	0.4	135
LL07NS01	3×10^5	0.01	11.5		GM131	0.9	310
LL08NS02	2.2×10^6	0.03	93	(20)	UV20HL21–27	0.7	210
LL09NS01	3.0×10^5	0.02	13		UV41HL4	0.4	120
LL10NS01	2.4×10^5	0.01	10.6		762–8A	0.5	150
LL11NS01	1.1×10^5	0.05	4.9		UV20HL4	0.2	50
LL12NS01	7.5×10^5	0.01	34	(20)	81P5D	0.4	120
LL13NS01	2.2×10^4	0.04	1.3		761	1.0	240
LL13NS02	8.5×10^5	0.03	47	(20)	GM131	0.5	120
LL14NS01	2.3×10^6	0.06	135		GM131	0.5	110
LL45NS01	2.6×10^6	0.02	152	(30)	811	1.0	210
LL15NS01	7.0×10^4	0.06	4.4		GM131	1.0	110
LL16NS03	7.6×10^5	0.02	51	(20)	HSF7	0.5	100
LL17NS02	3.4×10^5	0.02	24	(20)	HSF7	0.5	95
LL18NS01	8.9×10^5	0.13	72		761	1.0	170
LL19NS01	1.5×10^6	0.02	145	(10)	811	1.0	130
LL20NS01	3.9×10^6	0.01	354	(20)	811	1.0	140
LL21NS02	4.7×10^5	0.34	60	(20)	811	0.5	45
LL22NS01	6.1×10^5	0.05	71	(17)	811	0.5	50
LL0XNS01	2.1×10^6	0.33	84	(30)	UV24HL5	0.5	170
LL0YNS01	2.5×10^5	0.02	27		811	1.0	115

[a] The ID Code consists of eight alphanumeric items. The first two items indicate which laboratory made the library, i.e. LL = Lawrence Livermore National Laboratory. The next two items are underlined and indicate chromosome type (in the one case of a mixed 14/15 library, chromosome type is designated as 45). The fifth item is a letter indicating chromosome status, i.e. N for normal, T for translocation, etc. The sixth item is either S (for small-insert complete digest libraries). The final two items represent library construction number.

[b] Number of recombinants for one chromosome equivalent = $(3 \times 10^9) (0.65) (f)/4100$, where 3×10^9 bp is the size of the human haploid genome, 0.65 is the clonable fraction and f is the fraction of cellular DNA in particular chromosome; 4100 bp is the average fragment size. Numbers in parentheses refer to representation of amplified library; for very large libraries, only a fraction of the packaging reaction was amplified.

[c] Cell lines from which metaphase chromosomes were isolated):

(a) Normal diploid human fibroblast lines 761, 811, HSF7.

(b) Apparently normal human lymphoblastoid line GM131.

(c) CHOx human lymphocyte lines (human chromosome content:

     UV24HL10–12 (No. 1, 2 del, 3, 11, 13, 19)

     UV24HL6 (No. 1, 2 del, 3?, 11–13, 14?, 18?, 19)

     UV24HL5 (No. 2, X)

     314–1b (No. 3)

     UV20HL21–27 (No. 4, 8, 21)

     640–12 (No. 5, 9, 12)

     UV20HL15–33 (No. 6, 9, 13, 15, 17, 20, 21)

     UV41HL4 (No. 6, 9, 13, 16, 18, Y)

     762–8A (No. 10, Y)

     UV20HL4 (No. 1, 4–6, 11, 14–16, 19, 21)

     81P5D (No. 12, 15, X).

contributing factor is interphase nuclei. When using the EPICS V machine at LANL to sort chromosomes isolated from a suspension culture of human lymphoblastoid cells, it is not unusual to find that 10–20% of the sorted DNA results from interphase nuclei. This would produce an impurity level of 10–20%. Since the hybrid lines are grown in monolayer culture, the mitotic cells shaken off the monolayer should have less contamination by interphase nuclei than suspension cultures. Hence, this possible source of hamster contamination probably contributes less than 10–20% to the discrepancy between the two hybridization results.

Molecular analysis of library clones has been reported for about half the libraries. A very extensive analysis has been done on LL*04*NS01 by Gilliam *et al.* (*18*) in connection with studies of Huntington's disease, known to be due to a genetic defect near the tip of the short arm of chromosome 4. All 224 unique sequence human probes isolated from the library map to chromosome 4. This high purity is probably due to the fact that the hybrid line used (UV20HL21-27) contains only two other human chromosomes, 8 and 21. The chromosome 4 sort window cannot be contaminated by fragments of larger human chromosomes, since both chromosomes 8 and 21 are smaller. In addition, clumps of the two smaller human chromosomes or of fragments are very unlikely to contaminate the chromosome 4 sort window due to the high HO/CA fluorescence ratio of chromosome 4 compared to chromosomes 8 and 21. These characteristics make UV20HL21-27 a very favorable hybrid line. Other extensive molecular analyses have been done by Davis and Shows, in 1987, and Bruns, also in 1987, on LL*11*NS01; these independent investigators got identical results for purity (63–64%), each mapping about 200 unique sequence clones. Most other studies have been much less extensive, with results on much smaller numbers of unique sequence clones reported. An example is LL*20*NS01, with one group reporting five out of six clones mapping to chromosome 20 and another group reporting five out of eight clones mapping

to chromosome 20. There is probably no statistically significant difference between these two independent results, and this same conclusion holds for all the other mapping results on the same libraries shown in Table I.

Tables I and II list average insert size for several of the libraries, generally 2–3 kbp. This is fairly close to the 4.1 kbp to be expected from a complete digest of random sequence DNA by a restriction enzyme with a 6 bp recognition site. In addition to various factors which could produce this difference in average fragment size, the screening process selects against large fragment size, which is more likely to contain repetitive DNA.

Tables III and IV list all libraries constructed at both national laboratories and available from ATCC, and some of their salient characteristics. The input to the cloning procedure was generally about 10^6 chromosomes sorted from either human cell lines (for almost all the smaller chromosomes) or human–hamster cell lines (for almost all the larger chromosomes). This corresponds to about 0.3 μg of DNA for a chromosome of average size. Some libraries were constructed from as few as 0.2×10^6 chromosomes. The number of recombinant DNA molecules resulting from the ligation and packaging reactions varies by about two orders of magnitude among the various libraries, as does the frequency of nonrecombinants. This variability is probably inherent when cloning very small amounts of DNA. Since the statistical probability of any particular sequence of clonable size being present in a library increases with the size of a library, becoming 0.99 for a library size of five chromosome equivalents, libraries were constructed with this size as a goal. However, the smaller libraries still have a very large number of independent recombinants and are very useful. An example is LL*04*NS01 containing 0.8 chromosome equivalents $(2.3 \times 10^4$ independent recombinants). This library has provided over 200 single-copy chromosome 4 probes used in the search for restriction-fragment-length polymorphism markers for the Huntington's disease

Table IV

CHARACTERISTICS OF TWENTY-EIGHT CHROMOSOME-SPECIFIC DNA LIBRARIES CLONED INTO THE ECO R1 SITE OF CHARON 21A[a]

Library ID No.	Independent recombinants	Frequency of non-recombinants	Chromosome equivalents	Chromosome source[b]	Starting chromosomes ($\times 10^6$)	Starting DNA (ng)
LA01NS01	1.3×10^6	<0.01	31	UV24HL10–12	0.5	250
LA02NS01	7.0×10^4	0.04	1.8	UV24HL5	0.5	260
LA03NS01	2.7×10^4	0.04	0.4	314–1b	0.4	160
LA03NS02	2.1×10^5	0.08	6.4	314–1b	0.4	180
LA04NS01	5.1×10^4	<0.01	1.6	UV20HL21–27	0.9	370
LA04NS02	7.4×10^4	<0.01	2.3	UV20HL21–27	0.9	370
LA05NS01	1.3×10^6	<0.01	43	640–12	0.4	130
LA06NS01	4.8×10^4	0.06	1.7	UV20HL15–33	1.9	700
LA07NS01	2.4×10^5	0.06	9.2	MR3.316TG6	0.1	40
LA08NS04	3.6×10^4	0.10	1.5	UV20HL21–27	0.5	150
LA09NS01	1.6×10^5	0.07	7	HSF–7	0.3	90
LA10NS01	4.0×10^5	<0.01	18	762–8A	0.3	90
LA11NS02	6.2×10^4	0.17	2.8	80H10	0.4	100
LA12NS01	6.0×10^5	<0.01	27	81P5d	1.5	210
LA13NS03	7.5×10^4	0.14	4.2	HSF–7	0.4	80
LA14NS01	6.1×10^5	<0.01	36	1634	0.5	115
LA15NS02	4.0×10^5	0.06	20	81P5d	0.3	71
LA15NS03	6.8×10^4	<0.09	4	1634	0.4	82
LA16NS02	4.0×10^4	0.08	2	HSF–7	3.0	590
LA16NS03	3.3×10^5	0.03	21.7	HSF–7	0.5	1000
LA17NS03	1.1×10^5	0.01	7.9	GM130A	0.6	110
LA18NS04	2.5×10^5	0.02	19	HSF–7	0.5	81
LA19NS03	1.1×10^5	0.16	11	HSF–7	0.5	65
LA20NS01	1.6×10^4	0.24	1.5	HSF–7	0.5	72
LA21NS01	1.1×10^6	<0.01	137	HSF–7	0.9	87
LA22NS03	9.3×10^4	0.19	11	HSF–7	0.5	55
LA0XNS01	2.1×10^5	0.02	8.5	81P5d	1.2	380
LA0YNS01	1.1×10^5	0.10	11.5	HSF–7	0.5	64

[a] See footnotes to Table III for description of library ID No. and chromosome equivalent.

[b] Cell lines from which metaphase chromosomes were isolated:

(a) Normal diploid human fibroblast lines 761, 811, HSF7.

(b) Apparently normal human lymphoblastoid line GM130A and multiple X lymphoblastoid line 1634 (49,XXX).

(c) CHOx human lymphocyte lines (human chromosome content):

  UV24HL5 (No. 2, X)
  314–1b (No. 3)
  UV20HL21–27 (No. 4, 8, 21)
  640–12 (No. 5, 9, 12)
  UV20HL15–33 (No. 6, 9, 13, 15, 17, 20, 21)
  UV41HL4 (No. 6, 9)
  762–8A (No. 10)
  UV24HL10–12 (No. 1, 3, 11, 13).

(d) E36 X human lymphocyte lines (human chromosome content) 81P5d (X, 12, 15).

(e) V79 X human lymphocyte lines (human chromosome content) MR3.316TG6 (7, Xdel).

gene, on the tip of the short arm of chromosome 4 (*18*).

The efficiency of converting sorted chromosomal DNA into phage clones is low. For example, the LLNL Y-library was constructed from 10^6 sorted chromosomes and contains 27 chromosome equivalents, so that the conversion efficiency is $27/10^6 = 2.7 \times 10^{-5}$. This is the common experience in library construction, the ligation and packaging steps probably having the lowest efficiency. Recombinant phage growth on the *E. coli* lawn results in an amplification of 10^5–10^6, so that the final library ends up with about as much human DNA as was contained in the original sorted chromosomes.

The frequency of nonrecombinants should be low since these contain no inserts, are of no interest and waste time in clone screening procedures. Their frequency is usually measured indirectly in a parallel cloning procedure without insert DNA. An improved, direct method has been devised using a radiolabeled 18-base oligodeoxynucleotide whose sequence spans the cloning site and which hybridizes stably to nonrecombinants, but not to recombinants whose cloning site is interrupted by an insert (*13*). This method generally agrees with the indirect procedure, but for some libraries there are discrepancies, probably due to difficulties inherent in attempting to run the parallel reaction under conditions identical to the real cloning procedure.

Library usage has been predominantly by the medical genetics community in genome mapping and the search for RFLP markers linked to genes involved in such genetic diseases as Huntington's disease, cystic fibrosis and retinoblastoma, and in a variety of other studies, such as those on the origin of maleness and the location of the testis determining factor on the Y chromosome (*1,2,3,23*). There is also local usage of LL21NS02 at LLNL as a source of unique sequence probes for "painting" chromosome 21 by fluorescence *in situ* hybridization (*14*). The concept is that these probes can be used as a highly specific cytochemical stain for chromosome 21 if biotinylated, hybridized *in situ* to metaphase spreads, and made fluorescent by treatment with fluoresceinated avidin. This method should simplify the detection of trisomy 21, translocations involving this chromosome, and even the detection of trisomy 21 in interphase nuclei.

IV. FUTURE DIRECTIONS

Now that phase 1, the construction of two complete sets of small-insert complete digest libraries using the Hind III and Eco R1 cloning sites of Charon 21A has been accomplished; phase 2, the construction of large-insert partial digest libraries, is underway. Consideration of suitable vectors is important here. Phage insertional vectors have limited cloning capacity. For example, Charon 21A accepts insert sizes between 0 and 9 kbp. Phage replacement vectors accept insert sizes of about 9–23 kbp, whereas cosmid vectors accept insert sizes of about 35–45 kbp. The larger size is advantageous in studies involving gene structure and expression, chromosome walking and clone ordering. Phage libraries are easier to make and use, although this difference may diminish as cosmid vectors improve and experience with them increases. There has been considerable effort to simplify cosmid library construction via the development of double cos site cosmids (*6,26*), and to improve the representation of cosmid libraries with a new series of Lorist cosmids based on the lambda origin of replication to promote uniform cosmid replication (*16,17*). The phase 2 strategy is to use both cloning systems for the construction of large-insert libraries for each chromosomal type. We have decided to use the phage vector Charon 40 (*10*) which has several important characteristics, including features for ease of cloning and increase of representation, i.e. fraction of all chromosomal sequences actually present in the cloned library. However, new vectors (e.g. lambda GEMII (promega)) with advantageous features for chromosome walking and restriction mapping are appearing, and the decision for Charon 40 is not irrevocable.

Two cosmid vectors have been investigated at LLNL. The first is C2RB (6). The second is a modified version of Lorist X (18) which has double cos sites for ease of vector arm preparation. P. de Jong (LLNL) inserted the sequences necessary to convert Lorist X into a double cos site cosmid, and has also added a multicloning site. Current work at LLNL focuses on this modified Lorist X vector. LANL is concentrating on another double cos site cosmid, s cos-1 (18a) which contains Not1 sites flanking the BamH1 cloning site.

The genome is being divided up between the two laboratories, LLNL cloning 12 chromosomal types and LANL cloning the other 12. The first two chromosomal types selected for cloning are chromosomes 19 and 16. The large-insert libraries that result will be used for clone ordering (physical mapping) of chromosome 19 at LLNL and chromosome 16 at LANL. At present (June 1988), both have been cloned into Charon 40, and cosmid cloning experiments are underway.

ACKNOWLEDGEMENTS

This work was performed under the auspices of the US Department of Energy by the Lawrence Livermore National Laboratory under contract number W-7405-ENG-48 and by the Los Alamos National Laboratory under contract number W-7405-ENG-36.

REFERENCES

1. Affara, N. A., Ferguson-Smith, M. A., Tolmie, J., Kwok, K., Mitchell, M., Jamieson, D., Cooke, A., and Florentin, L. (1986). Variable transfer of Y-specific sequences in XX males. Nucl. Acids Res. 14 (13), 5375–5387.

2. Affara, N. A., Florentin, L., Morrison, N., Kwok, K., Mitchell, M., Cook, A., Jamieson, D., Glasgow, L., Meredith, L., Boyde, E., and Ferguson-Smith, M. A. (1986). Regional assignment of Y-linked DNA probes by deletion mapping and their homology with X-chromosome and autosomal sequences. Nucl. Acids Res. 14 (13), 5353–5373.

3. Andersson, M., Page, D. C., and de la Chapelle, A. (1986). Chromosome Y-specific DNA is transferred to the short arm of X chromosome in human XX males. Science 233, 786–788.

4. Athwal, R. S., Smarsh, M., Searle, B. M., and Surinder, S. D. (1985). Integration of a dominant selectable marker into human chromosomes and transfer of marked chromosomes to mouse cells by microcell fusion. Somat. Cell Molec. Genet. 11 (2), 177–187.

5. Bartholdi, M. F., Meyne, J., Albright, K., Luedemann, M., Campbell, E., Chritton, D., Deaven, L., and Cram, L. S. (1987). Meth. Enzymol. 151, 252–267.

6. Bates, P. F., and Swift, R. A. (1983). Double cos site vectors: simplified cosmid cloning. Gene 26, 137–146.

7. Dean, P. N., Kolla, S., and Van Dilla, M. A. (1989). Analysis of bivariate flow karyotypes. Cytometry 10, 109–123.

8. Deaven, L. L., Hildebrand, C. E., Fuscoe, J. C., and Van Dilla, M. A. (1986). Construction of the human chromosome specific DNA libraries: The National Laboratory Gene Library Project. Genet. Eng. 8, 317–332.

9. Deaven, L. L., Van Dilla, M. A., Bartholdi, M. F., Carrano, A. V., Cram, L. S., Fuscoe, J. C., Gray, J. W., Hildebrand, C. E., Moyzis, R. K., and Perlman, J. (1986). Construction of human chromosome-specific DNA libraries from flow-sorted chromosomes. Cold Spring Harbor Symp. Quant. Biol. 51, 159–167.

10. Dunn, I. S., and Blattner, F. R. (1987). Charons 367 to 40: multi enzyme, high capacity, recombination deficient replacement vectors with polylinkers and polystuffers. Nucleic Acids Res. 15, 6.

12. Fuscoe, J. C., Clark, L. M., and Van Dilla, M. A. (1986). Construction of fifteen human chromosome-specific DNA libraries from flow-purified chromosomes. Cytogenet. Cell Genet. 43, 79–86.

13. Fuscoe, J. C. (1987). Human chromosome-specific DNA libraries: use of an oligodeoxynucleotide probe to detect nonrecombinants. Gene 52, 291–296.

14. Fuscoe, J. C., Collins, C., Pinkel, D., and Gray, J. W. (1989). An efficient method for selecting unique sequence clones from DNA libraries and its application to fluorescent staining of human chromosome 21 using in situ hybridization. Genomics 5, 100–109.

15. Gerhard, D. S., Jones, C., Morse, H. G., Handelin, B., Weeks, V., and Housman, D. (1987). Analysis of human chromosome 11 by somatic cell genetics: Reexamination of derivatives of human–hamster cell line J1. Somat. Cell Molec. Genet. 13(4), 293–304.

16. Gibson, T. J., Rosenthal, A., and Waterston, R. H. (1987). Lorist 6, a cosmid vector with BamH1, Not1, Sca1, and Hind III cloning sites and altered neomycin phosphotransferase gene expression. Gene 53, 283–286.

17. Gibson, T. J., Coulson, A. R., Sulston, J. E., and Little, P. F. R. (1987). Lorist 2, a cosmid with transcriptional terminators insulating vector genes from interference by promoters within the insert; effect on DNA yield and cloned insert frequency. *Gene* **53**, 275–281.

18. Gilliam, T. C., Healey, S. T., MacDonald, M. E., Stewart, G. D., Wasmuth, J. J., Tanzi, R. E., Roy, J. C., and Gusella, J. F. (1987). Isolation of polymorphic DNA fragments from human chromosomes 4. *Nucl. Acids Res.* **15**, 4.

18a. Evans, G., Lewis, K. and Rothenberg, B. (1989). High efficiency vectors for cosmid microcloning and genomic analysis. *Gene* **79**, 9–20.

19. Gray, J. W., Dean, P. N., Fuscoe, J. C., Peters, D. C., Trask, B. J., van den Engh, G. J., and Van Dilla, M. A. (1987). High-speed chromosome sorting. *Science* **238**, 323–329.

20. Kunkel, L. M., Umadevi, T., Eisenhard, M., and Latt, S. A. (1982). Regional localization of the human X of DNA segments cloned from flow sorted chromosomes. *Nucl. Acids Res.* **10**, 1557–1578.

21. Lalande, M., Kunkel, L. M., Flint, A., and Latt, S. A. (1984). Development and use of metaphase chromosome flow-sorting methodology to obtain recombinant phage libraries enriched for parts of the human X chromosome. *Cytometry* **5**, 101–107.

22. Lebo, R. V., Gorin, F., Fletterick, R. J., Kao, F. T., Cheung, M. C., Bruce, B. D., and Kan, Y. W. (1984). High-resolution chromosome sorting and DNA spot–blot analysis assign McArdle's syndrome to chromosome 11. *Science* **225**, 57–59.

23. Page, D. C., Mosher, R., Simpson, E. M., Fisher, E. M. C., Graeme, M., Pollack, J., McGillivray, B., de la Chapelle, A., and Brown, L. G. (1987). The sex-determining region of the human Y chromosome encodes a finger protein. *Cell* **51** 1091–1104.

24. Perlman, J., and Fuscoe, J. C. (1986). Molecular characterization of the purity of seven human chromosome-specific DNA libraries. *Cytogenet. Cell Genet.* **43**, 87–96.

25. Pinkel, D., Straume, T., and Gray, J. W. (1986). Cytogenetic analysis using quantitative, high-sensitivity, fluorescence hybridization. *Proc. Natl Acad. Sci. USA* **83**, 2934–2938.

26. Poutstka, A., Rackwitz, H-R., Frischauf, A-M., Hohn, B., and Lehrach, H. (1984). Selective isolation of cosmid clones by homologous recombination in *Escherichia coli. Proc. Natl Acad. Sci. USA* **81**, 4129–4133.

27. Saxon, P. J., Strivatsan, E. S., Leipzig, G. V., Sameshima, J. H., and Stanbridge, E. J. (1985). Selective transfer of individual human chromosomes to recipient cells. *Molec. Cell. Biol.* **5**(1), 140–146.

28. Van Dilla, M. A., Dean, P. N., Laerum, O. D., and Melamed, M. R., eds (1985). "Flow Cytometry: Instrumentation and Data Analysis," Academic Press, London.

29. Van Dilla, M. A., Deaven, L. L., Albright, K. L., Allen, N. A., Aubuchon, M. R., Bartholdi, M. F., Browne, N. C., Campbell, E. W., Carrano, A. V., Clark, L. M., Cram, L. S., Fuscoe, J. C., Gray, J. W., Hildebrand, C. E., Jackson, P. J., Jett, J. H., Longmire, J. L., Lozes, C. R., Luedemann, M. L., Martin, J. C., McNinch, J. S., Meincke, L. J., Mendelsohn, M. L., Meyne, J., Moyzis, R. K., Munk, A. C., Perlman, J., Peters, D. C., Silva, A. J., and Trask, B. J. (1986). Human chromosome-specific DNA libraries: construction and availability. *Biotechnol.* **4**, 537–552.

Flow Cytometric Measurement of Specific DNA and RNA Sequences

J. G. J. BAUMAN

TNO Radiobiological Institute
Lange Kleiweg 151
2288 GJ Rijswijk
The Netherlands

D. PINKEL and B. J. TRASK

Lawrence Livermore National Laboratory
Biomedical Sciences Division
Livermore, CA 94550, USA

M. VAN DER PLOEG

Sylvius Laboratory
72 Wassenaarseweg
Leiden, The Netherlands

I. INTRODUCTION

One of the first applications of flow cytometry was the determination of the relative DNA and RNA contents of individual cells using chemical stains with specificity for nucleic acids. Later it was demonstrated that chromosomes and bacteria could also be measured. Certain of these dyes are capable of giving information on more subtle aspects of the state of the nucleic acids, such as the proportion of GC base pairs and whether the nucleic acid is single- or double-stranded. However, none of these stains can discriminate particular DNA or RNA nucleotide sequences.

Nucleic acid hybridization, which has been used for more than two decades, is capable of detecting specific nucleotide sequences. In this procedure, a labeled single-strand segment of "probe" nucleic acid is incubated under appropriate conditions with single-stranded target nucleic acid. The probe binds to the target and forms a double-stranded "hybrid" molecule at locations where the base sequences of the probe and target are complementary. Originally, the probes were radioactively labeled and the hybridization was detected with autoradiography or scintillation counters. More recently it has become possible to replace radioactivity with fluorescence, thus opening the opportunities for flow cytometric analysis. Since the first flow measurements of specific DNA sequences were made a few years ago (*105*), the sensitivity and specificity have improved two orders of magnitude, so that it is now possible to reliably detect hybridization to cells with probes for repetitive sequences which have 10^6 base pairs (Mbp) or more of target (*106*). The intensity of fluorescence following fluorescence *in situ* hybridization has also been shown to be proportional to the amount of target sequence. Thus, hybridization with a probe for a chromosome-specific repetitive sequence allows quantification of the number of chromosomes present per cell (assuming that each chromosome has the same amount of target sequences). Such sequences, most of which are located in the centromeric

regions, are known for approximately two-thirds of the human chromosomes. A partial list includes chromosomes 1, 3–11, 13, 15–20, 22, X and Y (*24,25,30,31,34,35,49,54,56,66, 67,81,83,112–117,121,122,124*). The amount of target sequence for these repetitive probes is frequently polymorphic. Flow measurements have demonstrated such variability (*107*). In addition to repetitive probes, composite probes made from collections of cloned sequences specific for any desired portion of the genome are now being produced (*89*). As these are further developed, they may become useful for flow measurements. Flow measurements using standard DNA stains are now used to determine the DNA index of the tumors, that is the ratio of the DNA content of tumor cells to that of normal diploid cells. This index gives an overall estimate of aneuploidy. Fluorescence hybridization with chromosome-specific repetitive DNA probes should allow determination of which chromosomes contribute to the aneuploidy. This may contribute to diagnosis and prognosis, and increase the rate at which information about tumor-specific aneuploidies is developed (*43*).

In this chapter we will present a brief introduction to nucleic acid hybridization, describe the major methods of nonradioactive probe labeling, and discuss the current status of flow cytometric detection of DNA and RNA sequences in cells, nuclei and chromosomes.

II. NUCLEIC ACID HYBRIDIZATION

A. Fundamental Concepts

The functional properties of nucleic acids are encoded in their nucleotide base sequences. For double-stranded molecules, which comprise most of the DNA and a portion of the RNA, the base sequences of the two strands on the double helix are complementary. The nucleotide adenine (A) on one strand is always paired with thymine (T) on the other. Similarly guanine (G) pairs with cytosine (C). In RNA, T is replaced by uridine (U). The hydrogen bonds between the complementary bases, which hold the strands together, are much weaker than the covalent bonds in the sugar phosphate backbone of each strand. Thus, if a double-stranded molecule is heated, the strands will separate, but the backbone will remain intact. The temperature at which strand separation occurs (the melting temperature, T_m) depends on the chemical environment, the base composition of the nucleic acid and the length of the molecule. For example the T_m of GC-rich nucleic acids is higher than that for AT-rich nucleic acids. The T_m (°C) of a DNA double helix of length d (base pairs) in an aqueous buffer containing a particular concentration of Na^+ (molarity) and percentage of formamide (v/v) can be estimated from the semi-empirical relation (18)

$$T_m = 81.5 + 0.41 \, (\% \, GC) + 16.6 \log [Na^+]$$
$$- 0.63 \, (\% \text{ formamide})$$
$$- (300 + 2000 [Na^+])/d. \quad (1)$$

Note that each 1% increase in the GC content of the DNA increases the melting temperature by 0.4°C. Reducing the sodium concentration a factor of two lowers T_m by 5°C. This formula includes the effect of formamide since it is one of the common ingredients in hybridization buffers. A 1% increase in the formamide concentration decreases T_m by 0.6°C. In other chemical environments (e.g. by addition of tetramethyl ammonium chloride), the dependence of T_m on base composition can be removed (123) and the T_m shifted more dramatically. The melting temperature of duplex RNA is higher than RNA·DNA hybrids, which is in turn higher than the DNA double helix. The melting relationships depend on the sequences involved (62). The binding energy between the two strands is reduced if the AT/GC base pairing rules are not followed perfectly. Thus the melting temperature will decrease as the fraction of mismatched base pairs in the double helix is increased.

Single-stranded nucleic acids with comple-mentary base sequences will form double-stranded hybrid molecules (hybridize) if the solution containing them is cooled to T_m. If the temperature is lowered further, duplexes with imperfect base pairing will have sufficient binding energy to form. The perfection of base pairing required for hybrid formation, which is called the "stringency" of the hybridization, is controlled by the difference between T_m and the hybridization temperature. The rate of the reaction depends on the concentrations of the probe and target sequences. The effective concentrations of DNA in solution can be increased by addition of dextran sulfate (or other polymers) to the hybridization mix (120). This discussion of hybridization is designed to give a qualitative overview and is greatly simplified. More information can be found in Cantor and Schimmel (27) and Beltz et al. (18). The quantitative information about hybridization has been determined for purified nucleic acids in solution. The actual behavior of probes in the complex environments of nuclei and metaphase chromosomes is not understood in detail, but is similar to what has just been described. Optimal conditions for hybridization to these targets are usually determined empirically.

B. IN SITU HYBRIDIZATION

Hybridization between the probe and target nucleic acid can be performed in solution, under conditions where the target is attached to a solid support such as a piece of filter paper, or where the target is still contained in a cell or chromosome. The latter case is termed in situ hybridization and is of most relevance for flow cytometry. In situ hybridization was introduced independently by three groups [Gall and Pardue (41), John et al. (58) and Buongiorno-Nardelli and Amaldi (23)]. These groups demonstrated the chromosomal localizations of the regions for satellite DNA and ribosomal RNA. Probes for these targets could be obtained from genomic DNA by bulk fractionation techniques (59,86). The broad applicability of the technique was originally limited by the difficulty in obtaining pure, highly radioactive probes for the desired sequences. This is reviewed in Coghlan *et al.* (*29*).

This situation changed dramatically with the advent of cloning (*78*), which permitted arbitrarily large amounts of selected nucleic acid sequences to be produced, and with the development of new methods of labeling probes to high specific activities. Around 1980, the ability to localize unique sequences on chromosomes by radioactive *in situ* hybridization became sufficiently reliable to be useful (*42,46,47*). [See Szabo and Ward (*101*) for a review of these developments]. Detection of nuclear DNA sequences, RNA, and viruses in interphase cells became practical as well (*19–21,44,57,75,77,97,111*). Bulk quantitation of mRNA expression in cells grown on coverslips or spotted onto nitrocellulose filters is possible using scintillation counting (*74,125*).

III. NONRADIOACTIVE HYBRIDIZATION

The central feature of nonradioactive hybridization is the replacement of radioactive labeling and detection of probes with cytochemical procedures. This approach is sometimes called hybridocytochemistry because it is the combination of the two techniques. The first successful system was that of Rudkin and Stollar (*94,95*). They hybridized RNA probes to DNA targets and detected the bound probe with an antibody directed against the RNA·DNA duplex. The remainder of the techniques that have been developed involve the use of chemically modified probes. These probes are either directly coupled to fluorochromes (*8,9*) or enzymes such as horseradish peroxidase (*93*), or they contain a modification, such as a hapten, that binds to an affinity reagent such as an antibody. The affinity reagent is labeled with fluorochromes or enzyme molecules. If an enzyme is used, the final step in the procedure requires supplying the enzyme substrate so that a visible product is deposited at the hybridization site.

The hybridization and fluorescence detection (fluorescence *in situ* hybridization: FISH) of a probe to a metaphase chromosome is schematically depicted in Fig. 1. It involves denaturation of the chromosomal DNA and probe so that both are single-stranded, incubation of the probe with the target and finally incubation of the hybridized target with an affinity reagent carrying a fluorochrome. Fluorescence hybridization of probes specific for human chromosomes 1, 6, 7, 9, 11, 15, 17, X and Y are shown in Plate 1 (p. 280). Note that the probes for the autosomes label two chromosomes in the metaphase spreads and that two yellow fluorescent spots are visible in the interphase nuclei. The sex chromosome probes show only one hybridization site in each cell. Hybridization stringency for many of these probes needs to be relatively high to achieve acceptable chromosome specificity. All of these hybridizations were done at 37°C in 55% formamide and 1XSSC (0.15 M NaCl and 0.015 M sodium citrate).

Nonisotopic hybridization methods have several advantages over radioactive techniques in many applications. These advantages stem from the high spatial resolution that can be obtained, the opportunity to distinguish the binding of multiple probes in the same target, the ability to easily quantify the amount of bound probe, the rapidity of the procedures and reduced complications arising from handling hazardous materials. High spatial resolution arises because the fluorochrome or enzyme precipitate is deposited close to the binding site of the probe. Probe position can be determined to the limit of optical resolution (*12,69*) or with electron microscopy (*45,55,64,79,118*). In many cases fluorescent techniques give higher resolution than do enzymes in optical microscopy since the enzyme precipitate tends to diffuse somewhat before it is fixed into the target (J. Landegent, private communication). The amount of fluorochrome can be measured

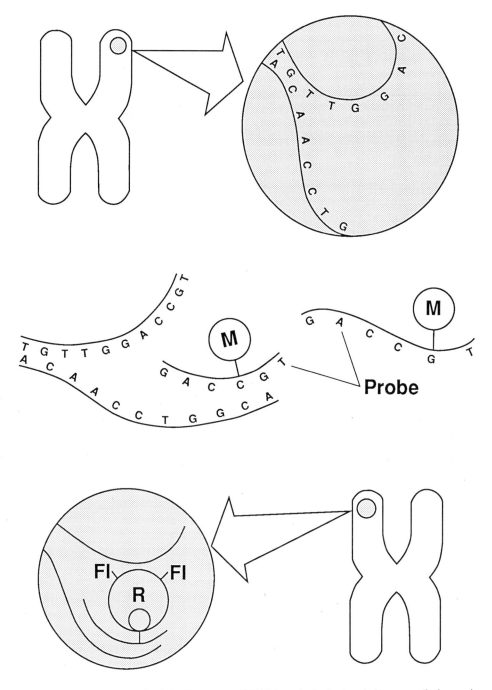

Fig. 1. Fluorescence hybridization. Chromosomal DNA is made single-stranded to expose the base pairs to permit hybridization. Probe molecules containing a chemical modification M diffuse into the target searching for places where their base sequence is complementary to those of the target. After binding, the excess probe is washed away, and the modification M is detected with an affinity reagent R carrying fluorochromes Fl.

with quantitative microscopy or flow cyto- metry. Autoradiographic resolution of the positions of ³H-labeled probes is limited to several micrometers because the autographic grain may occur at some distance from the bound probe. Nonradioactive procedures have all of the flexibility of immunochemistry at their disposal. Thus, several probes labeled with different procedures can be hybridized simultaneously (52,84,88,106), and the binding of each probe can be detected with a different fluorochrome or enzyme precipitate. While the length of the hybridization reaction is the same for both radioactive and nonradioactive pro- cedures, the cytochemical detection in the latter requires several hours, while autoradiography takes several days to weeks. Finally, the use of nonradioactive probes does not require special equipment or facilities for radioactive waste disposal. The main limitation of nonradioactive hybridization is the sensitivity, which has been less than for ³H probes. This has been changing rapidly. As will be described below, it is possible to detect the binding of probes of only a few kbp using standard fluorescence micro- scopy. We now discuss in detail several of the most widely used methods for nonradioactive probe labeling and probe detection.

A. IMMUNOCHEMICAL DETECTION OF RNA·DNA HYBRIDIZATION

Rudkin and Stollar were the first to report that injection of poly(rA.dT) complexed to methylated bovine serum albumin into rabbits could lead to the production of an antiserum specific for arbitrary RNA·DNA hybrids (94,95). However, such an antiserum is difficult to produce. Van Prooijen-Knegt *et al.* (90) were able to get one rabbit out of 12 to produce antibodies with sufficient avidity to be useful for detecting hybridization. Attempts at producing high-affinity monoclonal anti-RNA·DNA anti- bodies, using mice suffering from an auto- immune disease (91), or mice immunized with poly(rAdT), (100), have met with limited

success. Kitagawa and Stollar (63) reported anti-RNA·DNA antibodies raised in goats.

Anti-RNA·DNA antibodies have proven useful in detection of probe RNA bound to homologous DNA target sequences. With a good antiserum, 18S and/or 28S ribosomal RNA genes can be easily visualized on the short arms of the human acrocentric chromosomes from the D and G group (90) using purified ribosomal RNA as a probe. Since it is now possible to produce RNA probes comple- mentary to any cloned DNA sequence by using cloning vectors containing promoters for the RNA-polymerases from SP6 or T7 bacteriophages (32,82), this approach might be revived and developed further. Its inherent simplicity, notably the fact that the probe need not be modified, is very attractive. The antisera have also been applied to study the presence of endogenous RNA·DNA hybrids in cells or chromosomes (3).

B. DIRECT COUPLING OF A FLUOROPHORE TO THE 3'-TERMINUS OF A PROBE

Labeling of probes by direct chemical reactions with a fluorochrome is an elegant technique and was one of the initial probe modification systems to prove successful. The first published method made use of RNAs that were oxidized at the 3'-terminus with periodate in acetate buffer at pH 5.0. The oxidized *cis*-diol group of the ribose moiety resulted in two aldehyde groups which could react with compounds containing reactive amino groups. Fluorochromes such as fluorescein iso- thiocyanate (FITC) were reacted with hydra- zine and converted into thiosemicarbazides. The latter proved to have a high affinity for the aldehyde groups. By formation of a morpholine-like ring, one fluorochrome molecule was coupled to the end of each oxidized RNA molecule. The resulting labeled probe retains its specificity and efficiency for hybridization (8–10).

Among the successes of this approach has been the localization of tRNA genes in *D.*

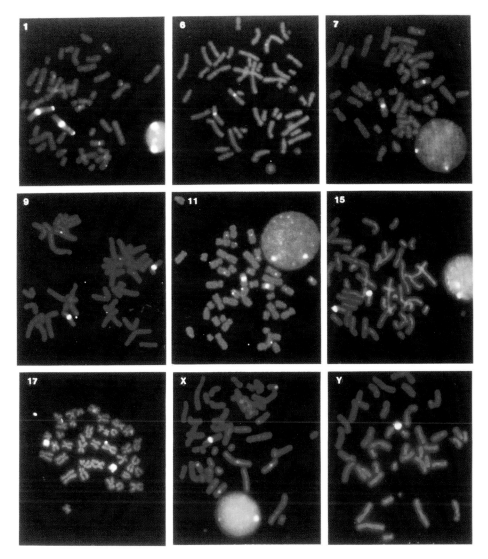

Plate 1. Fluorescence hybridization of human-chromosome-specific repetitive probes to metaphase spreads and interphase nuclei. These hybridizations were done with biotin-labeled probes at 37°C in 55% formamide, 1 x SSC. The bound probes were detected with avidin-fluorescein and the signals amplified using biotinylated anti-avidin antibodies and a second layer of avidin-FITC. The DNA has been counterstained with propidium iodide. Hybridization sites appear yellow in these photographs. The probes hybridize to the chromosome identified by the number in the upper left of each photograph. 1p (*25*) and para-centromeric region of chromosome 1 (*31*); 6 (*56*); 7 (*113*); 9 (*83*); 11 (*111*); 15 (*49*); 17 (*116,113*); X (*112*); Y (*24*); (D. Pinkel *et al.*, unpublished work).

melanogaster polytene chromosomes (*12*). This investigation also demonstrated the excellent topological resolution that can be obtained by fluorescence techniques resulting in the identification of two closely spaced chromosomal regions that have high homology to a probe for the heat shock genes. This double band could not be resolved unambiguously by autoradiography. A drawback of this approach is the relatively low sensitivity due to the fact that there is only one fluorescent molecule per probe molecule (average size 50 nucleotides). However, the signal can be amplified by the use of anti-FITC antibodies (*11*). Currently, only RNA probes can be labeled with this direct technique. However, the availability of cloning vectors supporting production of RNA transcripts minimizes this limitation. Other techniques are being developed for labeling DNA. Direct labeling has potential advantages for flow cytometric applications since no immunochemical steps are necessary after the hybridization to attach the fluorochromes. This might lower the amount of clumping and/or degradation of target nuclei, cells or chromosomes by reducing the number of wash steps they must go through.

C. CHEMICALLY MODIFIED PROBES

During the last 7 years several indirect hybridocytochemical procedures have been developed. These are called indirect because chemically modified probes are hybridized to the target, and the modification is subsequently detected with additional layers of affinity reagents. The four predominant methods involve biotinylation, AAF modification, sulfonation or mercuration of the probe. Biotinylation, AAF modification and sulfonation are similar in that relatively large molecules (molecular weight of several hundred) are attached to the probe before the hybridization procedure. These modifications slightly alter the hybridization reactions and result in a small but measurable decrease in the melting temperature (*40,68,72*). The fourth

procedure, developed by Hopman and coworkers (*13,50–53*) reduces this effect by attaching mercury atoms, which are much smaller than biotin and AAF, to the probe prior to hybridization. A larger ligand to which the fluorochrome can be attached is introduced after the probe has bound to the target.

a. Biotinylation

Ward and coworkers synthesized nucleotide triphosphate analogs of cytidine and uridine that contain biotin attached covalently to the C-5 position of the pyrimidine ring through an allylamine linker arm (*72*). These nucleotide analogs can be introduced into probes using polymerase enzymes in nick translation, primer extension, and polymerase chain reaction (*119*) procedures, in the same way probes are produced using radioactive nucleotides. Usually one of the nucleotides used in this reaction, for example dUTP, is biotinylated. Typically, 10–40% of the Ts in the probe are replaced by biotin-U during a nick translation. Techniques are available to produce both DNA and RNA probes in this manner. Biotin may also be introduced by photochemical reaction (*39*). The biotin-labeled probes readily form hybrids with complementary target sequences. The biotin can be detected by fluorochrome or enzyme-labeled avidin, streptavidin or anti-biotin antibodies. Biotinylation is the most widely used probe labeling scheme. The sensitivity is sufficient to visualize the binding to single-copy targets only a few kbp in length (*2,4,76,89*). Biotin-modified nucleotides with linker arms of various lengths are available from a number of sources including BRL (Bethesda, MD) ENZO (New York, NY) and Clontech (Palo Alto, CA).

b. AAF Modification

The carcinogen *N*-acetoxy-*N*-acetyl-2-aminofluorene (AAF) binds covalently to the C-8 position of the guanine residues when it is incubated with DNA or RNA at 37°C for 1 h (*68,102*). The extent of the reaction, which depends on the relative concentrations of DNA and the activated AAF, is easily controlled.

In most cases modification of ~5% of the total nucleotides is optimal for hybridization applications. The AAF–guanosine is a strong immunogen, and poly- and monoclonal antibodies are easily obtained (*6,102*). With the procedure, viral genomic sequences (*92*), human ribosomal DNA, mouse satellite DNA and even single-copy genes [15 kbp of the human thyroglobulin gene (*69*)] have been visualized *in situ*. The chromosomal location of a unique DNA fragment closely linked to Huntington's disease, the D4S10 locus, was assigned near the telomere of the short arm of chromosome 4 using the cosmid cloned genomic fragment C5.5 (*71*). *N*-Acetoxy AAF, is available from Chemsyn Laboratories, Lexana, Kansas. Anti-AAF antibodies are not yet commercially available.

c. Sulfonation

DNA probes can also be tagged by insertion of antigenic sulfone groups into the cytosine moieties of the denatured probe. The sulfonation is accomplished with sodium bisulfite at high molarity, and the resulting sulfone is stabilized by substitution of the function of the amine on C-4 with the aid of a nucleophilic reagent such as methylhydroxylamine. By these simple chemical reactions, 10–15% of the cytosine residues in the probe are transformed into 4*N*-methoxy-5,6-dihydrocytosine-6-sulfonate. After the hybridization procedure the bound probe is detected with a monoclonal antibody specific for the sulfone groups. Kits for this procedure are available from Orgenics Ltd (Yavne, Israel) and FMC Bioproducts (DK-2665 Vallensbaek Strand, Denmark, and Rockland, Maine, USA).

d. Mercuration

Nucleic acid probes can be mercurated using mercury acetate followed by isolation in the form of a cyanide complex. The mercury ions are covalently bound to the C-5 atom of the pyrimidine bases uracil and cytosine, and do not influence the hybridization behavior of the nucleic acid probes (*33*). After hybridization, the mercury atoms in the bound probe are detected using ligands that contain both a sulfhydryl group (to react with the mercury), a stabilizing positive charge to reduce the repulsion by the negatively charged DNA and a hapten (such as TNP) or a fluorochrome. The hapten is then detected with a fluorescently labeled antibody. Hybridization with mercurated probes seems to be as sensitive and specific as the other procedures. Single-copy probes as small as 7 kbp have been detected (J. Wiegant, private communication). It has been used to detect binding of the thyroglobulin and C5.5 cosmid probes (*50–53*; A. H. N. Hopman *et al.*, personal communication) by fluorescence detection of the TNP–hapten. The TNP ligand for this procedure is now commercially available from Euro-Diagnostics BV (Apeldoorn, The Netherlands) and Organon Technika (The Netherlands).

e. Other Modification Methods

Several other modification methods are available or have been described. They are not expected to be more sensitive than those discussed above, but they may offer advantages in specific applications. The greater the number of independent probe modification methods that are available, the higher the number of different target sequences that can be discriminated in simultaneous hybridizations. These modification methods include:

(1) Chemical coupling of TNP to DNA (*96*) and detection with anti-TNP antibodies.

(2) Covalent linking of the enzyme horseradish peroxidase to DNA (*93*).

(3) Polymerase mediated incorporation of allylamine-modified nucleotide. These nucleotides seem to be incorporated at higher efficiency than biotin modified nucleotides. A ligand (such as TNP or biotin) is then coupled to the allylamine in the probe. This method lends itself to the direct fluorochrome labeling of probes. However, nonspecific binding of fluoresceinated allylamine-U-RNA to cellular structures has been observed (J. G. J. Bauman, unpublished results). Allylamine (labeled nucleotides are available from Bethesda Research Laboratories).

(4) Incorporation of brominated nucleotides

(both the ribo- and deoxyribo-forms) followed by detection with anti-BrdUrd antibodies (103). This method is of limited use for hybridization because the available anti-BrdUrd antibodies do not recognise the BrdUrd in the context of double-stranded nucleic acids. Thus only single-strand probe tails not associated with the target DNA that are available bind the antibody (110). Anti-BrdUrd antibodies are available from several sources including Becton Dickinson (Mountain View, CA, USA) and Immunotech SA (France).

(5) Incorporation of digoxigenin-carrying nucleotides followed by detection with anti-digoxigenin antibodies. This method is offered as a kit for Southern analysis by Boehringer Mannheim (Mannheim, FRG). The commercially available antibody is directly conjugated to alkaline phosphatase, which currently prohibits its application to fluorescence techniques.

f. Modified Oligonucleotides

Several reports have been published on the use of chemically synthesized oligonucleotide probes into which ligands such as biotin have been incorporated (28). Detection of these probes after hybridization is as above. This approach is potentially very powerful because well-defined probes can be synthesized in milligram quantities when an automated DNA synthesizer is available. Probes are short, usually under 100 nucleotides, facilitating penetration into fixed chromosomes and tissues (5). However, many different short probes must be synthesized to cover enough of a potential target to give a useful signal. This might by the method of choice for detection of repetitive sequences.

D. ENZYMATIC DETECTION OF BOUND PROBE

Two enzyme systems have been used most often for the visualization of probe hybridization: horseradish peroxidase (PO) and alkaline phosphatase (AP). These are coupled, sometimes in large clusters, to antibodies or (strept)avidin. PO has a very high turnover number, so that each enzyme molecule produces a great deal of precipitate. 3'3-Diaminobenzidine tetrahydrochloride (DAB) is a commonly used substrate with PO. This brown precipitate gives a precise histochemical localization, is noncrystalline and can be stored permanently without fading or other alterations. It can be observed microscopically with trans-illumination or with the more sensitive reflection contrast microscopy (70). In flow cytometry it might be detected by light scatter. There are several procedures to enhance the signal by subsequent chemical reactions [imidazole, silver–gold amplification (60)]. Alkaline phosphatase is also a very stable enzyme. Tetranitroblue–tetrazoliumchloride (TNBT) and 5-bromo-4-chloro-3-indolyl phosphate (BCIP) are used in combination as a substrate for this enzyme. A purple-blue precipitate that remains well localized (60) is produced.

E. FLUORESCENCE DETECTION OF BOUND PROBE

Currently, the most effective fluorescence hybridization procedures involve the use of biotin-, AAF-, sulfone- and mercury-labeled probes. With the first three, the fluorochromes are added using affinity reagents. With the mercury method, special ligands carrying either a fluorochrome (53) or another hapten are used as the first step, followed by additional immunochemical steps if necessary. Among the fluorochromes that have proven useful to date are AMCA (ultraviolet excitation, blue fluorescence; BioCarb, Lund, Sweden), fluorescein (blue excitation, yellow-green fluorescence), rhodamine (blue-green excitation, red fluorescence) and Texas red (green excitation and red fluorescence). The phycobiliproteins, such as phycoerythrin (85), have not proven generally effective in hybridization work. Their large size may inhibit penetration into the complex environment of the nucleus, cell or metaphase chromosome. Additional fluorochromes are under development that may extend the useful spectral range to the near-infrared (*38*), allow more effective single-wavelength excitation of multiple dyes whose emissions are spectrally distinct (*48*), and have narrower excitation and emission bands to allow discrimination of more dyes in a single specimen.

The sensitivity of the fluorescence detection can be increased using multiple layers of affinity reagents as are used for immunofluorescence. Most of the detection procedures that use antibodies are inherently double layer. The first antibody is specific for the modification in the probe, and the second carries the fluorochrome. This may be amplified further, for example using an anti-fluorescein antibody to bind to the fluorescein carried by the second antibody (*11*). This is then detected using a fourth fluoresceinated antibody layer. With biotin-labeled probes, an anti-biotin antibody can be used for the first layer, followed by an amplification chain similar to the one just described. Alternately, avidin may be used for primary detection, followed by alternating layers of biotinylated anti-avidin antibody and avidin (*87,89*). (Avidin and biotinylated anti-avidin are available from Vector Laboratories, Burlingame, CA, USA.)

IV. FLOW CYTOMETRY

Use of FISH in conjunction with flow cytometry was discussed as soon as the initial nonautoradiographic techniques were developed (*7*). It took several years, however, before practical procedures were developed that permitted the application of FISH to nuclei (*105*) and cells (*15*) in liquid suspension so that they could be successfully measured using flow cytometry. Although information about the spatial localization of the hybridized probe is not in general obtained by flow measurements, the amount of probe that has hybridized to each cell can be quantified simply and rapidly. The fluorescence intensity of thousands of cells and/ or nuclei can be measured in a matter of minutes, giving an accurate measurement of the fluorescence intensity or target size in each

cell in the population. Measurement of other parameters can be obtained simultaneously.

Significant effort has been devoted to quantifications of specific DNA sequences in cells, nuclei, and chromosomes and specific RNA sequences in cells. Detection of DNA and RNA sequences in nuclei and cells has been successful (*16,105*; see Appendices I and II). However, hybridization to isolated chromosomes has been more difficult (*36,104*). Flow cytometric measurement requires that the *in situ* hybridization protocol be performed on objects that are either in liquid suspension throughout the procedure or that can be released into suspension after being stained. The objects must remain as single entities without clumping. Hybridization to targets on a solid support, such as a microscope slide, is simpler because attachment to the substrate maintains the integrity of the target. On the other hand, the accessibility of target sites to the probe and detection reagents may be higher for particles in suspension than to the same particles fixed on slides. Because fluorescence intensity is integrated over the entire object using conventional flow cytometry, nonspecific probe hybridization and binding of the detection reagents, and autofluorescence are greater problems than with microscopy. Weak emission distributed over the whole cell will not interfere with visual recognition of the localized high-intensity specific binding sites, but if the total intensity of the background approaches or exceeds that of the specific signal, it will reduce the ability to detect differences in the amount of bound probe in flow. Cellular autofluorescence is frequently enhanced in cells by the fixation and hybridization procedures, and is currently an important limiting factor in mRNA measurements.

A. MEASUREMENT OF NUCLEAR DNA SEQUENCES

Flow measurement of nuclear DNA sequences is undergoing rapid development since first being accomplished (*105*). This first work demonstrated the ability to detect hybridization of a probe to the highly repeated mouse satellite sequence. The size of the target detected by the probe was about 300 Mbp, or approximately 10% of the mouse genome. Since that work, the sensitivity and specificity of FISH has been improved so that targets on the order of 1 Mbp can now be measured (*106,107*). The presence of single unique sequences of a few kbp, or single genes, which may encompass several tens or hundreds of kbp, cannot currently be detected using flow cytometry. However, there are many important mammalian sequences with sufficient repetition so that 1 Mbp sensitivity is useful. Among these are the previously mentioned human-chromosome-specific sequences, ten of which are shown in Plate 1. Using these probes, flow measurements can be used to determine the relative number of copies of a particular chromosome present in each cell of a population (*106,107*). Polymorphism among individuals in the amounts of specific repetitive sequences they have also can be detected (*107*). Identification of aneuploid chromosomes, such as may exist in a tumor, should be feasible with this technique (*43*). Quantification of the number of copies of amplified sequences (e.g. dihydrofolate reductase or c-myc) also appears possible (B. Trask and M. Pallavicini, private communication). Likewise, it may be possible to assess the degree of viral sequence incorporation in the genome of cells.

The hybridization procedure for flow cytometry consists of several steps. One must isolate, stabilize and denature the target cells, prepare the single-stranded probe mixture, combine the two for the hybridization, wash out unbound probe and perform the fluorescence staining. Total cellular DNA can be stained using dyes like DAPI or propidium iodide after FISH. A detailed procedure for suspension hybridization is included in Appendix I. This procedure is largely that published previously by Trask *et al.* (*106*). Fluorescence hybridization and flow cytometric analysis can be performed with as few as 10^4 nuclei (*105*). The approach can be applied to interphase nuclei isolated from a variety of cultured human cell types including amniocyte-derived fibroblasts, skin fibroblasts, EBV-transformed lymphoblasts and peripheral blood lymphocytes. Thus, the procedure should be applicable to cell types for which little material is available.

The ability of flow cytometric measurements of nuclear DNA sequences to give quantitative information about the amount of target sequence is shown in Fig. 2. Fluorescence intensity is linear with target size over a 30-fold range. Measurements were made using human genomic DNA to probe a series of human–hamster hybrid cells containing various combinations of human chromosomes, as well as diploid and tetraploid human cells. Under the conditions of this hybridization, only the highly repeated sequences in the probe are present at high enough concentration for significant hybridization. The repetitive sequences in humans and hamsters have diverged sufficiently during evolution so that there is essentially no binding of human DNA to hamster DNA in these experiments.

Figure 3 shows a bivariate distribution of the measurement of the human Y-specific probe, pY3.4A, hybridized to a mixture of four different cell types [diploid male lymphoblast (46,XY), diploid female lymphoblast (46,XX), tetraploid female lymphoblast (92,XXXX) and aneuploid male fibroblast (47,XYY)]. This probe has a target of about 6 Mbp on the long arm of the Y chromosome, but has additional binding sites near the centromeres of chromosome 9 and, to a lesser extent other autosomes. Thus, female human cells contain about 15% as much target (1 Mbp) for this probe as do typical male cells (*67,73*; K. Smith, personal communication), although the amount of these sequences is polymorphic among individuals. Probe hybridization was detected using avidin FITC in this experiment and cellular DNA was stained with Hoechst 33258. The cells were analyzed in a dual-beam flow cytometer where the fluorescence following ultraviolet excitation was recorded as a measure of cellular DNA content and green fluorescence following blue excitation was recorded as a measure of probe binding.

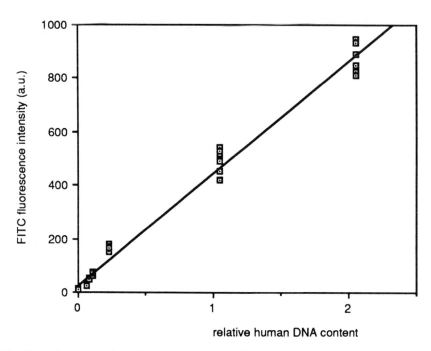

Fig. 2. Probe fluorescence intensity *vs* target sequence. Seven types of nuclei with differing amounts of human DNA were hybridized as mixtures with a human genomic DNA probe and measured on a flow cytometer. The relative fluorescence intensities of the G_1 peaks of each cell type due to probe binding are highly correlated ($R = 0.99$) with the amount of target sequence. [Data from Trask *et al.* (*106*).]

All four cell populations are visible in the bivariate distributions. Diploid female nuclei have the lowest probe fluorescence. Tetraploid female nuclei are more fluorescent; their G_0G_1 peak falls on top of the G_2/M peak of the diploid cells. The XY and XYY nuclei form two distinct distributions. A projection of the probe fluorescence of the G_0/G_1 peaks of the three diploid cell populations is shown at the left edge of the figure. The coefficient of variation (CV) of these peaks is about 15%. The CV of the Hoechst 33258 fluorescence for the G_0/G_1 peaks ranges from 5 to 10%. Note that the Hoechst 33258 intensity of the XYY nuclei is lower than that for the other diploid cells. This represents an increased loss or increased denaturation of the DNA in these cells. Despite this reduction in Hoechst 33258 fluorescence, the probe intensity of XYY cells is 2.3 times that of XY cells. In this histogram, the XY cells have a probe fluorescence 3.0 times brighter than XX

cells, and 1.7 times brighter than tetraploid XXXX cells.

When these measurements were repeated for multiple male and female cells to average for labeling variability and polymorphism in sequence content, the average XYY probe fluorescence intensity was 2.2 times greater than the XY intensity. Normal male XY cells were 2.9 times brighter than normal female cells and 2.2 times as intense as the tetraploid XXXX lines. These measurements indicate that the cell types can easily be distinguished based on the amount of target sequence that they contain. Background fluorescence from the nonspecific binding of the probe and/or fluorescence detection reagents is sufficient to reduce the expected difference between the male and female cells from about six-fold to three-fold. This result and measurements with a probe whose target size is approximately 2–3 Mbp indicate that background fluorescence is

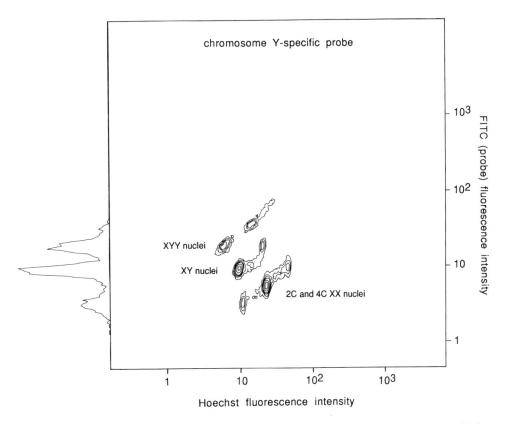

Fig. 3. Y probe fluorescence from four different human cell types. Nuclei from normal male (XY), aneuploid male (XYY), normal female (XX) and tetraploid female cells were mixed together and hybridized with a Y-specific probe, and bound probe was measured in flow. DNA in the nuclei was stained with Hoechst 33258. Contour lines represent the numbers of events in each channel. The distribution along the ordinate is a projection of the probe fluorescence intensities of the G_1 subpopulations of the XY, XYY and XX cells. [Data from Trask *et al.* (*106*).]

equivalent to a probe target of about 1 Mbp. Thus, with the current techniques, nuclei containing 1 Mbp of actual target have approximately twice the fluorescence intensity as those without any target (*106*).

The CV of probe fluorescence intensity distributions in 50 hybridizations using several genomic and chromosome-specific repetitive probes ranged from 12 to 31%, with an average of 21% (*106*). The ability to discriminate populations with different amounts of target sequence, for example, aneuploid cells, depends critically on the CV of the measurement. With the current capability, it is

estimated that aneuploid XYY cells occurring at a frequency of at least 5% among normal male cells could be detected with flow cytometry.

The simultaneous measurement of DNA content and probe fluorescence gives exciting additional information. The approximate time in S-phase when the target sequence is replicated can be derived from the shape of the distribution. Note that the probe fluorescence of XY cells (Fig. 3) does not increase as cells enter S-phase and their DNA content increases. Only in late S-phase does probe fluorescence increase. This is consistent with earlier studies

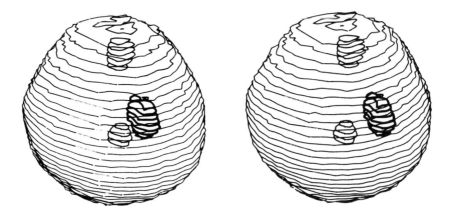

Fig. 4.   Stereo pair of a nucleus showing positions of the Y chromosome and both copies of chromosome 17. The Y-specific probe was modified with AAF and detected with FITC. The 17 probe was modified with biotin and detected with streptavidin–Texas Red. Total DNA was stained with DAPI. The spatial distribution of each fluorochrome was reconstructed from a series of images of focal planes separated by 0.5 μm. Additional contour lines were interpolated each 0.25 μm. The distributions of each fluorochrome were then superimposed to produce the figure. [Data from Trask *et al.* (*106*).]

showing that the long arm of the Y chromosome is one of the last areas of the genome to replicate (*26,61*).

Nuclei that have been hybridized in liquid suspension in preparation for flow cytometric measurement also are ideal subjects for the study of the binding distribution of probes within the nucleus using fluorescence microscopy (*1,80,88,106*; H. van Dekken *et al.*, unpublished work). If the nuclei are hybridized with chromosome-specific probes such as those shown in Plate 1, they appear much as they do in that figure except that the nuclei have retained their original three-dimensional morphology. Since the diameter of a nucleus is much larger than the depth of focus of microscope optics, the hybridization spots of the probes in general occur in different focal planes. A series of two-dimensional images at different focal planes can be collected and used to reconstruct a three-dimensional representation of probe binding in the nucleus. Figure 4 shows the reconstruction of the positions of chromosomes 17 and Y in a nucleus. The probes were modified with AAF (chromosome Y) and biotin (chromosome 17) and detected with FITC and Texas red,

respectively; the entire nucleus was counter-stained with DAPI. The determination of the three-dimensional positions of chromosomes in nuclei will be a research area of increasing importance.

Figure 5 shows a bivariate histogram of measurements of human lymphocytes hybridized with probes for the Y chromosome and the centromeric region of chromosome 1. The chromosome 1 probe was labeled with biotin and detected with avidin–TRITC. The Y probe was modified with AAF and detected with a FITC-labeled second antibody. Total DNA was stained with DAPI. The cell populations were hybridized, stained and run separately with the same instrument settings. The histogram shows the green *vs* red fluorescence distributions of DAPI-bright objects in each sample. After mathematical correction for the cross-talk between FITC and TRITC (*107*), the measurements were plotted together. The four populations indicate unhybridized cells and cells hybridized with either or both probes. Work is underway in a number of laboratories to extend these techniques to simultaneously detect DNA and

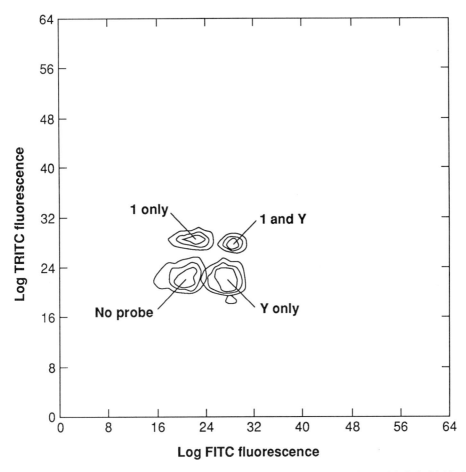

Fig. 5. Flow measurement of chromosomes 1 and Y. The chromosome 1 probe was labeled with biotin and detected with avidin–TRITC. The Y probe was modified with AAF and an FITC second antibody was used. Cell populations were hybridized with no probe, each of the probes singly, and both probes. They were stained and measured independently, and the four measurements displayed together after correction for cross-talk between the FITC and TRITC detection. [Data from van Dekken *et al.* (*107*).]

RNA sequences and proteins. These goals will drive instrument design to the addition of more excitation wavelengths to exploit four or more fluorochromes in single experiments.

B. MEASUREMENT OF CHROMOSOMES

One of the major goals of fluorescence hybridization has been the flow measurement of probes hybridized to isolated chromosomes as specific chromosome markers or for gene localization. This goal has so far proven elusive.

The measurement requires a protocol that produces single-stranded DNA in the chromosomes, while preserving their integrity and avoiding clumping. The first indication that this might be possible was the success in using anti-BrdUrd antibodies (which recognize BrdUrd only in single-stranded DNA) to detect BrdUrd in chromosomes, while still preserving the ability to discriminate independent chromosome types in a standard DNA-based flow karyotype (*104*). Fluorescence hybridization to chromosomes in liquid suspension has recently been demonstrated (*36*; B. Trask, unpublished observations). Unfortunately, only a small fraction of the hybridized chromosomes appear intact and free in suspension. The majority are clumped or fragmented and are not suitable for flow measurement. The only published report of flow measurement is by Dudin *et al.* (*36*). They show one slit-scan profile from a chromosome from a human–hamster hybrid cell that is the result of an interspecies translocation. It is hybridized with human genomic DNA, which binds to the human portion of the chromosome.

C. HYBRIDIZATION TO mRNA

Messenger RNA is transcribed from one strand of the DNA, is processed to produce an active mRNA sequence and moves to the ribosomes where the sequence is translated into a protein. *In situ* hybridization to RNA in cells is an important tool used to monitor gene expression. Specific mRNA levels are currently quantified by counting silver grains in autoradiographs or by scintillation counting of hybridized slides or filters to which 10^3–10^4 cells have been applied (*74,97,125*; J. Dunne *et al.*, unpublished work). Work aimed at extending this capability to flow cytometry is in its infancy. Hybridization to RNA in cells in suspension (or which can be released into suspension) presents several demands on cell fixation protocols not presented by DNA work. RNA can be lost from the cells, since it consists of small molecules that are not necessarily bound to cellular structures, and it is readily degraded by the ubiquitous RNase. On the positive side, mRNA is single-stranded and, therefore, it is not necessary to denature prior to hybridization. In addition, the cytoplasm of fixed cells seems to be more accessible than the nucleus for long probes and large detection molecules.

Probes can be made to bind selectively to DNA or RNA in cells. If one does not denature, only RNA targets are available. Probes complementary to cellular RNA can be produced by using double-stranded nucleic

acids, or single-stranded molecules in which the sequence is transcribed from the antisense DNA strand. These are called "antisense probes". If the DNA in the target is denatured, both DNA and RNA are available for hybridization. Hybridization to RNA can be prevented by using a single-stranded probe whose sequence is identical to the RNA. This is called a "sense" probe since its sequence is transcribed from the sense strand of the DNA. Thus, if the target is cellular RNA, the sense probe can be used as a control for the specificity of the hybridization obtained with the antisense probe (*32*). The sense probe will bind to the sense DNA strand if the DNA has been denatured. Single-stranded probes can be produced efficiently using cloning vectors containing bacteriophage DNA-dependent RNA polymerase promotors that make labeled transcripts from only one strand of the cloned DNA sequence.

Flow cytometric measurement of hybridization to RNA has recently been demonstrated (*16*). The target was the ribosomal RNA (rRNA), which is present at high concentration. Both sense and antisense probes made from a 2.1 kbp fragment representing most of the 3' half of the 28S rRNA molecule were used. This fragment was cloned in the pGEM plasmid which contains the bacteriophage T7 and SP6 RNA polymerase promoters on opposite sides of a multiple cloning site, permitting transcriptions of both sense and antisense probes as desired. The transcription reaction contained biotin–UTP (rather than the dUTP used for DNA probes). A detailed protocol for mRNA hybridization is given in Appendix II. Those aspects which are critical for success are prevention of target RNA degradation using diethylpyrocarbonate (DEPC) to inhibit endogeneous RNases, breaking the probe in fragments 100–200 nucleotides in length, and appropriately fixing the cells so that they maintain their RNA.

Figure 6 shows the results of the measurement of hybridization of the antisense probe to mouse bone marrow cells. DNA in the cells was stained with DAPI (excited with 250 mW of

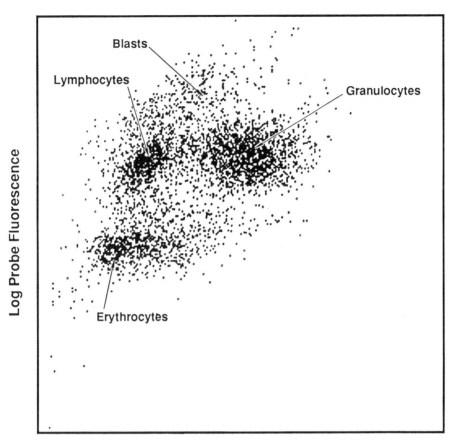

Fig. 6. Measurement of ribosomal RNA in mouse bone marrow cells. Cells were hybridized with an antisense probe, which was detected with FITC. DNA was stained with DAPI. Data was taken in list mode. Measurement of probe fluorescence *vs* forward light scatter (FLS) is shown. [Data from Bauman and Bentvelzen (*16*).]

ultraviolet light from an argon ion laser) and the probe was detected with streptavidin–FITC (excited with 500 mW of 488 nm light from a second laser). Forward (FLS) and perpendicular (PLS) light scatter and time of flight (TOF) also were measured to improve identification of single cells. The data was recorded in list mode so that it could be examined at leisure using many gating combinations. The relationship of probe fluorescence to FLS is shown. The measurements indicate that the concentration of rRNA in lymphocytes is much higher than in granulocytes since lymphocytes are substantially smaller cells (lower FLS). Hybridization with the sense control probe yielded much weaker signals (not shown). These could be reduced by RNase, indicating the probe was predominantly single stranded and was thus bound nonspecifically into the cell. Some cells with very low DNA content have appreciable FITC fluorescence. Microscopic observation shows that these are reticulocytes which have retained their rRNA.

Flow measurements have also been used to determine the total poly-A$^+$ RNA in normal and regenerating bone marrow cells, and in several leukemic cell lines (unpublished work). Expression of the murine mammary tumor virus long terminal repeat M-MTV-LTR transcripts in L1210 murine leukemia cells has also been accomplished (J. G. J. Bauman et al., unpublished work). The major limitation to the sensitivity of RNA measurements by flow is the generation of cellular autofluorescence during the in situ hybridization procedure. The work presented here indicates that 10^4 or greater target RNA molecules can be detected. Recent results indicate a factor of 10 increase in sensitivity. If autofluorescence remained at the level found in living cells (rather than increasing during the hybridization), it is estimated that as few as ten target molecules could be detected.

V. CONCLUSION

Flow cytometric measurement of specific DNA and RNA sequences shows great promise. The techniques described permit detection in cells of DNA and RNA targets that are larger than 1 and 10 Mbp, respectively. The CV of the DNA measurements is ~20%. Multicolor techniques have demonstrated the ability to measure the simultaneous binding of two probes, and given the number of different probe modification techniques and fluorochromes that are available or under development, this is likely to be extended to three or more in the near future. Among the applications are detection of aneuploidy in cells using chromosome-specific repetitive probes and the quantification of amplified genetic sequences. This has potential value for prenatal diagnosis, where the predominant cytogenetic abnormalities are aneuploidies of chromosomes 13, 18, 21, X and Y. In tumor cytogenetics, where aneuploidy and gene amplification occur commonly, flow measurements offer the potential to rapidly collect data on the distribution of particular DNA sequences over the cell population. As hybridization to isolated chromosomes is perfected, flow measurement will be able to detect structural abnormalities such as translocations. Since hybridization

techniques for staining the entire length of a selected chromosome have now been developed (89), different chromosome types could be stained different colors so that translocations would produce bicolor chromosomes.

The ability to measure RNA as well as DNA in cells, combined with the use of antibodies to detect specific proteins, offers the potential to study the whole gene expression pathway from DNA to transcription to translation on a cell by cell basis. The coming of age of fluorescent nucleic acid probes promises to have an impact on flow cytometry similar to the development of specific staining with monoclonal antibodies.

APPENDIX I

PROTOCOL FOR SUSPENSION FLUORESCENCE IN SITU HYBRIDIZATION TO NUCLEAR DNA (106)

Cell Preparation

Hybridization can be accomplished on either isolated nuclei or whole cells. Nuclei are isolated by a procedure developed for the isolation of chromosomes from mitotic cells (108). Cells are centrifuged at 150g for 10 min at room temperature and resuspended at a cell concentration of 5×10^6/ml of isolation buffer (50 mM KCl, 5 mM Hepes, 10 mM MgSO$_4$, 3 mM dithioerythritol, 0.15 mg/ml RNase, pH 8.0). After 10 min incubation at room temperature, Triton X-100 (2.5% v/v) is added to a final concentration of 0.25%. The cells are held on ice for 10 min and then vigorously vortexed to disperse the nuclei. This nuclear suspension is then incubated 30 min at 37°C for RNA digestion and put on ice. Recently, the nuclei isolation has been successfully omitted, and cells are simply washed in PBS before fixation (B. Trask, unpublished results). Cells or nuclei are then fixed to maintain their integrity by the addition of cold 100% ethanol while vortexing to achieve a final concentration of 70%. After 10 min on ice, they are pelleted at 150g for 10 min at 4°C. The supernate is removed and three times the original volume of fresh 100% ethanol is added while vortexing. Although nuclei additionally fixed with paraformaldehyde (1% PF in PBS + 50 mM MgSO$_4$, pH 7.6, 10 min, room temperature) are very suitable for three-dimensional localization of hybridized sequences (*106*; H. van Dekken *et al.*, unpublished work), they are unsuitable for flow measurements due to nonspecific binding of probe and detection reagents (B. Trask, unpublished observations).

Fixed cells or nuclei are centrifuged and resuspended in 0.1 N HCl/0.05% Triton X-100 at a concentration of 10^7/ml for 10 min to remove proteins. This step improves the accessibility and decreases nonspecific binding of probe and detection reagents (B. Trask, unpublished observations). The same procedure was developed for staining cells with antibodies to bromodeoxyuridine (*17*). The nuclei are then washed twice in IB (isolation buffer without dithioerythritol or RNase and containing 0.05% Triton X-100) and are resuspended in IB. Clumps are dispersed as necessary by syringing the suspension through a 22-gauge needle. Nuclei are counted using a hemacytometer after staining with Hoechst 33258 (2 μg/ml) and the concentration adjusted to 10^8/ml. The nuclei are stable and can be stored at 4°C prior to hybridization. Nuclei from different cell types are mixed together at this point for fluorescence intensity comparisons.

Hybridization

The hybridization reaction and immuno-fluorescent staining can be carried out in a 1.5 ml microcentrifuge tube. One microliter of fixed, acid-extracted nuclei or cells (10^5 particles) is added to 18 μl of a hybridization buffer consisting of five parts formamide to two parts 50% (w/v) dextran sulfate in water, one part 20 × SSC (1 × SSC is 0.15 M NaCl, 0.015 M sodium citrate) and one part sonicated herring sperm DNA (10 mg/ml). Finally, 1 μl of the labeled probe or probe mixture is added. Typical probe concentrations in the complete hybridization reaction are 1 ng/μl for biotin-labeled probes and 5 ng/μl for AAF-labeled probes. The mixture is heated to 70°C for 10 min to denature both the probe and the target DNA and then incubated overnight at constant temperature. This temperature is typically between 37 and 45°C, depending on the desired stringency.

Unbound probe is removed by adding 1.25 ml of 50% formamide/2 × SSC, pH 7, and incubating at 42°C for 10 min with occasional mixing. The mixture is brought to room temperature and 10^7 dimethyl-suberimidate-fixed human red blood cells (RBCs) are added as a carrier to improve recovery of nuclei in the numerous washing and immunofluorescence

labeling steps that follow. The nuclei are then centrifuged at 150g for 10 min, resuspended (tapping the tube to disperse the pellet) in 1.25 ml of 2 × SSC, pH 7.0, for 10–15 min, centrifuged and finally resuspended in 1.25 ml of PBS/0.05% Tween-20 in preparation for the immunofluorescent staining.

Fluorescent Staining

The nuclei and carrier RBCs are centrifuged and resuspended in buffers containing the appropriate affinity reagents. For example, when AAF-labeled probes are used, the mouse monoclonal anti-AAF antibody (*6*) can be added in PBS/0.05%, Tween-20/ 2% normal goat serum. For 10^5 cells, incubation is performed in 200 μl of buffer. The nuclei are incubated 45 min at 37°C. Nuclei are washed by incubation for 15 min in 1 ml PBS/Tween-20. The fluoresceinated second antibody, goat–anti-mouse immunoglobulin-FITC, is added in PBS/0.05% Tween-20/2% normal goat serum. Fluorescently labeled avidin or streptavidin can be used to detect hybridized biotinylated probes. Concentrations in the range of 1–5 μg/ml in 0.1 M phosphate buffer, pH 8.0, containing 0.05% NP-40 and 5% nonfat dry milk (PNM) are optimal. Avidin is usually found to give a brighter signal than streptavidin. (The solids are removed from PNM buffer by spinning the buffer at 12 000 rpm for 5 min prior to using it. The supernate is collected and stored at 4°C after the addition of 0.01% sodium azide.) When AAF-labeled and biotinylated probes are detected simultaneously, (strept)avidin and anti-AAF are added in PNM buffer and the anti-Ig-fluorophore in PBS/ Tween-20/normal goat serum. Incubations are 45 min at 37°C. Nuclei are washed after each step by incubation for 15 min in 1 ml phosphate buffer/ NP-40 or PBS/Tween-20 followed by centrifugation.

Flow Cytometric Measurements

Nuclei are resuspended in 1 ml PBS containing 0.05% Tween-20 and 2 μg/ml Hoechst 33258 (or DAPI). Nuclei are measured on a dual-beam flow cytometer so that nuclei can be identified on the basis of their Hoechst 33258 fluorescence (RBCs have no DNA and bind little Hoechst 33258). The nuclei and RBCs are first illuminated by 200 mW ultraviolet laser light for Hoechst 33258 excitation. Hoechst 33258 fluorescence is measured through a KV418 (Schott) filter.

Electronic thresholds are set to select nuclei and exclude erythrocytes and small debris from further analysis. One watt of 488 nm light from a second laser is used for FITC excitation. The fluorescence of each particle identified as a nucleus by its Hoechst 33258 fluorescence is collected through a 530 nm bandpass filter (DF 530/30 nm).

APPENDIX II

PROTOCOL FOR SUSPENSION FLUORESCENCE *IN SITU* HYBRIDIZATION TO CELLULAR mRNA (*16*)

Cell Preparation

Hybridization is best performed on cells that are cultured in media without phenol red. Freshly isolated cells from blood, bone marrow, spleen, etc., can also be used. Before fixation, the cells are washed once with Hank's balanced salt solution (without phenol red) with HEPES buffer (pH 7.2) and osmolarity 300 ± 5 mOsm (HH), and thoroughly resuspended by repeatedly pipetting with 0.5 ml HH using a 1 ml Eppendorf pipet tip, counted and finally adjusted to 2.5×10^7 cells/ml. Cells are fixed by the addition of an equal volume HH containing 2% formaline. It is important to adjust the osmolarity of the fixative to the cell type used in order to preserve cellular structure. Fixation is for 5 min at room temperature and continues during the subsequent centrifugation step (10 min, 300*g* at 4°C). Different cell types might require slightly different fixation times for optimal results. The cell pellet is resuspended as above in HH, and 100% ethanol is added to a final concentration of 70%. After centrifugation the cells are resuspended in 70% ethanol (in water). Fixed cells can be stored at $-20°C$ until used. In our experiments, the same stock of cells has been used for several months without significant loss in hybridization properties.

Probes

For the detection of specific RNA sequences, we prefer the use of single-stranded RNA (ssRNA) probes (*32*), prepared by transcription from a DNA fragment subcloned into plasmid pGEM2 (Promega Biotec, Madison, WI) using standard techniques. In our studies, a 2.1 kbp BglII–Eco R1 fragment was used that was obtained from the 7.3 kbp Eco R1 fragment (*37*) of the human 28S ribosomal RNA gene isolated from a sorted chromosome 21 library (*65*; clone pHR28.1, a gift from Dr B. D. Young, ICRF, Dept of Medical Oncology, St Bartholomews

Hospital, London). The RNA transcripts are prepared following the directions given by the manufacturers using either SP6 (Promega Biotec, or GIBCO-BRL, Gibco Europe, Breda, The Netherlands) or T7 (Promega, BRL or Boehringer Mannheim, FRG) DNA-dependent RNA polymerases. For the incorporation of 5-[*N*-(*N*-biotinyl-*e*-aminocaproyl)-3-aminoallyl] uridine tri-phosphate (biotin-11-UTP, BRL) we use enzyme concentrations which are two-fold higher for the T7 and three-fold higher for the SP6 enzyme than recommended by the manufacturer. The biotin-11-UTP concentration is increased two-fold. This ensures sufficient yield of full length transcripts. The template DNA must be removed by DNase digestion (DNase RQI, Promega Biotec), and the probes are purified by centrifugation chromatography through a 1 ml Sephadex-G25 column, eluted with $0.1 \times SSC$ containing 0.1% sodium dodecyl sulphate (SDS) ($1 \times SSC$ is $0.15 M$ NaCl, $15 mM$ trisodium citrate, pH 7.0). The RNA yield from $1 \mu g$ template is always recovered in a $75 \mu l$ volume. For a representative transcription reaction of the 2.1 kbp rDNA probe, the yield as measured by ultraviolet absorbance was $1.9 \mu g$ of transcript, resulting in a probe concentration in stock of $25 \mu g/ml$. Yield and quality of the transcripts are estimated by agarose gel electrophoresis and direct filterspot tests following directions as described for biotinylated DNA probes (GIBCO-BRL), except that RNA is denatured in $8 \times SSC/3.6 M$ formaldehyde, 10 min at 80°C before spotting on nitrocellulose membrane filters.

The ssRNA probes are degraded before use by controlled limited alkaline hydrolysis (*32*) to about 100–150 nucleotides in size. The size is checked by electrophoresis using 1.5% agarose gels. By direct filter spot test it was determined that the alkaline hydrolysis did not result in detectable loss of biotin moieties. When estimating transcript sizes it is necessary to take into account the fact that full biotinylation or probes decreases the electrophoretic mobility of the RNA molecules, resulting in an overestimation of the size by about 30–40%. Probes are stored at $-20°C$.

Hybridization

For reproducible ISH results it is essential to block endogenous RNase activity in fixed cells before rehydration. This is done with 0.2% diethylpyrocarbonate (DEPC, $1 \mu l$ of 10% DEPC in ethanol per $50 \mu l$ fixed cell suspension in 70%

ethanol) for 15 min at room temperature (RT). In addition, all reagents and equipment are kept free of RNase. After DEPC treatment, cells are centrifuged (5 min, 1500g) in 1.5 ml Eppendorf-type reaction tubes. The fixative is carefully removed. The cell pellets are resuspended to a final cell concentration of 1×10^7/ml in HH, 0.5% Tween-20 (HH–T, HH is DEPC-treated and autoclaved before addition of Tween-20). After 5 min at room temperature, one volume 20 × SSC is added, the tube content is mixed and two volumes deionized formamide (Merck, Darmstad, FRG; after deionization with AG50-X8, Bio-Rad, Richmond, CA, stored at $-20°$C) are added. Approximately $1–2 \times 10^5$ cells are then transferred from this suspension to the required number of Eppendorf tubes.

When highly concentrated cell suspensions are handled, the remaining liquids in the disposable pipet tips and tubes contribute significantly to the overall cell loss (up to 5% per step). Cell loss occurs mainly during the post hybridization washing and staining steps. The number of washings and centrifugations are therefore kept to a minimum. Recovery of cells varies and has been estimated to be 20–50%. Standard table-top centrifugation with swing-out buckets, with careful removal of supernates was found to give optimal results. Cell loss by sticking of cells to the tube wall is minimized by maintaining washing volumes at 200 μl or less. In our experiments, the number of cells is reduced to the absolute minimum in view of the large amounts of probes needed to saturate the target.

The cells, now in 50% formamide, 5 × SSC, are centrifuged as above and the cell pellet is resuspended by repeated pipetting in 10 μl hybridization mixture (50% formamide, 5 × SSC, 0.5 mg/ml E. coli tRNA, 0.5% SDS) containing the probe. In general, 10μl of probe stock is used, equivalent to about 0.25 μg of probe, dried in vacuo and dissolved in the hybridization mixture. The final probe concentration is then 25 μg/ml. Incubation is performed at 45°C for 17 h in an oven with forced ventilation. Recent experiments show that under the above conditions hybridization of antisense rRNA probe is virtually complete after 3 h. The tubes are agitated by placing the tube rack on a vibrating device. Excess probe is removed after hybridization by adding 100μl hybridization mixture and further incubation at 45°C for 45 min. The cells are pelleted as above, resuspended in 100μl 0.1 × SSC, 0.1% SDS, and incubated at 45°C for 30 min to remove mismatched hybrids.

Fluorescent Staining

Cells are centrifuged and resuspended in 100μl HH–T after which streptavidin–FITC (Amersham Int., Amersham, UK) is added (5μg/ml). After incubation for 30 min at 45°C, 400μl HH–T is added. The cells are either washed once more or directly analysed by flow cytometry. DAPI at 0.2μg/ml is added to counterstain the DNA in the cells. When single-beam flow cytometry must be performed, counterstaining is done with ethidium bromide (EB) at low concentrations. DNA histograms obtained with EB are not as good as for DAPI staining, since EB also binds to RNA in the cells.

Flow Cytometric Measurements

The fluorescence and the light scatter of the cells is measured using a FACS II with 0.5 W 488 nm Argon laser light as described earlier (14), or, when cells are counterstained with DAPI, a dual-beam flow cytometer; (RELACS-III) is used. Up to eight parameters from each individual cell can then be collected in list mode. Signals from 3000–10 000 cells are collected. When two lasers are used the forward (FLS) and perpendicular light scatter (PLS, 488 nm band-pass filter, Corion, Holliston, MA), green fluorescence (530/20 nm bandpass, Corion) and the time of flight (TOF) of the FLS are derived from the first laser (tuned at 488 nm, 0.5 W). The second laser (tuned to 351–363 nm, 250 mW) is used to induce the DAPI fluorescence (2 × KV408, Schott, Mainz, FRG, +SP450, Corion). Also, the second laser is used to derive autofluorescence signals in the green or red spectral range in order to perform autofluorescence corrections as described by Steinkamp and Stewart (99). Histograms are constructed using the signals from single cells selected by gating the FLS, PLS, TOF and DAPI fluorescence parameters. The green fluorescence is measured using a logarithmic amplifier. The mean linear fluorescence intensities are calculated from the histograms as the sum of the product of the linearized channel number times the number of events per channel, divided by the total number of events. Channel 1 on the log scale is taken to be 1.0 arbitrary units on the linear scale and 60 channels difference represents a ten-fold difference in signal intensity.

Microscopy

Fluorescence microscopy is used to confirm hybridization results after flow cytometry. Remaining cells are centrifuged and resuspended in antifade solution containing either DAPI or EB to counterstain the DNA. Antifade solution consists of 50 g/l 1,4-diazobicyclo[2.2.2]octane (22) in glycerol/ 0.1 M Tris–HCl, pH 8.0 (90:10 v/v). Kodak Ekta-chrome 200 ASA film is used for photography.

RNase Pretreatment

In order to determine whether the target is RNA or DNA the cells can be pretreated with RNase. For this the cells are rehydrated in $2 \times$ SSC at 1×10^7 cells/ml. RNase A (free of DNase, from pancreas, Boehringer Mannheim, FRG) is added to $10 \mu g/ml$ and incubation is for 15 min at 37°C. Cells are pelleted by centrifugation and resuspended at the original concentration in HH–T containing 1% DEPC to inactivate residual RNase. After a further 15 min at 37°C cells are pelleted by centrifugation and processed for hybridization as described above.

RNase Post Treatment

To assay specificity of the hybridization with sense control or antisense RNA probe the cells can be post treated with RNase. Specific double-stranded hybrids are resistant to RNase digestion. After hybridization the cells are pelleted by centrifugation and washed once with $0.1 \times$ SSC, 0.1% SDS and resuspended in the same buffer. The samples are incubated for 20 min at 46°C followed by addition of 1/10 volume of $20 \times$ SSC and RNase (10 $\mu g/ml$) and further incubation at 37°C for 15 min. The samples are cooled on ice, diluted with HH–T and fluorescence stained as above.

REFERENCES

1. Agard, D., and Sedat, J. (1983). Three dimensional architecture of a polytene nucleus. *Nature* **302**, 676–681.

2. Albertson, D. G. (1985). Mapping muscle protein genes by *in situ* hybridization using biotin-labeled probes. *EMBO* **4**, 2493–2498.

3. Alcover, A., Izquierdo, M., Stollar, D., Kitagawa, Y., Miranda, M., and Alonso, C. (1982). *In situ* immunofluorescent visualization of chromosomal transcripts in polytene chromosomes. *Chromosoma* **87**, 263–277.

4. Ambros, P., Matzke, M., and Matzke, A. (1986). Detection of a 17 kb unique sequence (T-DNA) in plant chromosomes by *in situ* hybridization. *Chromosoma* **94**, 11–18.

5. Arentzen, R., Manning, R. W., Wolfson, B., Davis, L. G., Higgins, G. A., Lin, Y., and Baldino Jr., F. (1985). Detection of neuropeptide mRNA in rat brain by *in situ* hybridization with radiolabeled synthetic oligonucleotide probes. *Nucleosides Nucleotides* **4**, 261–261.

6. Baan, R. A., Schoen, M. A., Zaalberg, O. B., and Lohman, P. H. M. (1982). In "Mutagens in Our Environment." (R. Sorsa, and H. Vaino, eds), pp. 111. Alan R. Liss, New York.

7. Bauman, J. G. J., and van Duijn, P. (1980). Hybrido-cytochemical detection of specific nucleic acids. Fifth Int. Symp. on Flow Cytometry, Rome, *Basic Appl. Histochem.* **24**, 255.

8. Bauman, J. G. J., Wiegant, J., Borst, P., and van Duijn, P. (1980). A new method for fluorescence microscopical localization of specific DNA sequences by *in situ* hybridization of fluorochrome-labeled RNA. *Exp. Cell Res.* **128**, 485–490.

9. Bauman, J. G. J., Wiegant, J., and van Duijn, P. (1981). Cytochemical hybridization with fluoro-chrome-labeled RNA. I. Development of a method using nucleic acids bound to agarose beads as a model. *J. Histochem. Cytochem.* **29**, 238–246.

10. Bauman, J. G. J., Wiegant, J., and van Duijn, P. (1981). Cytochemical hybridization with fluorochrome labeled RNA. II. Applications. *J. Histochem. Cytochem.* **29**, 238–246.

11. Bauman, J. G. J., Wiegant, J., and van Duijn, P. (1981). Cytochemical hybridization with fluoro-chrome labeled RNA. III. Increased sensitivity by the use of anti-fluorescein antibodies. *Histochem.* **73**, 181–193.

12. Bauman, L. G. J., Wiegant, J., van Duijn, P., Lubsen, N. H., Sondermeijer, P. J., Hennig, W., and Kubli, E. (1981). Rapid and high resolution detection of *in situ* hybridization to polytene chromosomes using fluorochrome-labeled RNA. *Chromosoma* **84**, 1–18.

13. Bauman, L. G. J., Wiegant, J., and van Duijn, P. (1983). The development, using poly (Hg-U) in a model system, of a new method to visualize cytochemical hybridization in fluorescence microscopy. *J. Histochem. Cytochem.* **31**, 571–578.

14. Bauman, J. G. J., Mulder, A. H., and van den Engh, G. (1985). Effect of surface antigen labeling on spleen colony formation, comparison of the indirect immunofluorescence and the biotin–avidin methods. *Exp. Hematol.* **13**, 780–787.

15. Bauman, L. G. J., Bentvelsen, P., and van Bekkum, D. W. (1986). Detection of mRNA by flow cytometry using fluorescent *in situ* hybridization. *Exp. Hematol.* **14**, 433–433.

16. Bauman, J. G. J., and Bentvelzen, P. (1988). Flow cytometric detection of ribosomal RNA in suspended cells by fluorescent *in situ* hybridization. *Cytometry* **9**, 517–524.

17. Beisker, W., Dolbeare, F., and Gray, J. (1987). An improved immunocytochemical procedure for high sensitivity detection of incorporated bromo-deoxyuridine. *Cytometry* **8**, 235–239.

18. Beltz, G. A., Jacobs, K. A., Eikbush, T. H., Cherbas, P. T., and Kafutos, F. C. (1983). Isolation of multigene families and determination of homologies by filter hybridization methods. *In* "Methods in Enzymology, Vol. 100, Recombinant DNA Part B" (R. Wu, L. Grossman and K. Moldavc, eds), pp. 261–285. Academic Press, London.

19. Brahic, M., and Haase, A. (1978). Detection of viral sequences of low reiteration frequency by *in situ* hybridization. *Proc. Natl Acad. Sci. USA* **75**, 6125–6129.

20. Brahic, M., Haase, A., and Cash, E. (1984). Simultaneous *in situ* detection of viral RNA and antigens. *Proc. Natl. Acad. Sci. USA* **91**, 5445–5448.

21. Brigati, D. J., Myerson, D., Leary, J. J., Spalholz, B., Travis, S. Z., Fong, C. K. Y., Hsiung, G. D., and Ward, D. C. (1983). Detection of viral genomes in cultured cells and paraffin-embedded tissue sections using biotin-labeled hybridization procedures. *Virology* **126**, 32–50.

22. Bock, G., Hilchenbach, M., Schauenstein, K., and Wick, G. (1985). Photometric analysis of antifade reagents for immunofluorescence with laser and conventional illumination sources. *J. Histochem. Cytochem.* **33**, 699–705.

23. Buongiorno-Nardelli, N., and Amaldi, F. (1970). Autoradiographic detection of molecular hybrids between RNA and DNA in tissue sections. *Nature* **2254**, 946–948.

24. Burk, R., Szabo, P., O'Brien, S., Nash, W., Yu, L., and Smith, K. (1985). Organization and chromosomal specificity of autosomal homologs of human Y chromosome repeated DNA. *Chromosoma* **92**, 225–233.

25. Buroker, N., Bestwick, R., Haight, G., Magenis, R. E., and Litt, M. (1987). A hypervariable repeated sequence on human chromosome 1p36. *Hum. Genet.* **77**, 175–181.

26. Camargo, M., and Cervenka, J. (1982). Patterns of DNA replication of human chromosomes. II Replication map and replication model. *Am. J. Hum. Genet.* **34**, 757–780.

27. Cantor, C. R., and Schimmel, P. R. (1980). "Biophysical chemistry Part III: The Behavior of Biological Macromolecules." Freeman, San Francisco.

28. Chollet, A., and Kawashima, E. H. (1985). Biotin-labeled synthetic oligodeoxyribonucleotides: chemical synthesis and uses as hybridization probes. *Nucl. Acids Res.* **13**, 1529–1541.

29. Coghlan, J. P., Aldred, P., Haralambidis, J., Niall, H. D., Penschow, J. D., and Tregear, G. W. (1985). Hybridization histochemistry. *Anal. Biochem.* **149**, 1–28.

30. Cooke, H. (1976). Repeated sequence specific to human males. *Nature* **262**, 182–186.

31. Cooke, J. J., and Hindley, J. (1979). Cloning of human satellite III DNA: Different components are on different chromosomes. *Nucl. Acids Res.* **6**, 3177–3197.

32. Cox, K. H., DeLeon, D. V., Angerer, L. M., and Angerer, R. C. (1984). Detection of mRNAs in sea urchin embryos by *in situ* hybridization using asymmetric RNA probes. *Dev. Biol.* **101**, 485–502.

33. Dale, R. M. K., and Ward, D. C. (1975). Mercurated polynucleotides: New probes for hybridization and selective polymer fractionation. *Biochem.* **14**, 2458–2469.

34. Devilee, P., Cremer, T., Slagboom, P., Bakker, E., Scholl, H., Hager, H., Stevenson, A., Cornelisse, C., and Pearson, P. (1986). Two subsets of human alphoid repetitive DNA show distinct preferential localization in the pericentric regions of chromosomes 13, 18, and 21. *Cytogenet. Cell Genet.* **41**, 193–201.

35. Donlon, T., Wyman, A. R., Mulholland, J., Berker, D., Bruns, G., Latt, S., and Botstein, D. (1986). Alpha satellite-like sequences at the centromere of chromosome 8. *Am. J. Hum. Genet.* **39**, A196.

36. Dudin, G., Cremer, T., Schardin, M., Hausmann, M., Bier, F., and Cremer, C. (1987). A method for nuclei acid hybridization to isolated chromosomes in suspension. *Hum. Genet.* **76**, 290–292.

37. Erickson, J. M., Rushford, C. L., Dorney, D. J., Wilson, G. N., and Schmickel, R. D. (1981). Structure and variation of human ribosomal DNA: molecular analysis of cloned fragments. *Gene* **16**, 1–9.

38. Ernst, L., Mujmdar, R., Southwick, P., Tauriello, E., Mujumdar, S., DeBinso, R., Taylor, R., and Waggoner, E. (1987). New red and infrared excited probes. *Cytometry*, Suppl. 1, 30.

39. Forster, A. C., McInnes, J. L., Skingle, D. C., and Symons, R. H. (1985). Non-radioactive hybridization probes prepared by the chemical labeling of DNA and RNA with a novel reagent, photobiotin. *Nucl. Acids Res.* **13**, 745–761.

40. Fuchs, R., and Daune, M. (1972). Physical studies of deoxyribonucleic acid after covalent binding of a carcinogen. *Biochemistry* **11**, 2659–2666.

41. Gall, J. G., and Pardue, M. L. (1969). Formation and detection of RNA·DNA hybrid molecules in cytological preparations. *Proc. Natl Acad. Sci. USA* **63**, 378–383.

42. Gerhard, D. S., Kawasake, E., Bancroft, F., and Szabo, P. (1981). Localization of a unique gene by direct hybridization *in situ*. *Proc. Natl Acad. Sci. USA* **78**, 3755–3759.

43. Gray, J. W., Pinkel, D., Trask, B., van den Engh, G., Pallavicini, M., Fuscoe, J., Mullikin, J., and van Dekken, H. (1989). Analytical cytology applied to detection of prognostically important cytogenetic aberrations: Current status and future directions. Proceedings of prediction of tumor treatment response, April 21–24, 1987, Banff, Canada. *In* "Prediction of Tumour Treatment Response." (S. D.

Chapman, C. J. Peters, and H. R. Withers, eds). Pergamon Press, 1989.

44. Haase, A., Walker, D., Stowring, L., Ventura, P., Geballe, A., Blum, H., Brahic, M., Goldberg, R., and O'Brien, K. (1985). Detection of two viral genomes in single cells by double-label hybridization *in situ* and color microradioautography. *Science* **227**, 189–191.

45. Hamkalo, B. A., Farnham, P. J., Johnston, R., and Schimke, R. T. (1985). Ultrastructural features of minute chromosomes in a methotrexate-resistant mouse 3T3 cell line. *Proc. Natl Acad. Sci. USA* **82**, 1126–1130.

46. Harper, M., Saunders, G. (1981). Localization of single copy DNA sequences on G-banded human chromosomes by *in situ* hybridization. *Chromosoma* **83**, 431–439.

47. Harper, M., Ullrich, A., and Saunders, G., (1981). Localization of the human insulin gene to the distal end of the short arm of chromosome 11. *Proc. Natl Acad. Sci. USA* **78**, 4458–4460.

48. Haugland, R. (1987). A variety of new fluorescent probes. *Cytometry*, supple. 1, 31.

49. Higgins, M. J., Wang, H., Shtromas, I., Haliotis, T., Roder, J. C., Holden, J. J. A., and White, B. N. (1985). Organization of a repetitive human 1.8 kb Kpnl sequence localized in the heterochromatin of chromosome 15. *Chromosoma* **93**, 77–86.

50. Hopman, A. H. N., Wiegant, J., and van Duijn, P. (1986). A new hybridocytochemical method based on mercurated nucleic acid probes and sulfhydryl–hapten ligands. I. Stability of the mercury–sulfhydryl bond and influence of the ligand structure on immunocytochemical detection of the hapten. *Histochem.* **84**, 169–178.

51. Hopman, A. H. N., Wiegant, J., and van Duijn, P. (1986). A new hybridocytochemical method based on mercurated nucleic acid probes and sylfhydryl–hapten ligands. II. Effect of variations in ligand structure on the *in situ* detection of mercurated probes. *Histochem.* **84**, 179–185.

52. Hopman, A. H. N., Weigant, J., Raap, A. K., Landegent, J. E., van der Ploeg, M., and van Duijn, P. (1986). Bi-color detection of two target DNAs by non-radioactive *in situ* hybridization. *Histochem.* **35**, 1–4.

53. Hopman, A. H. N., Weigant, J., Tesser, G. I., and van Duijn, P. (1986). A non-radioactive *in situ* hybridization method based on mercurated nucleic acid probes and sylfhydryl–hapten ligands. *Nucl. Acids Res.* **4**, 6471–6488.

54. Hulsebos, T., Schonk, D., van Dalen, I., Coerwinkel-Dressen, M., Schepens, J., Ropers, H. H., and Wieringa, B. (1988). Isolation and characterization of alphoid DNA sequences specific for the peri-centromeric regions of chromosomes 4, 5, 9 and 19. *Cytogenet. Cell Genet.* **47**, 144–148.

55. Hutchinson, N., Langer-Safer, P., Ward, D., and Hamkalo, B. (1982). *In situ* hybridization at the electron microscope level: hybrid detection by autoradiography and colloidal gold. *J. Cell. Biol.* **95**, 609–618.

56. Jabs, E., Wolf, S., and Migeon, B. (1984). Characterization of a cloned DNA sequence that is present at centromeres of all human autosomes and the X chromosome and shows polymorphic variation. *Proc. Natl Acad. Sci. USA* **81**, 4884–4888.

57. John, H., Patrinou-Georgoulas, M., and Jones, K. (1977). Detection of myosin heavy chain mRNA during myogenesis in tissue culture by *in vitro* and *in situ* hybridization. *Cell* **12**, 501–508.

58. John, H., Birnstiel, M. L., and Jones, K. W. (1969). RNA/DNA hybrids at the cytological level. *Nature* **223**, 582–587.

59. Jones, K. (1970). Chromosomal and nuclear location of mouse satellite DNA in individual cells. *Nature* **225**, 912–915.

60. de Jong, A. S. H., van Kessel-van Vark, M. and Raap, A. K. (1985). Sensitivity of various visualization methods for peroxidase and alkaline phosphatase activity in immunoenzyme histochemistry. *Histochem. J.* **17**, 1119–1130.

61. Kaluzewski, B. (1982). BrdUrd–Hoechst–Giemsa analysis of DNA replication in synchronized lymphocyte cultures. *Chromosoma* **85**, 553–569.

62. Kafatos, F. C., Jones, C. W., and Efstratiadis, A. (1979). Determination of nucleic acid sequence homologies and relative concentrations by a dot blot hybridization procedure. *Nucl. Acid. Res.* **7**, 1541–1552.

63. Kitagawa, Y., and Stollar, B. D. (1982). Comparison of poly(A)-poly(dT) and Poly(I)-poly(dC) as immunogens for the induction of antibodies to RNA/DNA hybrids. *Mol. Immunol.* **19**, 413–420.

64. Kress, H., Meyerowitz, E., and Davidson, N. (1985). High resolution mapping of *in situ* hybridized biotinylated DNA to surface-spread *Drosophila* polytene chromosomes. *Chromosoma* **93**, 113–122.

65. Krumlauf, R., Jeanpierre, M., and Young, B. D. (1982). Construction and characterization of genomic libraries from specific human chromosomes. *Proc. Natl Acad. Sci. USA* **79**, 2971–2975.

66. Kunkel, L. M., Smith, K. D., Boyer, S. H., Borgaonkar, D. S., Wachtel, S. S., Miller, O. J., Breg, W. R., Jones, H. W., and Rary, J. M. (1977). Analysis of human Y-chromosome-specific reiterated DNA in chromosome variants. *Proc. Natl Acad. Sci. USA* **74**, 1245–1249.

67. Kunkel, L. M., Smith, K. D., and Boyer, S. H. (1979). Organization and heterogeneity of sequences within a repeating unit of human Y chromosome deoxyribonucleic acid. *Biochem.* **18**, 3343–3353.

68. Landegent, J. E., Jansen in de Wal, N., Baan, R. A., Hoeijmakers, J. H. J., and van der Ploeg, M. (1984).

2-Acetylaminofluorene-modified probes for indirect hybridocytochemical detection of specific nucleic acid sequences. *Exp. Cell Res.* **153**, 61–72.

69. Landegent, J. E., Jansen in de Wal, N., van Ommen, G. J., Baas, F., de Vijlder, J. J. M., van Duijn, P., and van der Ploeg, M. (1985). Chromosomal localization of a unique gene by non-autoradiographic *in situ* hybridization. *Nature* **317**, 175–177.

70. Landegent, J. E., Jansen in de Wal, N., Ploem, J. S., and van der Ploeg, M. (1985). Sensitive detection of hybridocytochemical results by means of reflection contrast microscopy. *J. Histochem. Cytochem.* **33**, 1241–1246.

71. Landegent, J. E., Jansen in de Wal, N., Visser-Groen, Y. M., Bakker, E., van der Ploeg, M., and Pearson, P. L. (1986). Fine mapping of the Huntington disease linked D4S10 locus by non-radioactive *in situ* hybridization. *Hum. Genet.* **73**, 354–357.

72. Langer, P. R., Waldrop, A. A., and Ward, D. C. (1981). Enzymatic synthesis of biotin-labeled polynucleotides: novel nucleic acid affinity probes. *Proc. Natl Acad. Sci. USA* **78**, 6633–6637.

73. Lau, Y., and Schonberg, S. (1984). A male-specific DNA probe detects hetero chromatin sequences in a familial Y_q-chromosome. *Am. J. Hum. Genet.* **36**, 1394–1396.

74. Lawrence, J., and Singer, R. (1985). Quantitative analysis of *in situ* hybridization methods for the detection of actin gene expression. *Nucl. Acids Res.* **13**, 1777–1799.

75. Lawrence, J. B., and Singer, R. H. (1986). Intracellular localization of messenger RNAs for cytoskeletal proteins. *Cell* **45**, 407–415.

76. Lawrence, J. B., Villnave, C. A., and Singer, R. H. (1988). Sensitive, high-resolution chromatin and chromosome mapping *in situ*: presence and orientation of two closely integrated copies of EBV in a lymphoma line. *Cell* **52**, 51–61.

77. Lynn, D., Angerer, L., Bruskin, A., Klein, W., and Angerer, R. (1983). Localization of a family of mRNAs in a single cell type and its precursors in sea urchin embryos. *Proc. Natl Acad. Sci. USA* **80**, 2656–2660.

78. Maniatis, T., Fritsch, E. F., and Sambrook, J. (1982). Molecular cloning. "A laboratory Manual." Cold Spring Harbor Laboratory, New York.

79. Manuelidis, L., Langer-Safer, P., and Ward, D. (1982). High-resolution mapping of satellite DNA using biotin-labeled DNA probes. *J. Cell. Biol.* **95**, 619–625.

80. Manuelidis, L., and Borden, J. (1988). Reproducible compartmentalization of individual chromosome domains in human CNS cells revealed by *in situ* hybridization and three dimensional reconstruction. *Chromosoma* **96**, 397–410.

81. McDermid, H. E., Duncan, A. M. V., Higgins, M. J., Hamerton, R. L., Rector, E., Brasch, K. R., and White, B. N. (1986). Isolation and characterization of an alpha-satellite repeated sequence from human chromosome 22. *Chromosoma* **94**, 228–234.

82. Melton, D. A., Krieg, P. A., Rebagliati, M. R., Maniatis, T., Zinn, K., and Green, M. R. (1984). Efficient *in vitro* synthesis of biologically active RNA and RNA hybridization probes from plasmids containing a bacteriophage SPs promotor. *Nucl. Acids Res.* **12**, 7035–7056.

83. Moyzis, R., Albright, K., Bartholdi, M., Cram, S., Deaven, L., Hildebrand, E., Joste, N., Longmire, J., Meyne, J., and Schwarzacher-Robinson, T. (1987). Human chromosome-specific repetitive DNA sequences: Novel markers for genetic analysis. *Chromosoma* **95**, 375–386.

84. Nederlof, P. M., Robinson, D., Abuknesha, R., Weigant, J., Hopman, A. H. N., Tanke, H. J.. and Raap, A. K. (1989). Three color fluorescence *in situ* hybridization for the simultaneous detection of multiple nucleic acid sequences in interphase nuclei and chromosomes. *Cytometry* **10**, 20–27.

85. Oi, V. T., Glazer, A. N., and Stryer, I. (1982). Fluorescent phycobiliprotein conjugates for analyses of cells and molecules. *J. Cell Biol.* **93**, 981–986.

86. Pardue, M., and Gall, J. (1970). Chromosomal localization of mouse satellite DNA. *Science* **168**, 1356–1358.

87. Pinkel, D., Straume, T., and Gray, J. W. (1986). Cytogenetic analysis using quantitative, high sensitivity, fluorescence hybridization. *Proc. Natl Acad. Sci. USA* **83**, 2934–2938.

88. Pinkel, D., Gray, J. W., Trask, B., van den Engh, G., Fuscoe, J., and van Dekken, H. (1986). Cytogenetic analysis by *in situ* hybridization with fluorescently labeled nucleic acid probes. *Cold Spring Harbor Symp. Quantit. Biol.* **LI**, 151–157.

89. Pinkel, D., Landegent, J., Collins, C., Fuscoe, J., Seagraves, R., Lucas, J., and Gray, J. W. (1989). Fluorescence *in situ* hybridization with human chromosome specific libraries: Detection of trisomy 21 and translocations of chromosome 4. *Proc. Natl Acad. Sci. USA* **85**, 9138–9142.

90. van Prooijen-Knegt, A. C., van Hoek, J. F. M., Bauman, J. G. J., van Duijn, P., Wool, I. G., and van der Ploeg, M. (1982). *In situ* hybridization of DNA sequences in human metaphase chromosomes visualized by an indirect fluorescent immuno-cytochemical procedure. *Exp. Cell Res.* **141**, 397–407.

91. Raap, A. K., Marijnen, L. G. J., and van der Ploeg, M. (1984). Anti-DNA/RNA sera: specificity tests and application in quantitative *in situ* hybridization. *Histochem.* **81**, 517–520.

92. Raap, A. K., Geelen, J. L., van der Meer, J. W. M.,

van de Rijke, F. M., van den Boogaart, P., and van der Ploeg, M. (1988). Non-radioactive *in situ* hybridization for the detection of cytomegalovirus infections. *Histochem.* **88**, 367–373.

93. Renz, M., and Kurz, C. (1984). A colorimetric method for DNA hybridization. *Nucl. Acids Res.* **12**, 3435–3444.

94. Rudkin, G. T., and Stollar, B. D. (1977). Naturally occurring DNA/RNA hybrids. I. Normal patterns in polytene chromosomes. INCN-UCLA Symp. *Mol. Cell Biol.* **7**, 257–269.

95. Rudkin, G. T., and Stollar, B. D. (1977). High resolution detection of DNA/RNA hybrids *in situ* by indirect immunofluorescence. *Nature* **265**, 472–473.

96. Shroyer, K. R., and Nakane, P. K. (1983). Use of TNP-labeled cDNA for *in situ* hybridization. *J. Cell. Biol.* **97**, 377a.

97. Singer, R., and Ward, D. (1982). Actin gene expression visualized in chicken muscle tissue culture by using *in situ* hybridization with a biotinated nucleotide analog. *Proc. Natl Acad. Sci. USA* **79**, 7331–7335.

98. Singer, R. H., Lawrence, J. B., Langevin, G. L., Rashtchian, R. N., Villnave, C. A., Cremer, T., Tesin, D., Manuelidis, L., and Ward, D. C. (1987). Double labeling *in situ* hybridization using non-isotopic and isotopic detection. *Acta Histochem. Cytochem.* **20**, 589–599.

99. Steinkamp, J. A., and Stewart, C. C. (1986). Dual-laser, differential fluorescence correction method for reducing cellular background autofluorescence. *Cytometry* **7**, 568–574.

100. Stuart, W. D., Bishop, J. G., Carson, H. L., and Frank, M. B. (1981). Location of 18/28S ribosomal RNA genes in two Hawaiian *Drosophila* Species by monoclonal immunological identification of RNA/DNA hybrids *in situ*. *Proc. Natl Acad. Sci. USA* **78**, 3751–3754.

101. Szabo, P., and Ward, D. C. (1982). What's new with hybridization *in situ*? *Trends Biochem. Sci. Dec.*, 425–427.

102. Tchen, P., Fuchs, R. P. P., Sage, E., and Leng, M. (1984). Chemically modified nucleic acids as immunodetectable probes in hybridization experiments. *Proc. Natl Acad. Sci. USA* **81**, 3466–3470.

103. Traincard, F., Ternynck, T., Danchin, A., and Avrameas, S. (1983). An immunoenzymatic procedure for the detection of nucleic acid hybridization. *Ann. Immunol. (Inst. Pasteur)* **134**, 399–405.

104. Trask, B. (1985). Studies of chromosomes and nuclei using flow cytometry. PhD. Thesis, Leiden University, Leiden, The Netherlands.

105. Trask, B., van den Engh, G., Landegent, J., Jansen in de Wal, N., and van der Ploeg, M. (1985). Detection of DNA sequences in nuclei in suspension by *in situ* hybridization and dual beam flow cytometry. *Science* **230**, 1401–1403.

106. Trask, B., van den Engh, G., Pinkel, D., Mullikin, J., Waldeman, F., Dekken, H., and Gray, J. (1988). Fluorescence *in situ* hybridization to interphase cell nuclei in suspension allows flow cytometric analysis of chromosome content and microscopic analysis of nuclear organization. *Hum. Genet.* **78**, 251–259.

107. van Dekken, H., Arkeskeün, G. J. A., Visser, J. W. H. and Bauman, J. C. J. (1990). Flow cytometric quantitation of chromosome specific repetitive DNA sequences by single and bicalor fluorescent *in situ* hybridization to human lymphocyte interphase nuclei. *Cytometry* (in press).

108. van den Engh, G., Trask, B. J., Gray, J. W., Langlois, R., and Yu, L.-C. (1985). Preparation of chromosome suspensions for flow cytometry II. Bivariate analysis of human chromosomes. *Cytometry* **6**, 92–100.

109. van den Engh, G., Stokdijk, W., and Peters, D. (1987). A fast data-acquisition system for multi-parameter flow analysis and sorting. *Cytometry*, supple. 1, 7.

110. Vanderlaan, M., and Thomas, C. B. (1985). Characterization of monoclonal antibodies to bromodeoxyuridine. *Cytometry* **6**, 501–505.

111. Venezky, D., Angerer, L., and Angerer, R. (1981). Accumulation of histone repeat transcripts in the sea urchin egg pronucleus. *Cell* **24**, 385–391.

112. Waye, J., and Willard, H. (1985). Chromosome-specific alpha satellite DNA: nucleotide sequence analysis of the 2.0 kilobase pair repeat from the human X chromosome. *Nucl. Acids Res.* **13**, 2731–2743.

113. Waye, J., and Willard, H. (1986). Structural organization and sequence of alpha-satellite DNA from human chromosome 17: evidence of evolution by unequal crossing over and an ancestral pentamer repeat shared with the human X chromosome. *Mol. Cell Biol.* **6**, 3156–3165.

114. Waye, J. S., Creeper, L. A., and Willard, H. F. (1987). Organization and evolution of alpha satellite DNA from human chromosome 11. *Chromosoma* **95**, 182–188.

115. Waye, J. S., Durfy, S. J., Pinkel, D., Davies, K. E., and Willard, H. F. (1989). Chromosome-specific alpha satellite DNA from human chromosome 1: hierarchical structure and genomic organization of a polymorphic domain spanning several hundred kilobase pairs of centromeric DNA (1 centromere). *Genomics* **1**, 43–51.

116. Waye, J. S., England, S. B. and Willard, H. F. (1987). Genomic organization of alpha satelite DNA on human chromosome 7: evidence for two distinct alphoid domains on a single chromosome. *Mol. Cell Biol.* **7**, 349–356.

117. Waye, J. S., and Willard, H. F. (1987). Nucleotide

sequence heterogenicity of alpha satellite repetitive DNA: a survey of alphoid sequences from different human chromosomes. *Nucl Acids Res.* **15**, 7549–7569.

118. Webster, H., Lamperth, L., Favilla, J., Lemke, G., Tesin, D., and Manuelidis, L. (1987). Use of a biotinylated probe and *in situ* hybridization for light and electron microscopic localization of PO mRNA in myelin-forming Schwann cells. *Histochem.* **86**, 441–444.

119. Weier, H. A., Pinkel, D., Segraves, R., and Gray, J. W. (1989). Synthesis of Y-chromosome specific, labeled DNA probes by *in vitro* DNA amplification. *J. Histochem. Cytochem.* (submitted).

120. Wetmur, J. G. (1975). Acceleration of DNA renaturation rates. *Biopolymers* **14**, 2517–2524.

121. Willard, H. (1985). Chromosome-specific organization of the human alpha satellite DNA. *Am. J. Hum. Genet.* **37**, 524–532.

122. Wolfe, J., Darling, S. M., Erickson, R. P., Craig, I. W., Buckle, V. J., Rigby, P. W. J., Willard, H. F., and Goodfellow, P. N. (1985). Isolation and characterization of an alphoid centromeric repeat family from the human Y chromosome. *J. Mol. Biol.* **182**, 477–485.

123. Wood, W. I., Gitschier, J., Lasky, L. A., and Lawn, R. (1985). Base composition-independent hybridization in tetramethylammoniumchloride: a method for oligonucleotide screening of highly complex gene libraries. *Proc. Natl Acad. Sci. USA* **81**, 1585–1588.

124. Yang, T., Hansen, S., Oishi, K., Ryder, O., and Hamkalo, B. (1982). Characterization of a cloned repetitive DNA sequence concentrated on the human X chromosome. *Proc. Natl Acad. Sci. USA* **79**, 6593–6597.

125. Yu, S., and Gorovsky, M. (1986). *In situ* dot blots: quantitation of mRNA in intact cells. *Nucl. Acids Res.* **14**, 7597–7615.

Index

303

X

X-inactivation, 251–252
X-linked genetic diseases, 237–238
X-rays
  flow karyotypes affected by,
    152–154
  see also Radiation damage

Y

Y-linked genes, transmission of, 238

Z

Zapper technique, 132